Texts in Applied Mathematics

Volume 51

For further volumes:
http://www.springer.com/series/1214

Thomas Rylander • Pär Ingelström
Anders Bondeson

Computational Electromagnetics

Second Edition

Thomas Rylander
Department of Signals and Systems
Chalmers University of Technology
Hörsalsvägen 11
SE-41296 Göteborg, Sweden

Pär Ingelström
Department of Electromagnetics
Chalmers University of Technology
Hörsalsvagen 11
Göteborg, Sweden

Anders Bondeson‡
‡Deceased

ISSN 0939-2475
ISBN 978-1-4899-8602-3 ISBN 978-1-4614-5351-2 (eBook)
DOI 10.1007/978-1-4614-5351-2
Springer New York Heidelberg Dordrecht London

Mathematics Subject Classification (2010): 01-08, 35Q61, 65M06, 65M12, 65M38, 65M60, 65M80, 65N06, 65N12, 65N25, 65N30, 65N38, 65N80

Printed on acid-free paper

Springer is part of Springer Science+Business Media (www.springer.com)

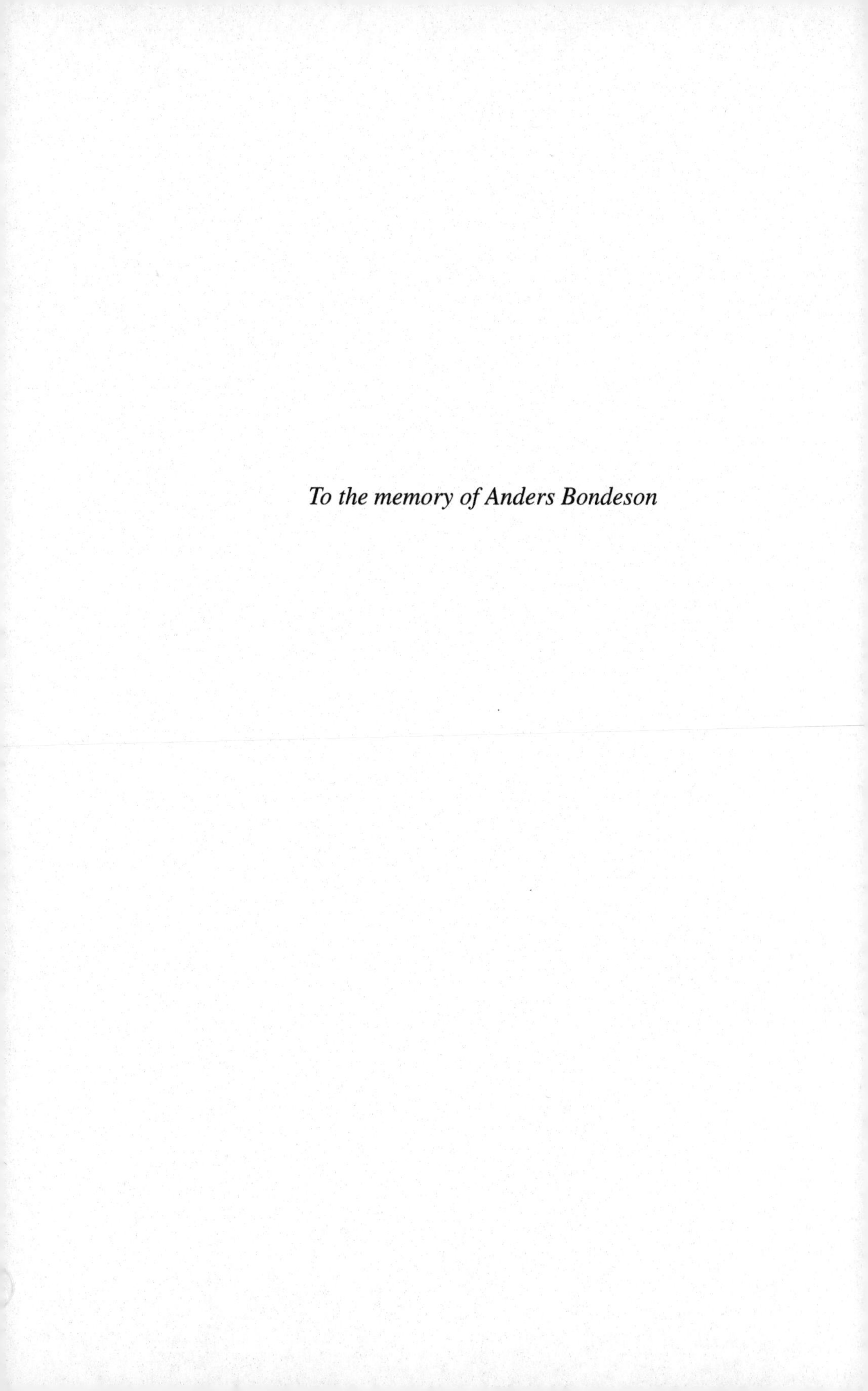

To the memory of Anders Bondeson

Foreword

The material in this book was developed for an undergraduate course in computational electromagnetics, initially given by Professor Anders Bondeson at Chalmers University of Technology, Göteborg, Sweden. It was used in various forms for almost a decade and it fostered a number of engineers who today work in academia and industry with computational electromagnetics as a main tool. During this time, we never managed to find any textbooks suitable for an introductory course on the subject matter, so we eventually decided on compiling our material into a book. The first edition of the book was published in 2005 and now, seven years later, the second edition is completed with new material that hopefully makes the book even more useful. In particular, we wish to thank students, readers and reviewers around the world for contributing with important feedback on the book.

On the 20th of March 2004, before the first edition of the book was completed, Professor Anders Bondeson passed away, suddenly and unexpectedly. This caused shock and deep sorrow to all of us who worked with him, and naturally interrupted the creation of the book. Nevertheless, we felt we should complete the first edition of the book, and we would like to thank the publisher Springer for all the understanding, support, and encouragement during the difficult time after the passing of our coauthor, colleague, friend and former advisor. In particular, we would like to thank Donna Chernyk, Achi Dosanjh, Jamie Ehrlich, Yana Mermel, Brian Halm, and Frank McGuckin at Springer. We are also grateful to the anonymous reviewers who contributed with valuable comments on the manuscript.

We dedicate this book to the memory of Anders Bondeson.

Göteborg

Thomas Rylander
Pär Ingelström

Preface to the Second Edition

The first edition of this book was published 7 years ago, and since then it has been used for courses and self-study around the world. We have received many useful comments on the presentation of the material and the contents of the book from students, lecturers, researchers, and other readers. In addition, we have used the book ourselves for various courses that we teach on computational electromagnetics (CEM), where we received direct feedback from our students on the book as a source of information and its pedagogical development of the topic. Given the collection of all comments that we received in combination with our own experience, we have complemented the second edition of the book with material that we believe will help the reader to learn the subject matter. In particular, we have strengthened the discussion concerning numerical techniques for the first-order system of Maxwell's equations and, as a consequence, their relation to the corresponding second-order differential equations. For finite-difference schemes, this is manifested in terms of the staggered grids that are used to represent the electric and magnetic fields, where particular emphasis is placed on analysis with complex exponentials. In the context of finite element methods (FEMs), the first-order system is treated by means of expanding the electric field in terms of curl-conforming elements and the magnetic field in terms of divergence-conforming elements. This representation associates the electric field with edges (referred to by some authors as a primal grid), and, similarly, the magnetic field is associated with faces (which may be thought of as a dual grid). In particular, for brick-shaped elements it is rather apparent that the primal grid and the dual grid together make up the staggered grids that are used for finite-difference schemes. Furthermore, we have incorporated a new Appendix B that features the lowest-order curl-conforming and divergence-conforming basis functions on the most common finite element shapes: the triangle and the quadrilateral, the tetrahedron, the prism, and the hexahedron. The automatic generation of symbolic expressions and vector field visualizations for these basis functions is provided in terms of a MATLAB implementation that can be downloaded from a URL provided at the end of this preface. The presentation of the material is further improved by cross references between the description of the finite difference schemes and the FEMs. In addition, the presentation of both finite differences and finite elements references the material in the new Appendix B, which may be useful for obtaining a unified perspective on CEM. In a similar fashion, the chapter on the method of moments (MoM) exploits this new Appendix B.

Appendix B on the lowest-order curl-conforming and divergence-conforming basis functions also contains basis functions for nodal elements (i.e., gradient-conforming basis functions), a definition of reference elements, and suitable quadrature rules. The concept of the reference element is useful for the implementation of the FEM (and MoM), where the reference element is related to the physical element by means of a mapping. The mapping is expressed in terms of the nodal basis functions, and the evaluation of discrete operators is formulated in terms of numerical integration (or quadrature) on the reference element. This type of approach is at the heart of most finite element programs and is particularly useful for problems that involve inhomogeneous materials or nonlinear materials. The second edition of the book contains a new Sect. 6.6 that describes in detail the concept of the reference element and its relation to the physical element by means of a mapping, where integration on the reference element is also featured. These concepts are presented with the triangle as an example, and references to the new Appendix B are used extensively to demonstrate how the techniques generalize to other element shapes. In addition, the new Appendix B is very useful as a source of information for computer implementations, where the reader can select and combine different discretization techniques that are suitable for the specific situation at hand. We have also added a new advanced real-world problem in Sect. 6.6.4 that demonstrates the use of all these techniques, where we solve a 3D eigenvalue problem formulated in terms of the FEM on tetrahedrons applied to the first-order system of Maxwell's equations with inhomogeneous and lossy media. The theoretical formulation is described in detail, and the complete computer implementation is included in the material that can be downloaded from the Internet, at the URL provided at the end of this preface. Also, we present some results in terms of 3D visualizations for a circular cylindrical cavity of finite length and discuss their relation to the conventional analysis. In addition, we solve a more challenging problem with inhomogeneous and lossy media in a cavity. The reader can easily modify the computer implementation to study, e.g., driven problems in the frequency domain or exploit the computer implementation provided here as a platform for more advanced programs with other features such as radiation boundary conditions, transient analysis, and nonlinear media. The book also contains material on numerical methods that exploit higher-order approximations, and this is demonstrated in particular in Sect. 3.3, which provides a method for deriving higher-order finite-difference approximations.

The second edition of this book also features a new Appendix A with five computer projects: (1) convergence and extrapolation, (2) finite differences in the frequency domain, (3) finite-difference time-domain scheme, (4) FEM, and (5) MoM. This collection of projects covers the material presented in the book, and the projects are designed to progress the knowledge and skills of the reader. They are useful for learning and understanding the material in the book and could be used as assignments in a course. Such assignments could be examined in several different ways such as (1) oral examination in a computer-laboratory setting, (2) presentation in a classroom setting where students and teacher ask questions, or (3) student-authored written reports. Experience shows that it is useful to let students work in small groups. This approach gives the students the possibility to present, challenge, and discuss different ideas on how to work on the assignments.

To our knowledge, the open literature is sparse on books that present contemporary CEM in an introductory manner that is appropriate for use at the undergraduate level. This book is intended to provide such material and prepare the reader for the more advanced literature on CEM. Apart from this text, the book by Davidson [22] and the book by Sadiku [68] are some of the very few other examples with such an aim. Additional books that present the main computational techniques in CEM have been published in recent years. Those by Garg [29] and Jin [42] contain a fair amount of classical analytical calculations for Maxwell's equations apart from the basic techniques in CEM. Warnick [88] and Sheng [72] have a stronger focus on CEM; in addition, Warnick [88] also discusses numerical techniques in somewhat more general settings.

The MATLAB implementations given in this book, together with the MATLAB implementations that support the projects in Appendix A and some other useful programs, are available for download at http://www.springer.com. We would appreciate it if errors found were brought to our attention. These will be posted on the aforementioned Web site.

Contents

List of Symbols

Symbol	Quantity	SI Unit
$\boldsymbol{E}$	Electric field strength	V/m
$\boldsymbol{D}$	Displacement field strength	C/m^2
$\boldsymbol{H}$	Magnetic field strength	A/m
$\boldsymbol{B}$	Magnetic flux	T = Vs/m^2
$\boldsymbol{J}$	Electric current density	A/m^2
ρ	Electric charge density	C/m^3
ϵ	Electric permittivity	As/Vm
μ	Magnetic permeability	Vs/Am
σ	Electric conductivity	S/m = A/Vm
t	time	s
ω	angular frequency	rad/s

List of Acronyms

ABC	Absorbing Boundary Conditions
BEM	Boundary Element Method
CEM	Computational Electromagnetics
CFIE	Combined Field Integral Equation
CG	Conjugate Gradient
DWP	Discrete Wavelet Packet
EMC	Electromagnetic Compatibility
FD	Finite Difference
FDTD	Finite-Difference Time-Domain
FEM	Finite Element Method
FMM	Fast Multipole Method
FIT	Finite Integration Technique
FV	Finite Volume
IML	Impedance Matrix Localization (technique)
MFIE	Magnetic Field Integral Equation
MLFMA	Multilevel Fast Multipole Algorithm
MoM	Method of Moments
NTF	Near-to-far field transformation
ODE	Ordinary Differential Equation
PEC	Perfect electric conductor
PML	Perfectly Matched Layer
RCS	Radar Cross Section
RWG	Rao-Wilton-Glisson (boundary elements)
TDIE	Time-Domain Integral-Equation
TE	Transverse Electric
TLM	Transmission Line Method
TM	Transverse Magnetic
TEM	Transverse Electromagnetic

Chapter 1
Introduction

Our modern society relies on electromagnetic devices and systems: television, radio, internet, microwave ovens, mobile telephones, satellite communication systems, radar systems, electrical motors, electrical generators, computers, microwave filters, lasers, industrial heating devices, medical imaging systems, electrical power networks, transformers and many more. Each of these examples is used in a broad range of situations. Radar, for example, is employed for fire-control, weather detection, airport traffic-control, missile tracking, missile guidance, speed control/enforcement, and traffic safety. Undoubtedly, electromagnetic phenomena have a profound impact on contemporary society.

The understanding of electromagnetic phenomena is treated by electromagnetic field theory: the study of interactions between electric charges at rest and in motion. (Electric charges in motion are often referred to as electric currents.) Electromagnetic field theory, or electromagnetics for short, describes the interactions between electric charges by Maxwell's equations: a system of coupled partial differential equations that relate sources (charges and currents) to the electromagnetic fields and fluxes. Analytical solutions in closed form are known for only a very limited number of special cases, which hardly ever are directly applicable to real-world applications. Instead, more or less crude approximations have been employed in various attempts to bridge the gap between theory and advanced applications.

The advent of computers has changed our ability to solve Maxwell's equations in a profound way. Ahead of the computer's time, it was advantageous to use considerable effort to avoid computations, often at the price of lengthy analytical manipulations and severely reduced applicability. With powerful computers at hand, however, it is more attractive to use analytically simple methods that may require large amounts of computation. Such computational methods can treat large classes of problems without modifications of the computer algorithms or programs. The part of electromagnetics that deals with computational methods is known as computational electromagnetics (CEM).

It is of significant importance for modern engineers and scientists who work in the area of electromagnetics to have a good command of the computational tools

T. Rylander et al., *Computational Electromagnetics*, Texts in Applied Mathematics 51, DOI 10.1007/978-1-4614-5351-2_1,

developed for electromagnetics problems. CEM allows for a faster and cheaper design process, where the use of expensive and time-consuming prototypes is minimized. These tools can also provide crucial information and understanding of a device's electromagnetic operation, which may be difficult or even impossible to achieve by means of experiments or analytical calculations. Automation of computations allows for extensive parametric studies. It is only relatively recently that optimization by computation has been used for electromagnetic design problems. In times of a rapid pace of development, analysis and optimization of electromagnetic devices by CEM tools may be crucial for maintaining competitiveness.

Today, there is a broad selection of commercially available computer programs that provide implementations of popular and powerful CEM algorithms. These programs can handle many engineering and research problems. However, a well-informed choice and correct use of software for reliable results and conclusions require good knowledge of CEM. Furthermore, problems that extend beyond the applicability of commercially available software packages demand modifications or additions that again rely on a good command of CEM.

1.1 Computational Electromagnetics

CEM is a young discipline. It is still growing, in response to the steadily increasing demand for software for the design and analysis of electrical devices. Twenty years ago, most electrical devices were designed by building and testing prototypes, a process that is both costly and slow. Today the design can be made faster and cheaper by means of numerical computation. CEM has become a main design tool in both industrial and academic research.

There are numerous application areas for CEM, and here we mention a few. In electric power engineering, computation is well established for the analysis and design of electrical machines, generators, transformers, and shields. In applications to microwaves, CEM is a more recent tool, but it is now used for designing microwave networks and antennas, and even microwave ovens. The analysis and optimization of radar cross sections (RCS) for stealth devices has been the driving force for the development of many new techniques in CEM. The clock frequencies of modern microprocessors are approaching the region where circuits occupy a large fraction of a wavelength. Then ordinary circuit theory no longer applies and it may be necessary to solve Maxwell's equations to design smaller and faster processors. The increased demand for electromagnetic compatibility (EMC) also poses new computational problems.

The performance of CEM tools is increasing rapidly. One reason for this is the steady growth of computer capacity over half a century. Another equally important reason is improvements in algorithms. The purpose of this book is to give an introduction to the most frequently used algorithms in CEM. These are finite differences (FD) (usually in the time domain), the finite element method

(FEM), and the boundary element method (BEM), which is usually referred to, for historical reasons, as the method of moments (MoM). Finite difference methods are more or less straightforward discretizations of Maxwell's equations in differential form, using the field components, or the potentials, on a structured grid of points as unknowns. Finite differences in general, and the finite-difference time-domain (FDTD) method in particular, are very efficient and require few operations per grid point. The FDTD is one of the most widespread methods in CEM, and it can be applied to a large variety of microwave problems. One drawback of finite difference methods is that they work well only on uniform Cartesian (structured) grids, and typically use the so-called staircase approximation of boundaries not aligned with the grid. Finite element methods in which the computational region is divided into unstructured grids (typically triangles in two dimensions and tetrahedra in three dimensions) can approximate complex boundaries much better, but are considerably slower in time-domain calculations. The FEM is mainly used for time-harmonic problems, and it is the standard method for eddy current calculations. The MoM discretizes Maxwell's equations in integral form, and the unknowns are sources such as currents or charges on the surfaces of conductors and dielectrics. This method is advantageous for problems involving open regions, and when the current-carrying surfaces are small. The MoM is often applied to scattering problems. We will discuss how the three types of methods, FD, FEM, and MoM, can be applied to different electromagnetics problems, in both the time domain and the frequency domain (time-harmonic fields and currents). Some other methods will be mentioned in Chap. 8.

1.2 Maxwell's Equations

Before discussing how to solve electromagnetics problems, we will first write down Maxwell's equations in the form in which they can be found in most textbooks on electromagnetics, see e.g. [5, 19, 32]. They are usually stated as Ampère's law

$$\nabla \times \boldsymbol{H} = \boldsymbol{J} + \frac{\partial \boldsymbol{D}}{\partial t}, \tag{1.1}$$

Faraday's law

$$\nabla \times \boldsymbol{E} = -\frac{\partial \boldsymbol{B}}{\partial t}, \tag{1.2}$$

Poisson's equation

$$\nabla \cdot \boldsymbol{D} = \rho, \tag{1.3}$$

and the condition of solenoidal magnetic flux density

$$\nabla \cdot \boldsymbol{B} = 0. \tag{1.4}$$

Here $\boldsymbol{H}$ is the magnetic field, $\boldsymbol{J}$ is the current density, $\boldsymbol{D}$ is the electric displacement, $\boldsymbol{E}$ is the electric field, $\boldsymbol{B}$ is the magnetic flux density, ρ is the electric charge density, and t denotes the time variable. Moreover, we have

$$\boldsymbol{H} = \frac{\boldsymbol{B}}{\mu_0} - \boldsymbol{M}, \quad \boldsymbol{D} = \epsilon_0 \boldsymbol{E} + \boldsymbol{P},$$

where $\mu_0 = 4\pi \cdot 10^{-7}$ Vs/Am is the free-space magnetic permeability, $\epsilon_0 = 1/(c_0^2 \mu_0) \approx 8.854 \cdot 10^{-12}$ As/Vm is the free-space electric permittivity, $\boldsymbol{M}$ is the magnetization and $\boldsymbol{P}$ is the polarization. In vacuum, the speed of light is $c_0 = 299\,792\,458$ m/s.

In this book, we will restrict attention to linear, isotropic and nondispersive materials for which the constitutive relations

$$\boldsymbol{B} = \mu \boldsymbol{H}, \quad \boldsymbol{D} = \epsilon \boldsymbol{E}$$

hold with frequency-independent electric permittivity ϵ and magnetic permeability μ. The permittivity is often written as $\epsilon = \epsilon_0 \epsilon_r$, where ϵ_r is called the relative permittivity. Similarly, the permeability is often written $\mu = \mu_0 \mu_r$ where μ_r is called the relative permeability.

For electrically conductive materials, an electric field causes a current density

$$\boldsymbol{J} = \sigma \boldsymbol{E}$$

where σ is the electric conductivity.

1.2.1 Boundary Conditions

Consider the situation in which one medium, characterized by ϵ_1 and μ_1, shares an interface with another medium, characterized by ϵ_2 and μ_2. We use the subindices 1 and 2 to denote quantities that are associated with media 1 and 2, respectively. At the interface, the tangential and normal fields must satisfy so-called boundary conditions, which are consequences of Maxwell's equations. For example, (1.4) states the condition of solenoidal magnetic flux density, and Gauss's theorem

$$\int_V \nabla \cdot \boldsymbol{B} dV = \oint_{\partial V} \boldsymbol{B} \cdot \hat{\boldsymbol{n}} dS, \tag{1.5}$$

where ∂V is the surface enclosing the volume V, applied to this conservation law yields the boundary condition

$$\hat{\boldsymbol{n}} \cdot (\boldsymbol{B}_2 - \boldsymbol{B}_1) = 0,$$

where $\hat{\boldsymbol{n}}$ is a unit normal to the interface that points into medium 2. Similarly, Poisson's equation (1.3) gives

$$\hat{\boldsymbol{n}} \cdot (\boldsymbol{D}_2 - \boldsymbol{D}_1) = \rho_s,$$

where ρ_s is the surface charge density on the interface. Stokes's theorem

$$\int_S (\nabla \times \boldsymbol{E}) \cdot d\boldsymbol{S} = \oint_{\partial S} \boldsymbol{E} \cdot d\boldsymbol{l}, \tag{1.6}$$

where ∂S is the curve enclosing the surface S, applied to Faraday's law (1.2) yields

$$\hat{\boldsymbol{n}} \times (\boldsymbol{E}_2 - \boldsymbol{E}_1) = \boldsymbol{0}$$

and, analogously, Ampère's law (1.1) gives

$$\hat{\boldsymbol{n}} \times (\boldsymbol{H}_2 - \boldsymbol{H}_1) = \boldsymbol{J}_s,$$

where $\boldsymbol{J}_s$ is the surface current on the interface between the two media.

The electric field inside a perfect electric conductor (PEC) is zero and, consequently, also the electric displacement. We get the boundary conditions $\hat{\boldsymbol{n}} \cdot \boldsymbol{D}_2 = \rho_s$ and $\hat{\boldsymbol{n}} \times \boldsymbol{E}_2 = \boldsymbol{0}$ when medium 1 is a PEC. At finite frequencies, Faraday's law yields that the magnetic flux density is zero inside a PEC (which also applies to the magnetic field) and we get the boundary conditions $\hat{\boldsymbol{n}} \cdot \boldsymbol{B}_2 = 0$ and $\hat{\boldsymbol{n}} \times \boldsymbol{H}_2 = \boldsymbol{J}_s$ when medium 1 is a PEC.

Another kind of boundary conditions, which do not correspond to any physical boundary, are *absorbing boundary conditions* (ABC). These are used to truncate the computational domain in case of open region problems and can be implemented using a variety of techniques. The most popular ABC is the *perfectly matched layer* (PML), which will be described in Sect. 5.3.1.

For a more detailed discussion on boundary conditions, the reader is referred to a textbook on electromagnetics; see, e.g., [5, 19, 32].

1.2.2 Energy Relations

For Maxwell's equations, it is useful (and in some cases essential) to regard the energy as being stored in the fields. For electrostatics, we have the energy density $w_\mathrm{e} = \epsilon|\boldsymbol{E}|^2/2$ and the work to assemble a static charge distribution is

$$W = \frac{1}{2}\int \epsilon|\boldsymbol{E}|^2 dV. \tag{1.7}$$

There are alternative expressions for the evaluation of W in terms of the charge distribution and the electrostatic potential. In magnetostatics, the corresponding energy density is $w_{\mathrm{m}} = |\boldsymbol{B}|^2/(2\mu)$. For a time-varying electromagnetic field, we have the energy density $w_{\mathrm{e}} + w_{\mathrm{m}}$ and this quantity is often used to form energy conservation expressions that involve the electromagnetic phenomena.

1.2.3 Time Evolution

Before discussing schemes for evolving Maxwell's equations (1.1)–(1.4) in time, we must note that they are not all independent. For example, Poisson's equation (1.3) is best viewed as an initial condition for the charge density. To see this, take the divergence of Ampère's law, which gives

$$\frac{\partial}{\partial t}\nabla \cdot \boldsymbol{D} + \nabla \cdot \boldsymbol{J} = 0. \tag{1.8}$$

Replacing $\nabla \cdot \boldsymbol{J}$ from the equation of continuity for electric charge

$$\frac{\partial \rho}{\partial t} + \nabla \cdot \boldsymbol{J} = 0,$$

we see that the divergence of Ampère's law (1.8) is the time derivative of Poisson's equation $\nabla \cdot \boldsymbol{D} = \rho$. Therefore, if the initial fields satisfy Poisson's equation, time advancement of Ampère's law together with the conservation of charge will ensure that Poisson's equation holds at later times. Similarly, the divergence of Faraday's law shows that the time derivative of $\nabla \cdot \boldsymbol{B}$ vanishes, so $\nabla \cdot \boldsymbol{B} = 0$ need only be given as an initial condition. Thus, $\nabla \cdot \boldsymbol{B} = 0$ can be seen as a restriction on valid initial conditions for Faraday's law.

We conclude that the time evolution of the fields is completely specified by

$$\epsilon \frac{\partial \boldsymbol{E}}{\partial t} = \nabla \times \boldsymbol{H} - \boldsymbol{J}, \tag{1.9}$$

$$\mu \frac{\partial \boldsymbol{H}}{\partial t} = -\nabla \times \boldsymbol{E}. \tag{1.10}$$

This form is used in the FDTD method to advance $\boldsymbol{E}$ and $\boldsymbol{H}$ in time, as will be described in Chap. 5. The initial conditions for this set of equations are the electric and magnetic fields $\boldsymbol{E}$ and $\boldsymbol{H}$, and they must satisfy (1.3) and (1.4).

The system of two first-order equations can be combined to a single second-order equation for $\boldsymbol{E}$:

$$\epsilon \frac{\partial^2 \boldsymbol{E}}{\partial t^2} + \nabla \times \frac{1}{\mu} \nabla \times \boldsymbol{E} = -\frac{\partial \boldsymbol{J}}{\partial t}, \tag{1.11}$$

which is often referred to as the curl-curl equation or the vector wave equation. We will use Maxwell's equations in this form in Chap. 6 on the FEM. The initial conditions that need to be specified for (1.11) are the electric field and its time derivative. In particular, FEM is generally used to solve the frequency domain form of the curl-curl equation, sometimes referred to as the vector Helmholtz equation, where $\exp(j\omega t)$ time dependence is assumed, so that the time derivative $\partial/\partial t$ is replaced by $j\omega$, where j is the imaginary unit and ω is the angular frequency.

The full Maxwell equations (1.9)–(1.10) or (1.11) are commonly used for microwave problems, such as antennas and microwave circuits. One of the difficulties one has to face in solving these equations is that the computational domain may extend over many wavelengths in all three coordinate directions, and that consequently the required number of unknowns needed for an accurate computation may be very large. To complicate matters, one may have to deal with complex three-dimensional (3D) geometry, including details, such as wires, that are much smaller than a wavelength. Moreover, microwave problems often involve open regions, and to model this, the computational domain has to be truncated by means of absorbing boundary conditions.

1.2.4 Dispersion Relation and Wave Velocities

The propagation of electromagnetic waves is often characterized in terms of the *dispersion relation*, which relates spatial and temporal variation of a monochromatic solution by means of its wavevector $\boldsymbol{k}$ and frequency ω, respectively. Often, we deal with nondispersive situations where the frequency is directly proportional to the wavenumber k. When the frequency is not proportional to the wavenumber, we have dispersion and this occurs physically for wave propagation in some media and waveguides. However, the discretization process may also cause dispersion, which is often referred to as numerical dispersion. In general, dispersion implies that a wave packet containing several different spatial frequencies will change shape as it propagates. Naturally, it is important that the numerical dispersion is small in comparison to the physical dispersion of interest.

To provide a brief introduction to dispersion and related issues, we use (1.11) to deduce the corresponding 1D wave equation:

$$\frac{\partial^2}{\partial t^2}E(z,t) = c^2\frac{\partial^2}{\partial z^2}E(z,t), \tag{1.12}$$

where the transverse electric field is denoted $E(z,t)$. Here, the speed of light c in the medium is constant. The exact solutions of (1.12) on an infinite interval have the form

$$E(z,t) = E^+(z-ct) + E^-(z+ct), \tag{1.13}$$

where E^+ and E^- represent waves traveling in the positive and negative z-directions, respectively. This solution typically involves a range of frequencies and, next, we consider one of these, i.e. the monochromatic case.

To obtain the dispersion relation for the 1D wave equation, we substitute $E = \exp(j\omega t - jkz)$ in (1.12), and then divide both sides by $\exp(j\omega t - jkz)$, which gives $\omega^2 = c^2k^2$. Consequently, the dispersion relation for the 1D wave equation is

$$\omega = ck. \tag{1.14}$$

The angular frequency ω is a *linear* function of the wavenumber k and this implies that all frequency components of a transient wave propagate with the same velocity. The *phase velocity* v_p, defined as the velocity of a constant phase surface, satisfies $(d/dt)(\omega t - kz) = \omega - kv_p = 0$, which gives

$$v_p = \omega/k. \tag{1.15}$$

Next, we consider the superposition of the two signals $E_\mathrm{A} = \exp[j(\omega - \Delta\omega)t - j(k - \Delta k)z]$ and $E_\mathrm{B} = \exp[j(\omega + \Delta\omega)t - j(k + \Delta k)z]$. The sum wave $E_\mathrm{A} + E_\mathrm{B}$ can be written as a carrier wave $\exp(j\omega t - jkz)$ times a slowly varying envelope which is $2\cos(t\Delta\omega - z\Delta k)$. We see that the propagation speed of the envelope is $\Delta\omega/\Delta k$ and, in the limit where $\Delta\omega$ and Δk become small, this is called the *group velocity*

$$v_g = \frac{\partial\omega}{\partial k}. \tag{1.16}$$

The envelope can be identified with a wave-packet and, if an energy density is associated with the magnitude of the wave, the transportation of energy occurs with the group velocity.

For the wave equation (1.12), both the phase and group velocities are constant and equal to the speed of light $v_p = v_g = c$. This is also evident from the explicit solution (1.13). Given this analytical treatment, all waves propagate with the same speed, independent of their wavenumber k. Therefore we say that there is no dispersion. However, a numerical treatment of (1.12) will, in almost all cases, suffer from numerical dispersion and this is discussed in Chap. 3, 4, and 5.

1.2.5 Low-Frequency Approximation

A special case of (1.11) is the "low-frequency approximation," used for instance for electrical machines, generators, and transformers. The low-frequency approximation consists in setting $\epsilon_0 = 0$, that is, one neglects the displacement current in (1.11):

$$\nabla \times \frac{1}{\mu} \nabla \times \boldsymbol{E} + \sigma \frac{\partial \boldsymbol{E}}{\partial t} = -\frac{\partial \boldsymbol{J}_\text{external}}{\partial t}, \tag{1.17}$$

where the electrical current density was taken as $\boldsymbol{J} = \sigma \boldsymbol{E} + \boldsymbol{J}_{\text{external}}$, and σ is the electrical conductivity. The low-frequency approximation gets rid of the electromagnetic waves present in the full Maxwell equations (1.9)–(1.10) and makes it possible to take time steps on the much longer time scale associated with the penetration of eddy currents in conductors. However, the low-frequency approximation is mathematically more complicated, because in regions where $\sigma = 0$, the time derivative of $\boldsymbol{E}$ drops out of (1.17). As a consequence, (1.17) gives no information about $\nabla \cdot \boldsymbol{E}$ in the nonconducting regions, so that $\boldsymbol{E}$ itself is not actually known. Since the low-frequency equations are important in the area of both electric power engineering and electromagnetic compatibility, we will discuss, briefly, some methods used to solve these equations in Sect. 6.8. Some challenges that frequently occur in eddy current problems come from extremely complicated 3D geometry and thin layers of currents caused by the skin effect.

1.2.6 Integral Formulation

A simple special case is electrostatics, where there is no time-dependence. For static conditions, Faraday's law implies $\nabla \times \boldsymbol{E} = \boldsymbol{0}$, so that $\boldsymbol{E} = -\nabla \phi$, where ϕ is the electrostatic potential. Poisson's equation then becomes

$$\nabla \cdot (\epsilon \nabla \phi) = -\rho. \tag{1.18}$$

The formulations mentioned so far are all differential equations. However, sometimes integral equations are useful. In three dimensions, the "solution" to Poisson's equation in free space is

$$\phi(\boldsymbol{r}) = \int \frac{\rho(\boldsymbol{r}')dV'}{4\pi\epsilon_0|\boldsymbol{r} - \boldsymbol{r}'|}. \tag{1.19}$$

This formulation is used in the MoM to solve for the charges on conductors needed to produce specified potential distributions, as discussed in Chap. 7.

Similar reformulations in terms of surface integrals exist also for the time-dependent Maxwell system. The integral equations are called the electric field integral equation (EFIE), the magnetic field integral equation (MFIE), and the combined field integral equation (CFIE). We will derive and employ the EFIE for a scattering problem in Chap. 7, which also contain discussions on the MFIE and CFIE.

Chapter 2
Convergence

The usage of computational electromagnetics in engineering and science more or less always originates from a physical situation that features a particular problem. Here, some examples of such situations could be to determine the radiation pattern of an antenna, the transfer function of a frequency selective surface, the scattering from small particles or the influence of a cell phone on its user. The physical problem is then described as a mathematical problem that involves Maxwell's equations. In a very limited number of cases, the mathematical problem can be solved analytically such that we have an *exact* solution in closed form. If there exists a solution to the problem that can not be calculated analytically, we can approximate the mathematical problem and pursue an *approximate* solution. In the context of CEM, such an approximate solution is often referred to as a numerical solution, since it typically involves extensive numerical computations in combination with relatively simple analytical expressions. These simple analytical expressions are normally applied to small subdomains of the original problem-domain, where the subdomain solutions are related to each other such that they collectively correspond to the solution to the original problem. The difference between an approximate solution and the exact solution is referred to as the error. It is desirable that the error can be reduced to an arbitrary low level such that the approximate solution *converge* to the exact solution, i.e. the accuracy of the numerical solution improves.

Thus, one must keep in mind that numerical tools never give the exact answer. The accuracy of the numerical result depends on the so-called resolution. Resolution may mean the number of grid points per wavelength in microwave problems, or how well the geometry of an electrical motor is represented by a finite element mesh. If the method works correctly, the computed answer will converge to the exact result as the resolution increases. However, with finite resolution, the error is nonzero, and one must estimate it to ensure that its magnitude is acceptable. This is particularly true for large systems, where it may be hard to resolve details of the geometry or to afford a sufficient number of points per wavelength. Examples of this state of affairs are found in 3D-modeling of electrical motors and generators, large array antennas, and computation of the radar cross sections of aircrafts.

T. Rylander et al., *Computational Electromagnetics*, Texts in Applied Mathematics 51, DOI 10.1007/978-1-4614-5351-2_2,

Applied mathematicians have derived a posteriori error estimates, which can be evaluated after an approximate numerical solution has been computed. However, such error estimates are only beginning to be established for Maxwell's equations, and discussion of these would take us far beyond an introductory course. For further information on this topic, see, e.g., [48, 69]. Nevertheless, error estimates are useful because they can be exploited for adaptive mesh refinement in regions that give large contributions to the error. A simpler method to estimate the error of a given computation is to do a convergence test by increasing the resolution uniformly, finding out the order of convergence, and then extrapolating the computed results to infinite resolution. That is the approach we will follow.

In general, one does not know the order of convergence of a computational method for a given problem a priori. Even though standard centered finite differences or linear finite elements converge with an error of order h^2 (where h is the grid spacing or the cell size) for regular problems, singular behavior of the solution decreases the order of convergence in most application problems. Singularities are introduced by sharp edges and tips of objects such as metallic conductors, dielectrics, and magnetic materials.

2.1 Extrapolation to Zero Cell Size

We will use a very simple problem, namely to calculate the electrostatic potential on the symmetry axis of a uniformly charged square, to illustrate how computed results can be extrapolated to zero cell size. The square is the region $-a < x < a$, $-a < y < a$, $z = 0$, the surface charge density $\rho_s(x, y) = \rho_{s0}$ is constant, and we seek the potential ϕ at two points on the symmetry axis: $(0, 0, a)$ and $(0, 0, 0)$. Using the symmetry, we can write the potential from this charge distribution as

$$\phi(0,0,z) = \frac{\rho_{s0}}{4\pi\epsilon_0} \int_{x'=-a}^{a} dx' \int_{y'=-a}^{a} \frac{dy'}{(x'^2 + y'^2 + z^2)^{1/2}} = \frac{\rho_{s0}}{\pi\epsilon_0} I(z,a),$$

with

$$I(z,a) \equiv \int_{x'=0}^{a} dx' \int_{y'=0}^{a} \frac{dy'}{(x'^2 + y'^2 + z^2)^{1/2}}. \tag{2.1}$$

To do the integral $I(z, a)$ numerically, we split the square into n^2 smaller squares of side $h = a/n$, and on each square, apply a simple integration rule such as midpoint integration

$$\int_{x}^{x+h} f(x)dx \approx hf\left(x + \frac{h}{2}\right) \tag{2.2}$$

or Simpson's rule

$$\int_x^{x+h} f(x)dx \approx \frac{h}{6}\left[f(x) + 4f\left(x + \frac{h}{2}\right) + f(x+h)\right] \tag{2.3}$$

in two dimensions. The integration can be written as a MATLAB function.

```
% ---------------------------------------------------------------
% Compute potential on symmetry axis of square plate
% ---------------------------------------------------------------
function pot = integr(z, a, n, rule)

% Arguments:
%    z    = the height over the plate
%    a    = the side of the square
%    n    = the number of elements along each side of the plate
%    rule = a string 'midpoint' or 'simpson' that specifies
%           the integration rule
% Returns:
%    pot = the potential at the point (0,0,z)

x  = linspace(0, a, n+1);
y  = linspace(0, a, n+1);
h  = a/n;
zs = z^2;

if (strcmp(rule, 'midpoint'))

  % Midpoint integration
  xs(1:n) =  (x(1:n) + h/2).^2;
  ys(1:n) =  (y(1:n) + h/2).^2;
  [xxs, yys] = meshgrid(xs,ys);

  int = sum(sum(1./sqrt(xxs + yys + zs)));

elseif (strcmp(rule, 'simpson'))

  % Simpson's rule
  int = 0;
  for i = 1:n
    x1 = x(i)^2; x2 = (x(i) + h/2)^2; x3 = (x(i) + h)^2;
    y1(1:n) = y(1:n).^2;
    y2(1:n) = (y(1:n) + h/2).^2;
    y3(1:n) = (y(1:n) + h).^2;
    int = int + sum( 1./sqrt(x1+y1+zs) + 1./sqrt(x1+y3+zs) ...
                   + 1./sqrt(x3+y1+zs) + 1./sqrt(x3+y3+zs)...
                   + 4./sqrt(x2+y1+zs) + 4./sqrt(x2+y3+zs)...
                   + 4./sqrt(x1+y2+zs) + 4./sqrt(x3+y2+zs)...
                   + 16./sqrt(x2+y2+zs))/36;
  end
```

Table 2.1 Integral $I(1,1)$ from numerical integration with different cell sizes

n [-]	h [m]	$I_{\text{midp}}(1,1)$ [m]	$I_{\text{Simpson}}(1,1)$ [m]
5	0.20000	0.79432 30171	0.79335 94378
7	0.14286	0.79385 04952	0.79335 92042
10	0.10000	0.79359 97873	0.79335 91413
15	0.06667	0.79346 60584	0.79335 91252
20	0.05000	0.79341 92684	0.79335 91225

```
else
  error(['Only midpoint integration and Simpson''s rule are ' ...
      'implemented'])
end
pot = int*h^2;
```

We call this function with $z = a = 1$ [`integr(1,1,n,rule)`] and different numbers of grid points `n` for `rule = 'simpson'` and `'midpoint'`, and then extrapolate the results to zero cell size to get as accurate an answer as possible. The first step is to establish the order of convergence. Table 2.1 shows some results of calling the function for different cell sizes $h = 1/n$.

We can carry out the extrapolation using MATLAB routines, by collecting the values of h, I_{midp}, and I_{Simpson} in vectors. Plotting I_{midp} versus h to some power p, we find an almost straight line for $p = 2$, as shown in Fig. 2.1. This indicates that the midpoint rule gives quadratic convergence, i.e., $I_{\text{midp}}(h) = I_0 + I_2h^2 + \cdots$ where I_0 is the extrapolated result. The term I_2h^2 in the Taylor expansion of I_{midp} is the dominant contribution to the error when h is sufficiently small, and for such resolutions the higher-order terms in the Taylor expansion can be neglected.

We extrapolate the computed results as a polynomial fit in h^2 using the MATLAB command

```
pfit = polyfit(h.^2,I,m)
```

Here, m is the order of the polynomial, and the extrapolated value of the integral is the coefficient for h^0. [With the MATLAB convention for storing polynomials, this is the $(m+1)$th component of `pfit`]. A first-order fit ($m = 1$) gives the extrapolation $I(1,1) \approx 0.79335\,88818$, second-order ($m = 2$) gives $0.79335\,91208$, and a third-order fit gives $0.79335\,91213$.

The results from the Simpson integration fall on an almost straight line when plotted against h^4, and we conclude that the dominant error scales as h^4. A fit of $I_{\text{Simpson}}(1,1)$ to a linear polynomial in h^4 gives the extrapolation 0.79335 91207, and quadratic and cubic fits give 0.79335 91202.

The correct answer to eight digits is $0.79335\,912$. Extrapolation allows us to establish this degree of accuracy with a rather moderate effort: a second-order fit of the low-order midpoint rule versus h^2, using data computed for rather coarse grids $h \geq 0.05$. This gives eight-digit accuracy of the extrapolation even though the

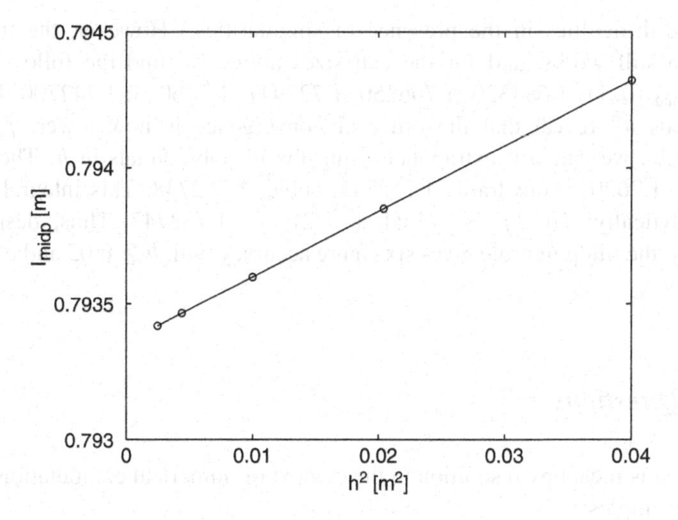

Fig. 2.1 Values of the integral $I(1,1)$ computed by the midpoint rule, plotted versus h^2

computed data has only three to four correct digits. Thus, extrapolation can bring very significant improvements of accuracy. Another advantage of extrapolation is that it makes us aware of how good the accuracy is. The example shows that good accuracy can also be obtained by using the higher-order Simpson integration, even without extrapolation, on a grid of moderate size.

A simple way to estimate the order of convergence is to carry out computations for a geometric sequence of cell sizes such that $h_i/h_{i+1} = h_{i+1}/h_{i+2}$. Assuming that the lowest-order term in the expansion of the error is sufficient, i.e. $I(h) = I_0 + I_p h^p$, and that the cell sizes form a geometric series, one can then estimate the order of convergence as

$$p = \ln\left[\frac{I(h_i) - I(h_{i+1})}{I(h_{i+1}) - I(h_{i+2})}\right] \Bigg/ \ln\left[\frac{h_i}{h_{i+1}}\right]. \tag{2.4}$$

When applied to the computed results for $h = 0.2, 0.1$ and 0.05, this formula gives $p = 2.002$ for the midpoint rule and $p = 3.985$ for Simpson, indicating that the convergence is quadratic and quartic, respectively, for the two methods.

2.1.1 A Singular Problem

It is instructive to consider a more singular problem, such as the potential on the midpoint of the plate, $z = 0$. Now, the integrand is singular, but the integral is nevertheless convergent. For this problem, Simpson integration gives a divergent result and cannot be used. (This illustrates the fact that high-order methods often experience difficulties in the presence of singularities.) However, the midpoint integration still works, and for the cell sizes above we find the following values for $I_{\text{midp}}(0, 1)$: 1.684320, 1.706250, 1.722947, 1.736083, 1.742700. Plots of I_{midp} versus h^p reveal that the order of convergence is now lower, $p = 1$. Nevertheless, we can still extrapolate using fits to polynomials in h. The results are linear, 1.762015; quadratic, 1.762745; cubic, 1.762748. This integral can be done analytically: $I(0, 1) = 2\ln(1 + \sqrt{2}) \approx 1.762747$. Thus, despite the singularity, the midpoint rule gives six-figure accuracy with $h \geq 0.05$ and quadratic extrapolation.

Review Questions

2.1-1 What is meant by resolution in the context of numerical computations? Give some examples.
2.1-2 How can the error in a computation be estimated?
2.1-3 What influences the error and the order of convergence?
2.1-4 Give a couple of examples of numerical integration rules and provide a simple comparison. Especially consider the differences for smooth and singular integrands.

2.2 Practical Procedures

The example we have just studied is very simple. Real application problems have more complex geometry than a square, but on the other hand, six-digit accuracy is very rarely needed, or even possible to achieve. Furthermore, numerical results converge in the very regular way we found here only if the grid can be refined uniformly over the whole computational region. When this is not possible, the convergence may be oscillatory, and the extrapolation to zero cell size becomes more difficult. In practice, it is often possible to extract a main power of convergence with the number of grid cells, but the remainder is too oscillatory to be convincingly fit by higher-order polynomials. A more robust and practical procedure for such cases is to use a linear fit of the computed results to h^p, where p is the estimated order of convergence. When the converged answer is not known, but the convergence is sufficiently regular, the order of convergence can be estimated from results for three different resolutions. To ascertain that the estimated order of convergence is not accidental, at least four different resolutions should be used. Once the order of convergence is established, extrapolation to zero cell size can be made by fitting a lowest-order expansion

$$I(h) = I_0 + I_p h^p \tag{2.5}$$

to the computed results.

Review Question

2.2-1 Why can extrapolation to zero cell size be difficult for nonuniformly refined grids?

Summary

- The accuracy of a numerical result depends on resolution. For example, a domain of integration can be divided into segments of size h, and a numerical evaluation of the integral I is then expressed as $I(h) = I_0 + I_p h^p + \cdots$, where I_0 is the exact result, $I_p h^p$ is the dominant error term (provided that h is sufficiently small), and p is the order of convergence.
- The order of convergence p can be estimated from

$$p = \ln\left[\frac{I(h_i) - I(h_{i+1})}{I(h_{i+1}) - I(h_{i+2})}\right] \Big/ \ln\left[\frac{h_i}{h_{i+1}}\right],$$

  which requires at least three computations and where $h_i/h_{i+1} = h_{i+1}/h_{i+2}$. The result should preferably be verified for at least four resolutions to ascertain that the estimated p is not accidental.
- A simple method to estimate the error of a given computation is to (i) do a convergence test by uniform grid refinement, (ii) find the order of convergence, and (iii) extrapolate the computed results to zero cell size.
- The order of convergence depend on the method *and* the regularity of the solution. Singular behavior of the solution decreases the order of convergence p in many real-world problems.

Problems

P.2-1 Derive the order of convergence for midpoint integration (2.2) and Simpson's rule (2.3) under the assumption that the integrand is regular. How does a singular integrand influence your derivation?

P.2-2 Show that (2.4) gives an estimate for p. Under what conditions is this estimate accurate?

Computer Projects

C.2-1 Repeat the calculations of $I(1,1)$ and $I(0,1)$, where $I(z,a)$ is defined in (2.1), using two-point Gaussian integration

$$\int_x^{x+h} f(x)dx = \frac{h}{2}\left[f\left(x+\frac{h}{2}\left(1-\frac{1}{\sqrt{3}}\right)\right)+f\left(x+\frac{h}{2}\left(1+\frac{1}{\sqrt{3}}\right)\right)\right]$$

and find the order of convergence.

C.2-2 Calculate the integral $\int_0^1 x^{-\alpha}dx$, with a singular integrand, numerically by dividing the interval into equal elements and applying midpoint integration on each. Investigate the cases $\alpha = 0.5$ and 0.8, find the order of convergence, and extrapolate to zero cell size. The exact integral is $1/(1-\alpha)$.

Chapter 3
Finite Differences

Maxwell's equations are usually formulated as differential equations (1.1)–(1.4), where we have derivatives with respect to space (represented by the curl and divergence operators) and derivatives with respect to time. The sought electromagnetic field is a function of space and time that must satisfy Maxwell's equations. A rather simple way to pursue a numerical solution is to represent the electromagnetic field by its function values at a (finite) set of discrete grid points, where we would normally use a uniform Cartesian grid with respect to space and time. Then, it is quite natural to approximate the derivatives of the electromagnetic field in Maxwell's equations by finite-difference approximations that involve the fields at neighboring points on the grid. This type of approach is referred to as a finite-difference method.

In a one-dimensional (1D) problem on the x-axis, for example, a finite-difference method consequently introduces a set of grid points $x_1, x_2, \ldots, x_N$ where a sought function $f(x)$ takes the values $f(x_1), f(x_2), \ldots, f(x_N)$. In practice, it is often convenient to use the notation $f_1, f_2, \ldots, f_N$ instead of $f(x_1), f(x_2), \ldots, f(x_N)$, respectively. Thus, the objective with the finite-difference method is to determine the unknown values $f_1, f_2, \ldots, f_N$ at the known grid points $x_1, x_2, \ldots, x_N$. If the finite-difference method works correctly, the approximate numerical solution converges to the exact solution as the number of grid points tends to infinity. (Naturally, the required computational resources increase as the number of grid points increase.) It should be noted that a finite-difference method does not normally incorporate an intrinsic representation of the field solution between the grid points.

We will first recapitulate expressions for first- and second-order differences on a uniform grid with grid points $x_{n+i} = x_n + ih$, where i is an integer and h is the distance between the grid points (often referred to as cell size). The basis for this is the Taylor expansion

$$f(x+\delta) = f(x) + \delta f'(x) + \frac{\delta^2}{2} f''(x) + \frac{\delta^3}{6} f'''(x) + \cdots \tag{3.1}$$

T. Rylander et al., *Computational Electromagnetics*, Texts in Applied Mathematics 51, DOI 10.1007/978-1-4614-5351-2_3,

To get the first derivative on a grid point x, we could use the noncentered difference $[f(x+h) - f(x)]/h = f'(x) + O(h)$, but the error here is of first order in h. One way to increase the order of approximation is to take the difference across two cells, which gives

$$\frac{f(x+h) - f(x-h)}{2h} = f'(x) + O(h^2). \tag{3.2}$$

As we shall see shortly, this becomes very inaccurate for short wavelengths, in particular, when the wavelength is less than four grid cells. A better alternative is to use "staggered grids" and compute the first-order derivative on the "half-grid" $x_{i+\frac{1}{2}} = x_i + h/2$:

$$\frac{f(x+h) - f(x)}{h} = f'\left(x + \frac{h}{2}\right) + O(h^2). \tag{3.3}$$

A difference formula for the second derivative on an equidistant grid can be developed by applying (3.3) repeatedly, which gives

$$\frac{f(x+h) - 2f(x) + f(x-h)}{h^2} = f''(x) + O(h^2). \tag{3.4}$$

We note that the $O(h^2)$ errors in (3.2)–(3.3) are achieved only if the solution is sufficiently regular (for example, if $f''(x)$, $f'''(x)$, etc are bounded).

3.1 A 2D Capacitance Problem

As an application of finite differences to an electrostatic potential problem, we will compute the capacitance of a coaxial transmission line. The two-dimensional (2D) geometry shown in Fig. 3.1 consists of an inner conductor of rectangular cross section $a \times b$, placed coaxially with an outer waveguide of rectangular cross section $c \times d$.

In the vacuum region between the inner and outer conductors, the electrostatic potential ϕ satisfies Laplace's equation

$$\nabla^2\phi = \frac{\partial^2\phi}{\partial x^2} + \frac{\partial^2\phi}{\partial y^2} = 0, \tag{3.5}$$

where the potential is constant on the conductors. We let ϕ_1 denote the value for the potential on the inner conductor, and correspondingly, the potential on the outer conductor is denoted by ϕ_2.

We assume that the geometry can be fitted on a grid of squares. (It is possible to use nonsquare, and even nonuniform, finite difference grids. However, finite

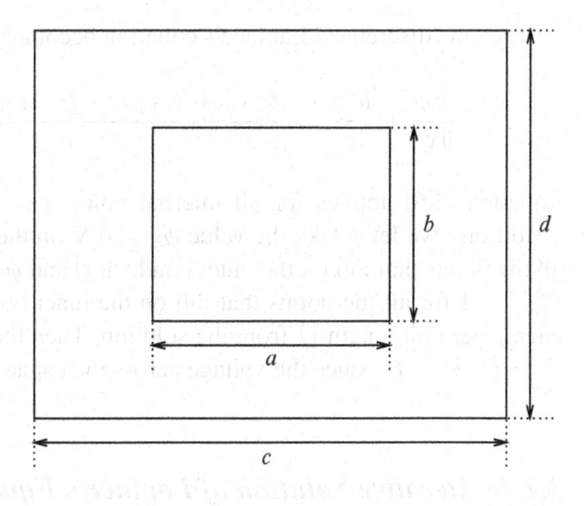

Fig. 3.1 Geometry of the coaxial transmission line

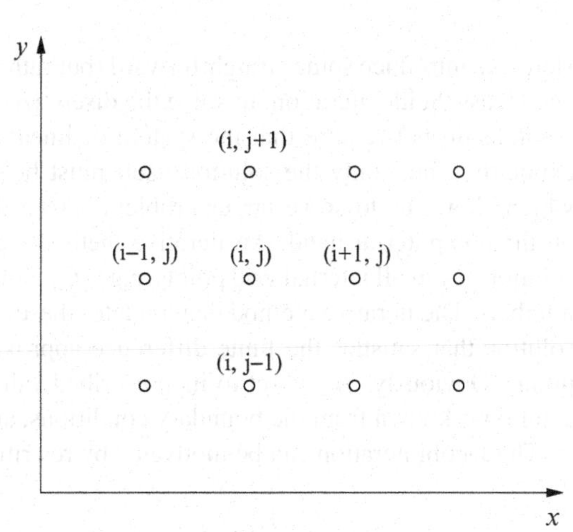

Fig. 3.2 2D finite difference grid

difference grids are often uniform and square, and we will not go beyond that. Nonuniformities are better treated by finite elements.)

We use the square grid

$$x_i = ih, \; i = \ldots, -1, 0, 1, 2, \ldots,$$
$$y_j = jh, \; j = \ldots, -1, 0, 1, 2, \ldots,$$

illustrated in Fig. 3.2, and introduce the potential at the grid points

$$f_{i,j} = \phi(ih, jh)$$

as unknowns.

Then the discretized Laplace's equation becomes

$$\frac{\partial^2 \phi}{\partial x^2} + \frac{\partial^2 \phi}{\partial y^2} \approx \frac{f_{i-1,j} + f_{i+1,j} + f_{i,j-1} + f_{i,j+1} - 4f_{i,j}}{h^2} = 0. \qquad (3.6)$$

Equation (3.6) applies for all internal points (x_i, y_j) on the grid. As boundary conditions, we let ϕ take the value $\phi_2 = 0$ V on the outer conductor ($f_{i,j} = 0$ for all the points that fall on the outer conductor) and $\phi_1 = 1$ V on the inner conductor ($f_{i,j} = 1$ for all the points that fall on the inner conductor). We will compute the charge per unit length Q from the solution. Then the capacitance per unit length is $C = Q/V = Q$, since the voltage across the capacitor is $V = 1$ V.

3.1.1 *Iterative Solution of Laplace's Equation*

Here, we introduce some straightforward (but rather old) methods, known as Jacobi and Gauss–Seidel iteration, to solve the discretized Laplace's equation (3.6). These methods do not require that the system of linear equations be formed and stored explicitly. Thus, only the solution itself must be stored in the computer memory, which allows us to solve larger problems given the amount of memory available on the computer at hand. An iterative method starts with an initial guess for the solution $f_{i,j}$ at all internal grid points, e.g., $f_{i,j} = 0$ or some other arbitrarily chosen numbers. The iterative method then updates these values until we reach a converged solution that satisfies the finite difference approximation (3.6) at all internal grid points. Obviously, $f_{i,j}$ is set to its prescribed values on the boundaries, where the solution is known from the boundary conditions, and these values are kept fixed.

The Jacobi iteration can be motivated by rewriting (3.6) as

$$f_{i,j} = \frac{1}{4}\left(f_{i-1,j} + f_{i+1,j} + f_{i,j-1} + f_{i,j+1}\right),$$

which states that at every grid point, the potential is the average of the potential at the four nearest neighbors. The Jacobi scheme uses this as the prescription for assigning new values

$$f_{i,j}^{(n+1)} = \frac{1}{4}\left(f_{i-1,j}^{(n)} + f_{i+1,j}^{(n)} + f_{i,j-1}^{(n)} + f_{i,j+1}^{(n)}\right),$$

where superscripts denote the iteration count. This scheme gives very slow convergence, but one can do better by simple modifications. One modification is the so-called Gauss–Seidel iteration, where the "old" values of f are immediately overwritten by new ones, as soon as they are computed. If f is updated in the order of increasing i and j, the Gauss–Seidel scheme is

$$f_{i,j}^{(n+1)} = \frac{1}{4}\left(f_{i-1,j}^{(n+1)} + f_{i+1,j}^{(n)} + f_{i,j-1}^{(n+1)} + f_{i,j+1}^{(n)}\right).$$

The other improvement is "overrelaxation," which means that the correction in going from iteration level n to $n+1$ is multiplied by a relaxation parameter R:

$$f_{i,j}^{(n+1)} = f_{i,j}^{(n)} + R\left(\frac{f_{i-1,j}^{(n+1)} + f_{i+1,j}^{(n)} + f_{i,j-1}^{(n+1)} + f_{i,j+1}^{(n)}}{4} - f_{i,j}^{(n)}\right); \tag{3.7}$$

$R > 1$ greatly improves the convergence, but for stability reasons R must be less than 2. For the Laplace's equation, a heuristic estimate for the optimal value of R varies with the number of grid points in one direction, N, as

$$R_{\text{opt}} = 2 - c/N,$$

where c is an N-independent number that depends on the geometry [7].

3.1.2 Computing the Capacitance

We now have all the elements needed to compute the capacitance between the two conductors. The computation can be broken down into the following parts:

1. Generate a grid such that the conducting boundaries fall on the grid points. For the particular problem here, we can exploit the symmetry and compute only on the upper right quarter, to reduce the number of unknowns. (Around a line of symmetry with a constant i, we enforce the symmetry by $f_{i+n,j} = f_{i-n,j}$, where n is a positive integer. Symmetry lines with a constant j are treated analogously)
2. Introduce the boundary conditions by setting $f = 0$ on the outer conductor and $f = V = 1$ on the inner conductor.
3. Set up an array to identify whether a grid point is inside the region where the potential is computed from Laplace's equation.
4. Iterate with the Gauss–Seidel scheme over the internal points to solve for the potential.
5. The capacitance per unit length is $C = Q/V = Q$. The charge on the inner conductor Q can be computed from Gauss's law

$$Q = \epsilon_0 \oint \boldsymbol{E} \cdot \hat{\boldsymbol{n}}\, dl = -\epsilon_0 \oint \frac{\partial \phi}{\partial n} dl, \tag{3.8}$$

   where the closed integration contour encircles the inner conductor.
6. If the change of the capacitance in the last iteration is small enough, stop iterating.
7. Once the calculation is finished, refine the grid several times and extrapolate the result to zero cell size.

3.1.3 MATLAB: Capacitance of Coaxial Cable

We will compute the capacitance for the geometry shown in Fig. 3.1 with $a = b = 1$ cm and $c = d = 2$ cm. Here, the capacitance is expressed in terms of the charge on the inner conductor. As an alternative to the Gauss–Seidel iteration, we could use MATLAB routines for solving linear systems of equations. However, we take this opportunity to introduce a simple, yet quite efficient, iterative method. More advanced iterative methods are discussed in Appendices C and D.

The following MATLAB function computes the capacitance following the outline in Sect. 3.1.2.

```
% -----------------------------------------------------------
% Compute capacitance per unit length of
% a coaxial pair of rectangles
% -----------------------------------------------------------
function cap = capacitor(a, b, c, d, n, tol, rel)

% Arguments:
%     a   =  width of inner conductor
%     b   = height of inner conductor
%     c   =  width of outer conductor
%     d   = height of outer conductor
%     n   = number of points in the x-direction (horizontal)
%     tol = relative tolerance for capacitance
%     rel = relaxation parameter
%           (a good choice is 2-c/n, where c is about pi)
% Returns:
%     cap = capacitance per unit length [pF/m]

% Make grids
h  = 0.5*c/n;                      % Grid size
na = round(0.5*a/h);               % Number of segments on 'a'
x  = linspace(0,0.5*c,n+1);        % Grid points along x-axis
m  = round(0.5*d/h);               % Number of segments on 'd'
mb = round(0.5*b/h);               % Number of segments on 'b'
y  = linspace(0,0.5*d,m+1);        % Grid points along y-axis

% Initialize potential and mask array
f = zeros(n+1,m+1);                % 2D-array with solution
mask = ones(n+1,m+1)*rel;          % 2D-array with relaxation
                                   % [mask(i,j) = 0 implies
                                   %  unchanged f(i,j)]

for i = 1:na+1
  for j = 1:mb+1
    mask(i,j) = 0;
    f(i,j)    = 1;
  end
end

% Gauss Seidel iteration
oldcap = 0;
for iter = 1:1000                  % Maximum number of iterations
```

```
  f = seidel(f,mask,n,m);       % Perform Gauss-Seidel iteration
  cap = gauss(n,m,h,f);         % Compute the capacitance
  if (abs(cap-oldcap)/cap<tol)
    break                       % Stop if change in capacitance
                                % is sufficiently small
  else
    oldcap = cap;               % Contiue until converged
  end
end
str = sprintf('Number of iterations = %4i',iter); disp(str)

% ----------------------------------------------------------------
% Make one Seidel iteration
% ----------------------------------------------------------------
function f = seidel(f, mask, n, m)

% Arguments:
%    f    = 2D-array with solution
%    mask = 2D-array with relaxation
%    n    = number of points in the x-direction (horizontal)
%    m    = number of points in the y-direction (vertical)
% Returns:
%    f    = 2D-array with solution after Gauss-Seidel iteration

% Gauss seidel iteration
for i = 2:n
  for j = 2:m
    f(i,j) = f(i,j) + mask(i,j)* ...
             (0.25*( f(i-1,j) + f(i+1,j) ...
                   + f(i,j-1) + f(i,j+1)) - f(i,j));
  end
end

% Symmetry on left boundary i-1 -> i+1
i = 1;
for j = 2:m
  f(i,j) = f(i,j) + mask(i,j)* ...
           (0.25*( f(i+1,j) + f(i+1,j) ...
                 + f(i,j-1) + f(i,j+1)) - f(i,j));
end

% Symmetry on lower boundary j-1 -> j+1
j = 1;
for i = 2:n
  f(i,j) = f(i,j) + mask(i,j)* ...
           (0.25*( f(i-1,j) + f(i+1,j) ...
                 + f(i,j+1) + f(i,j+1)) - f(i,j));
end
% ----------------------------------------------------------------
% Compute capacitance from the potential
% ----------------------------------------------------------------
function cap = gauss(n, m, h, f)

% Arguments:
```

Table 3.1 Capacitance vs. cell size for finite difference solution

n [-]	h [m]	C [pF/m]
10	0.1000	92.09715
20	0.0500	91.18849
30	0.0333	90.94575
40	0.0250	90.83912
50	0.0200	90.78080

```
%    n    = number of points in the x-direction (horizontal)
%    m    = number of points in the y-direction (vertical)
%    h    = cell size
%    f    = 2D-array with solution
% Returns:
%    cap = capacitance per unit length [pF/m]

q = 0;

for i = 1:n
  q = q + (f(i,m)+f(i+1,m))*0.5; % integrate along upper boundary
end

for j = 1:m
  q = q + (f(n,j)+f(n,j+1))*0.5; % integrate along right boundary
end

cap = q*4;              % 4 quadrants
cap = cap*8.854187;     % epsilon0*1e12 gives answer in pF/m
```

Table 3.1 shows some results of calling the function with different grid sizes and $a = b = 1$ cm, $c = d = 2$ cm, the tolerance 10^{-9}, and the relaxation paramter 1.9. When the results are plotted against h^p, they appear to fall on a straight line for $p \approx 1.5$. If we had the patience to wait for longer runs, write more efficient MATLAB code, or program the calculation in a language such as Fortran or C, the resolution could be improved, and we would find that the asymptotic order of convergence is 4/3. An important thing to learn from this example is that the convergence is slower than the normal $O(h^2)$ convergence for the difference formula (3.4). In fact, the $O(h^2)$ convergence occurs only when the solution is sufficiently regular, and the decreased order of convergence in this example is the result of the singular behavior of the solution at the corners of the inner conductor. As will be shown in Chap. 7, the potential at such a "reentrant" corner, where the angle in the solution region is 270°, varies as the distance r to the corner to the power 2/3. This implies that the electric field is singular, $E \propto r^{-1/3}$. With the computed results in Table 3.1, and assuming that the order of convergence is 1.5, a second- or higher-order polynomial fit of the data versus $h^{1.5}$ gives an extrapolated answer for the capacitance as $C = 90.6$ pF/m.

Appendix C contains some information on more efficient algorithms for the solution of linear systems. Many of these algorithms are also available in MATLAB. Thus, we could use some of these routines to solve larger problems and get better resolution. Another way to improve the convergence when the solution is singular is adaptive grid refinement. However, this is more easily done with finite elements than with finite differences.

Review Questions

3.1-1 What are the constituents of a finite-difference method?
3.1-2 Derive (3.2)–(3.4) given (3.1). When are $O(h^2)$ errors achieved?
3.1-3 Use (3.4) to deduce (3.6). What will the corresponding discrete Laplace operator look like in three dimensions?
3.1-4 How can a known potential distribution be used to compute the capacitance of a coaxial cable?
3.1-5 What is the order of convergence for the problem shown in Fig. 3.1?

3.2 Finite Difference Derivatives of Complex Exponentials

For Laplace's equation, straightforward application of finite differences works well. However, when derivatives of odd order are involved, a different technique is required to get good results. For demonstration purposes, we consider a simple case that involves the important aspects of this problem. For Ampère's law (1.1) and Faraday's law (1.2) in frequency domain, we consider a one-dimensional wave on the form $\boldsymbol{E} = \hat{x}E_x(z)$ and $\boldsymbol{H} = \hat{y}H_y(z)$, which propagates in vacuum. Given these circumstances, only the x-component is nonzero for Ampère's law and only the y-component is nonzero for Faraday's law. Consequently, we have Faraday's law and Ampère's law, respectively, on the form

$$\frac{dE_x(z)}{dz} = -j\omega\mu_0 H_y(z)$$

$$-\frac{dH_y(z)}{dz} = j\omega\epsilon_0 E_x(z)$$

Clearly, these equations involve the first-order derivative, where it operates on the electric field in Faraday's law and on the magnetic field in Ampère's law. For this problem, we wish to construct a lowest-order finite-difference scheme with a truncation error that is proportional to h^2. Thus, we have essentially two different alternatives for the representation of the first-order derivative with respect to the z-coordinate: (i) the finite-difference approximation (3.2) that extends over two cells; and (ii) the finite-difference approximation (3.3) that extends over only one cell. In addition, we must use the approximation for the first-order derivative at the midpoint of the finite-difference stencil in order to achieve an error that is proportional to h^2.

To summarize, it appears as if we have two solutions to this problem and that these solutions are equally good. However, a more detailed analysis shows that one of the two options must be abandoned since it features unphysical behavior. In order to be more specific, the two options are listed below.

- Introduce a computational grid with the grid points $z_i = ih$, where the electric field is represented by $E_x(z_i)$ and the magnetic field is represented by $H_y(z_i)$. Thus, we can use finite-difference approximations of the first-order derivative that extends over two cells, which yields the construction

$$\frac{E_x(z_{i+1}) - E_x(z_{i-1})}{2h} = -j\omega\mu_0 H_y(z_i)$$
$$-\frac{H_y(z_{i+1}) - H_y(z_{i-1})}{2h} = j\omega\epsilon_0 E_x(z_i)$$

  This approximation of Faraday's law and Ampère's law has a truncation error that is proportional to h^2. At this point, this construction appears to be satisfactory but, as shown later in this section, it actually is disastrous.
- The alternative solution is to use the finite-difference approximation (3.3) that extends over only one cell. If we introduce the grid points $z_i = ih$ for the electric field, we could write Faraday's law as

$$\frac{E_x(z_{i+1}) - E_x(z_i)}{h} = -j\omega\mu_0 H_y\left(z_{i+\frac{1}{2}}\right)$$

  where we are forced to evaluate the magnetic field at the midpoint between z_{i+1} and z_i in order to have a truncation error that is proportional to h^2. As a consequence, we use the grid points $z_{i+\frac{1}{2}} = (i + \frac{1}{2})h$ for the magnetic field $H_y\left(z_{i+\frac{1}{2}}\right)$, where i is still an integer. Thus, the unknowns for the magnetic field are shifted half a cell with respect to the unknowns for the electric field and this is referred to as staggered grids. With this construction, we discretize Ampère's law as

$$\frac{H_y\left(z_{i+\frac{1}{2}}\right) - H_y\left(z_{i-\frac{1}{2}}\right)}{h} = j\omega\epsilon_0 E_x(z_i)$$

  where the electric field is located at the midpoint z_i between the grid points $z_{i+\frac{1}{2}}$ and $z_{i-\frac{1}{2}}$ for the magnetic field.

The heart of the problem (associated with the finite-difference approximation that extends over two cells) can be described and understood by means the dispersion relation (1.14), which states that $\omega = ck$. Given the ansatz $E_x(z) = E_0 \exp(-jkz)$ and $H_y(z) = H_0 \exp(-jkz)$, we recall that the dispersion relation is a consequence of Faraday's law and Ampère's law. At this point, it's important to recall that the frequency ω represents the time variation of the electromagnetic field. As the frequency ω increases, the time variation becomes more and more rapid. According to the dispersion relation $\omega = ck$, the wavenumber k is proportional to the frequency given that the speed of light in the medium of propagation is constant with respect to frequency and, consequently, an increase in frequency implies an increase in the wavenumber. Here, the wavenumber $k = 2\pi/\lambda$ represents the spatial variation of the electromagnetic field, where λ is the wavelength. Thus, the spatial variation becomes more and more rapid as the wavenumber increases. However, and this is the crucial point, the spatial variation is not explicitly represented by the wavenumber in our finite-difference scheme, but instead it is the finite-difference approximation of the first-order derivatives that *evaluates* the spatial variation. Nevertheless, a function that varies rapidly yields a large derivative and, according to Faraday's law and Ampère's law, this must be associated with a high frequency ω. Similarly, a slow variation with respect to space yields a low frequency. We note that this is all in agreement with the dispersion relation (1.14), which is derived from Faraday's law and Ampère's law. Indeed, the dispersion relation reflects important physical properties of an electromagnetic wave problem. These aspect are assessed quantitatively in the rest of this section.

It is instructive to consider how the difference approximations (3.2)–(3.4) act on complex exponentials. Two reasons for studying complex exponentials are these:

- All functions can be decomposed as sums over complex exponentials (the Fourier transform).
- The complex exponentials $\exp(jkx)$, where j is the imaginary unit and k is the wavenumber ($k = 2\pi/\lambda$, where λ is the wavelength) are eigenfunctions of the derivative operator, $(\partial/\partial x)\exp(jkx) = jk\exp(jkx)$.

We consider a uniform 1D grid with grid points

$$x_i = ih, \quad i = \ldots, -2, -1, 0, 1, 2, \ldots,$$

and we will examine the difference approximations by evaluating them for complex exponentials, $f = \exp(jkx)$. The wavenumbers can be restricted so that $|kh| \leq \pi$. This is because, when any harmonic function is represented on a grid of points with spacing h, one can always shift kh by any integer multiple of 2π so that $kh \in [-\pi, \pi]$, without changing the value of f at any grid point.

Derivative operators can be defined as

$$D_x = f'/f, \quad D_{xx} = f''/f, \tag{3.9}$$

and for $f = \exp(jkx)$, the exact analytical results are

$$D_x = jk\ , \ D_{xx} = D_x^2 = -k^2. \tag{3.10}$$

3.2.1 First-Order Derivative

For the first derivative, the numerical difference formulas applied to $f(x) = \mathrm{e}^{jkx}$ give the results shown as functions of kh in Fig. 3.3:

- Equation (3.2), derivative across two cells, f' on the "integer grid":

$$D_x = \frac{f'(x_i)}{f(x_i)} = \frac{f(x_i + h) - f(x_i - h)}{2h\,f(x_i)} = \frac{\mathrm{e}^{jkh} - \mathrm{e}^{-jkh}}{2h} = \frac{j}{h}\sin(kh). \tag{3.11}$$

  This gives an effective numerical wavenumber

$$k_{\mathrm{num}}^{\text{two-cell}} = \frac{\sin kh}{h} = k\left(1 - \frac{k^2h^2}{6} + \cdots\right). \tag{3.12}$$

  The leading term in the expansion is correct, and the relative error is $-k^2h^2/6$, so the error increases with decreasing wavelength.

- Equation (3.3), derivative across one cell, f' on the half-grid:

$$D_x = \frac{f'(x_{i+\frac{1}{2}})}{f(x_{i+\frac{1}{2}})} = \frac{f(x_i + h) - f(x_i)}{hf(x_i + h/2)} = \frac{\mathrm{e}^{jkh/2} - \mathrm{e}^{-jkh/2}}{h} = \frac{2j}{h}\sin\left(\frac{kh}{2}\right) \tag{3.13}$$

  This gives an approximation with a smaller error

$$k_{\mathrm{num}}^{\text{staggered}} = \frac{2}{h}\sin\frac{kh}{2} = k\left(1 - \frac{k^2h^2}{24} + \cdots\right). \tag{3.14}$$

  Such an arrangement, where the first derivative is computed on the half-grid, is called *staggered grids*.

The difference formula across two cells gives very poor results when $kh > \pi/2$. In particular, for $kh = \pi$, it gives the rather strange result $f' = 0$ and $k_{\mathrm{num}}^{\text{two-cell}} = 0$. Fig. 3.4 illustrates how this comes about: when $kh = \pi$, $f(x_i)$ jumps between plus and minus the same value between neighboring points. Points at the distance of $2h$ have the same value of f, and therefore $f' = 0$ at every point on the integer grid. Thus, the most rapidly oscillating function has the derivative equal to zero everywhere on the integer grid. Notice also that the two-cell difference formula gives $\partial k_{\mathrm{num}}^{\text{two-cell}}/\partial k < 0$ for $\pi/2 < kh < \pi$. In a wave-propagation problem, this would have the consequence that the group velocity ($v_g = \partial\omega/\partial k$) changes sign, and signals propagate in the wrong direction.

The expression (3.14) for the more compact derivative on the staggered grid is clearly better at the shortest wavelength $kh = \pi$. Although the result $k_{\mathrm{num}}^{\text{staggered}} h = 2$ for $kh = \pi$ is not very accurate, it is at least nonzero and this arrangement gives no negative group velocity.

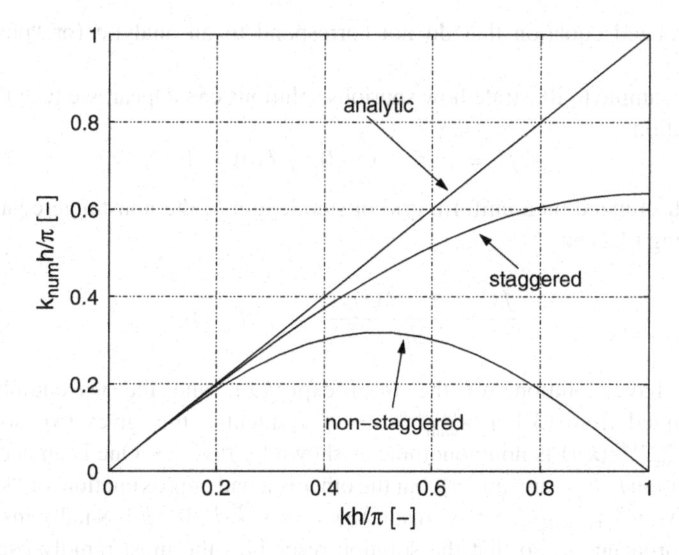

Fig. 3.3 Finite difference approximation of wavenumber from first derivative $k = -jf'/f$, with $f = \exp(jkx)$ for staggered and nonstaggered grids. Note the bad approximation of the nonstaggered form when $kh \to \pi$

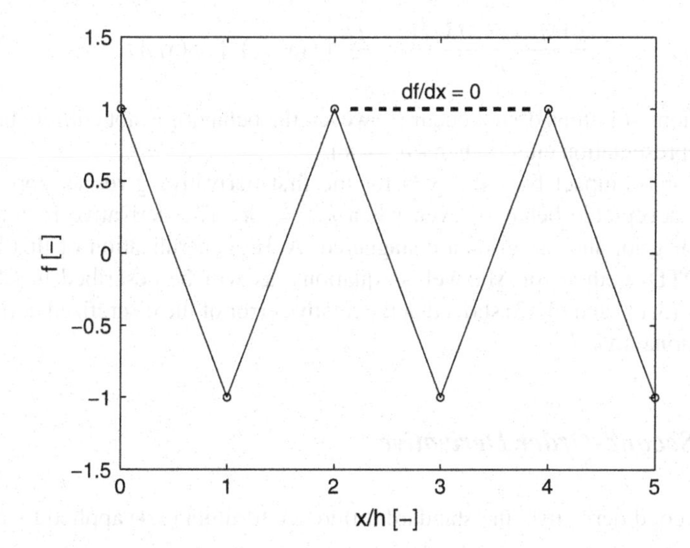

Fig. 3.4 Fastest oscillating function on a finite difference grid with $kh = \pi$ has the derivative equal to zero at all integer points on the grid

3.2.2 Spurious Solutions and Staggered Grids

The inability of the difference formula across two cells to see rapid oscillations can cause difficulties known as "spurious modes." By spurious modes we mean solutions of a discretized equation that do not correspond to an analytic (or "physical") solution.

As an example to illustrate how spurious solutions can appear, we take the first-order equation

$$f' = j\lambda f, \quad x > 0, \quad f(0) = 1$$

If this is discretized on a uniform grid of step length h, the nonstaggered approximation using (3.2) is

$$\frac{f(x_{i+1}) - f(x_{i-1})}{2h} = j\lambda f(x_i). \tag{3.15}$$

This will have solutions of the form $\exp(jkx)$, and the wavenumber can be determined from (3.12): $k_{\mathrm{num}}^{\text{two-cell}} = \lambda$. Evidently, this gives two solutions, because $k_{\mathrm{num}}^{\text{two-cell}}(kh)$ is nonmonotonic as shown by Fig. 3.3. One is an acceptable approximation $k_1 h = \arcsin(\lambda h)$, but the other is a bad approximation, or "spurious mode," having $k_{\mathrm{spurious}} h = \pi - \arcsin(\lambda h) = \pi - k_1 h$. If λh is small, this branch for kh approaches π, so that the solution resembles the most rapidly oscillating function shown in Fig. 3.4, even though the correct solution varies slowly on the scale of the grid. If we use the approximation on a staggered grid, with the stencil

$$\frac{f(x_{i+1}) - f(x_i)}{h} = \frac{j\lambda}{2}\left[f(x_{i+1}) + f(x_i)\right], \tag{3.16}$$

such spurious solutions do not occur (however, the behavior is not entirely physical for this representation either when $kh \to \pi$).

The more compact formula (3.3) for the first derivative gives an approximation with acceptable behavior even when $kh = \pi$. The derivative is computed on the half-grid, and the grids are staggered. A 3D generalization of this is used in the FDTD method for Maxwell's equations, as will be described in Chap. 5. Equations (3.11) and (3.13) show that the relative error of the discretized derivatives is proportional to k^2h^2.

3.2.3 *Second-Order Derivative*

For the second derivative, the standard difference formula (3.4) applied to $f(x) = \mathrm{e}^{jkx}$ gives

$$D_{xx} = \frac{\mathrm{e}^{jkh} - 2 + \mathrm{e}^{-jkh}}{h^2} = -\frac{4}{h^2}\sin^2\left(\frac{kh}{2}\right). \tag{3.17}$$

Therefore,

$$k_{\mathrm{num}}^2 = (k_{\mathrm{num}}^{\mathrm{staggered}})^2 = \frac{4}{h^2}\sin^2\frac{kh}{2} = k^2\left(1 - \frac{k^2h^2}{12} + \cdots\right), \tag{3.18}$$

which is illustrated in Fig. 3.5.

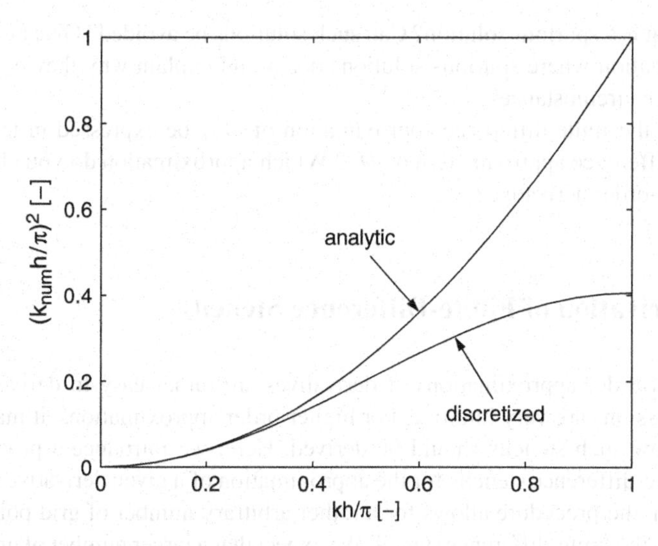

Fig. 3.5 Finite difference approximation of $k^2 = -f''/f$, with $f = \exp(jkx)$ analytically and with standard three-point difference formula

The result is only moderately accurate at the shortest wavelength ($-D_{xx} = 4/h^2$, when $kh = \pi$, to be compared with the analytic result π^2/h^2). But at least $-D_{xx}$ grows monotonically with k, so this approximation does not introduce spurious solutions. To achieve 1% accuracy in computed frequencies (which means 2% accuracy in D_{xx}), one needs $k^2h^2 < 0.24$, or 13 grid points per wavelength. If we consider the problem of calculating the fields from a mobile telephone, at 900 MHz with $\lambda = 33$ cm, in a car of length 5 m, we see that the number of cells in one direction required to get 1% phase (or frequency) error is at least $13 \times 5/0.33 \approx 200$. Evidently, a 3D computation for mobile phones in cars requires several million cells. We emphasize the absolute error will accumulate as the wave propagates. When the wave has propagated 15 wavelengths with 1% relative phase error, the absolute phase error is $15 \cdot 360/100 = 54$ degrees.

Review Questions

3.2-1 Why is it useful to study finite difference derivatives of complex exponentials?

3.2-2 Why is the wavenumber restricted by $|kh| \leq \pi$ on a grid with cell size h?

3.2-3 Derive the results in (3.12) and (3.14). Establish a value for kh when the first two terms in the expansions give 0.5% error of the numerical wavenumber. Repeat this analysis for (3.18).

3.2-4 What is a staggered grid and why is it useful?

3.2-5 What is a spurious solution? Can such solutions be avoided? Give an example of a situation where spurious solutions occur and explain why they exist under the given circumstances.

3.2-6 Can the finite difference approximation of D_{xx} be expressed in terms of a finite difference approximation of D_x? Which approximation do you choose for the first-order derivative?

3.3 Derivation of Finite-Difference Stencils

The lowest-order approximations of derivatives are rather easy to derive and the final expressions are very intuitive. For higher-order approximations, it may not be obvious how such stencils should be derived. Here, we introduce a procedure to derive finite-difference stencils for the approximation of a given derivative operator. In addition, the procedure allows for a rather arbitrary number of grid points to be involved in the finite-difference stencil. We expect that a larger number of grid points yields a leading error term that decays faster as the grid is refined, i.e. we achieve a higher order of convergence. Clearly, a larger number of grid points also implies that a larger number of floating point operations must be executed to evaluate the finite-difference approximation.

The procedure is based on the Taylor expansion (3.1), which is repeated here for convenience

$$f(x+\delta) = f(x) + \delta f'(x) + \frac{\delta^2}{2} f''(x) + \frac{\delta^3}{6} f'''(x) + \cdots$$

In order to demonstrate the procedure, we consider some examples where we derive new and useful finite-difference stencils.

3.3.1 *Higher-Order Approximation of First-Order Derivative*

First, we intend to derive a finite-difference stencil that resembles the one-cell finite-difference approximation, where the main difference is that we include one additional grid point on each side of the lowest-order stencil. Consequently, we consider an approximation to the first-order derivative that involves the four grid points $-3h/2$, $-h/2$, $h/2$ and $3h/2$.

Thus, we have

$$f_{i-\frac{3}{2}} = f\left(x - \frac{3h}{2}\right) = f(x) - \frac{3h}{2} f'(x) + \frac{9h^2}{8} f''(x) - \frac{9h^3}{16} f'''(x) + \cdots$$

$$f_{i-\frac{1}{2}} = f\left(x - \frac{h}{2}\right) = f(x) - \frac{h}{2} f'(x) + \frac{h^2}{8} f''(x) - \frac{h^3}{48} f'''(x) + \cdots$$

$$f_{i+\frac{1}{2}} = f\left(x+\frac{h}{2}\right) = f(x) + \frac{h}{2}f'(x) + \frac{h^2}{8}f''(x) + \frac{h^3}{48}f'''(x) + \cdots$$

$$f_{i+\frac{3}{2}} = f\left(x+\frac{3h}{2}\right) = f(x) + \frac{3h}{2}f'(x) + \frac{9h^2}{8}f''(x) + \frac{9h^3}{16}f'''(x) + \cdots$$

We now introduce the unknown constants $a_{i-\frac{3}{2}}$, $a_{i-\frac{1}{2}}$, $a_{i+\frac{1}{2}}$ and $a_{i+\frac{3}{2}}$, which are used as coefficients in the finite-difference approximation. Thus, we have the following finite-difference approximation of the first order derivative

$$\begin{aligned}
&a_{i-\frac{3}{2}}f_{i-\frac{3}{2}} + a_{i-\frac{1}{2}}f_{i-\frac{1}{2}} + a_{i+\frac{1}{2}}f_{i+\frac{1}{2}} + a_{i+\frac{3}{2}}f_{i+\frac{3}{2}} = \\
&\quad = \left(a_{i-\frac{3}{2}} + a_{i-\frac{1}{2}} + a_{i+\frac{1}{2}} + a_{i+\frac{3}{2}}\right)f(x) \\
&\quad + \left(-\frac{3h}{2}a_{i-\frac{3}{2}} - \frac{h}{2}a_{i-\frac{1}{2}} + \frac{h}{2}a_{i+\frac{1}{2}} + \frac{3h}{2}a_{i+\frac{3}{2}}\right)f'(x) \\
&\quad + \left(\frac{9h^2}{8}a_{i-\frac{3}{2}} + \frac{h^2}{8}a_{i-\frac{1}{2}} + \frac{h^2}{8}a_{i+\frac{1}{2}} + \frac{9h^2}{8}a_{i+\frac{3}{2}}\right)f''(x) \\
&\quad + \left(\frac{-9h^3}{16}a_{i-\frac{3}{2}} - \frac{h^3}{48}a_{i-\frac{1}{2}} + \frac{h^3}{48}a_{i+\frac{1}{2}} + \frac{9h^3}{16}a_{i+\frac{3}{2}}\right)f'''(x) + \cdots
\end{aligned}$$

In order to determine the coefficients, we impose the following requirements

$$0 = a_{i-\frac{3}{2}} + a_{i-\frac{1}{2}} + a_{i+\frac{1}{2}} + a_{i+\frac{3}{2}} \tag{3.19}$$

$$1 = -\frac{3h}{2}a_{i-\frac{3}{2}} - \frac{h}{2}a_{i-\frac{1}{2}} + \frac{h}{2}a_{i+\frac{1}{2}} + \frac{3h}{2}a_{i+\frac{3}{2}} \tag{3.20}$$

$$0 = \frac{9h^2}{8}a_{i-\frac{3}{2}} + \frac{h^2}{8}a_{i-\frac{1}{2}} + \frac{h^2}{8}a_{i+\frac{1}{2}} + \frac{9h^2}{8}a_{i+\frac{3}{2}} \tag{3.21}$$

$$0 = \frac{-9h^3}{16}a_{i-\frac{3}{2}} - \frac{h^3}{48}a_{i-\frac{1}{2}} + \frac{h^3}{48}a_{i+\frac{1}{2}} + \frac{9h^3}{16}a_{i+\frac{3}{2}} \tag{3.22}$$

that essentially state that the coefficients $a_{i-\frac{3}{2}}, a_{i-\frac{1}{2}}, a_{i+\frac{1}{2}}$ and $a_{i+\frac{3}{2}}$ are chosen such that $a_{i-\frac{3}{2}}f_{i-\frac{3}{2}} + a_{i-\frac{1}{2}}f_{i-\frac{1}{2}} + a_{i+\frac{1}{2}}f_{i+\frac{1}{2}} + a_{i+\frac{3}{2}}f_{i+\frac{3}{2}} \approx f'(x)$. The solution to the system of linear equations (3.19)-(3.22) is

$$a_{i-\frac{3}{2}} = \frac{1}{24h} \qquad a_{i-\frac{1}{2}} = -\frac{9}{8h} \qquad a_{i+\frac{1}{2}} = \frac{9}{8h} \qquad a_{i+\frac{3}{2}} = -\frac{1}{24h}$$

and we have the approximation

$$f'(x_i) \approx \frac{f_{i-\frac{3}{2}} - 27f_{i-\frac{1}{2}} + 27f_{i+\frac{1}{2}} - f_{i+\frac{3}{2}}}{24h}$$

Expressed in terms of the notation used previously, this stencil is given by

$$f'_{i+\frac{1}{2}} \approx \frac{f_{i-1} - 27f_i + 27f_{i+1} - f_{i+2}}{24h}$$

This finite-difference approximation differentiates fourth-order polynomials exactly.

3.3.1.1 Complex Exponentials

Given this higher-order approximation of the first-order derivative, we perform the analysis with complex exponentials, which yields

$$\begin{aligned} D_x &= \frac{\mathrm{e}^{-j3kh/2} - 27\mathrm{e}^{-jkh/2} + 27\mathrm{e}^{jkh/2} - \mathrm{e}^{j3kh/2}}{24h} \\ &= \frac{j}{3h}\left(7 - \cos^2\left(\frac{kh}{2}\right)\right)\sin\left(\frac{kh}{2}\right). \end{aligned}$$

A Taylor expansion around $kh = 0$ gives the numerical wavenumber

$$k_{\text{num}}^{\text{4th order}} = k\left(1 - \frac{3k^4h^4}{640} + \cdots\right)$$

which has a leading error term that is proportional to h^4. Thus, a reduction of the cell size h by a factor of two reduces the error in the wavenumber by a factor 16, which should be compared with the corresponding lowest-order approximation that reduces the error by only a factor 4.

3.3.1.2 Difficulties for Higher-Order Approximations

It is non-trivial to impose boundary conditions for finite-difference schemes that exploit higher-order stencils. For a problem with a boundary that is aligned with the Cartesian coordinate axes and coincides with the grid points used in the finite-difference discretization, the lowest-order approximation stencil can be applied to every interior point of the domain without difficulties. For the lowest-order approximation stencils that are located closest to the boundary, the outermost points of the stencil do not normally extend beyond the boundary but, instead, they would be located on the boundary which makes it possible to incorporate information from the boundary conditions in a natural manner. For a higher-order finite-difference stencil, however, such a situation would imply that some points of the stencil are located outside of the computational domain. A common solution to this problem is to use the lowest-order stencil in the very vicinity of the boundary in combination with the higher-order stencil applied to internal grid-points that are at a sufficiently

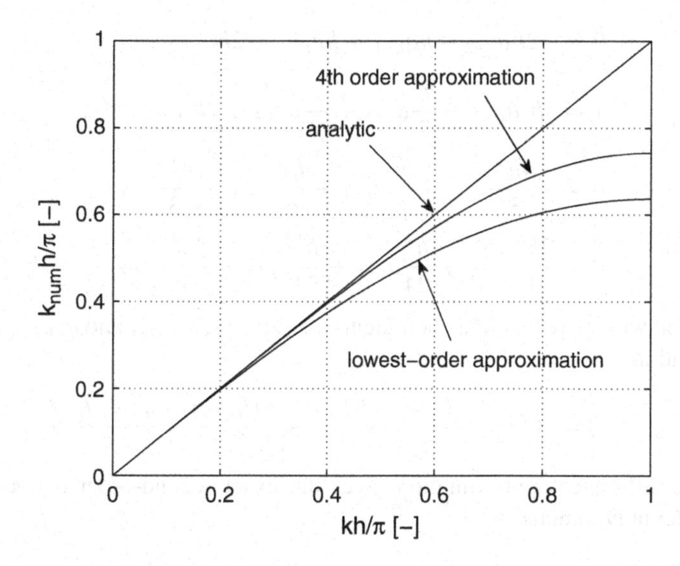

Fig. 3.6 Finite difference approximation of wavenumber from first derivative $k = -jf'/f$, with $f = \exp(jkx)$, for the fourth order finite-difference stencil on staggered grids, as compared to the lowest-order approximation and the analytical result. Note that the fourth order finite-difference approximation yields a more accurate dispersion curve for the entire interval $0 \leq kh \leq \pi$, as compared to the lowest-order approximation

large distance from the boundary. However, this yields a numerical scheme that mixes different types of approximations and, as a consequence, the error analysis becomes more difficult. Furthermore, the order of convergence is reduced since it is determined by the poorest approximation present the numerical scheme.

A considerably better solution is to use higher-order finite element methods. Such methods can exploit the higher-order approximation also in the very vicinity of the boundary and, in addition, the boundary can be curved and does not have to coincide with the Cartesian coordinate axes.

3.3.2 Higher-Order Approximation of Second-Order Derivative

We exploit the same technique to derive the corresponding result for finite-difference approximation of the second-order derivative. The lowest-order stencil involves three grid points and, again, we include one grid point on each side of the lowest-order stencil. Thus, we use the five grid points $-2h$, $-h$, 0, h and $2h$. This yields five equations that can be used to compute the five unknown coefficients a_{i-2}, a_{i-1}, a_i, a_{i+1} and a_{i+2}. Following the same procedure as above, we get the five equations

$$0 = a_{i-2} + a_{i-1} + a_i + a_{i+1} + a_{i+2}$$

$$
\begin{aligned}
0 &= -2ha_{i-2} - ha_{i-1} + ha_{i+1} + 2ha_{i+2} \\
1 &= 2h^2a_{i-2} + \frac{h^2}{2}a_{i-1} + \frac{h^2}{2}a_{i+1} + 2h^2a_{i+2} \\
0 &= -\frac{4h^3}{3}a_{i-2} - \frac{h^3}{6}a_{i-1} + \frac{h^3}{6}a_{i+1} + \frac{4h^3}{3}a_{i+2} \\
0 &= \frac{2h^4}{3}a_{i-2} + \frac{h^4}{24}a_{i-1} + \frac{h^4}{24}a_{i+1} + \frac{2h^4}{3}a_{i+2}
\end{aligned}
$$

The solution with respect to the coefficients $a_{i-2}, a_{i-1}, a_i, a_{i+1}$ and a_{i+2} yields the approximation

$$
f''(x_i) = f_i'' = \frac{-f_{i-2} + 16f_{i-1} - 30f_i + 16f_{i+1} - f_{i+2}}{12h^2}
$$

This finite-difference approximation gives the exact second-order derivative for fourth-order polynomials.

3.3.2.1 Complex Exponentials

The analysis with complex exponentials for the higher-order approximation of the second-order derivative yields

$$
\begin{aligned}
D_{xx} &= \frac{-\mathrm{e}^{-j2kh} + 16\mathrm{e}^{-jkh} - 30 + 16\mathrm{e}^{jkh} - \mathrm{e}^{j2kh}}{12h^2} \\
&= -\frac{4}{3h^2}\left(4 - \cos^2\left(\frac{kh}{2}\right)\right)\sin^2\left(\frac{kh}{2}\right)
\end{aligned}
$$

A Taylor expansion around $kh = 0$ gives the numerical wavenumber

$$
\left(k_{\text{num}}^{\text{4th order}}\right)^2 = k^2\left(1 - \frac{k^4h^4}{90} + \cdots\right)
$$

which has a leading error term that is proportional to h^4. Thus, a reduction of the cell size h by a factor of two reduces the error in the wavenumber by a factor 16, which should be compared with the corresponding lowest-order approximation that reduces the error by only a factor 4.

Summary

- Derivatives can be approximated by differences between neighboring points on a grid. A so-called uniform grid uses a constant grid point spacing h; i.e., the grid points are given by $x_{n+i} = x_n + ih$, where i is an integer.
- The first-order derivative of a function f on a staggered grid (evaluated at the midpoint $(x_{i+1} + x_i)/2$) is

$$f'_{i+\frac{1}{2}} \approx \frac{f_{i+1} - f_i}{h},$$

and that across two cells (evaluated at the center grid point x_i) is

$$f'_i \approx \frac{f_{i+1} - f_{i-1}}{2h}.$$

The second-order derivative (evaluated at the center grid point x_i) is

$$f''_i \approx \frac{f_{i+1} - 2f_i + f_{i-1}}{h^2}.$$

- The discretized Laplacian operator is

$$\nabla^2 f = \frac{\partial^2 f}{\partial x^2} + \frac{\partial^2 f}{\partial y^2} \approx \frac{f_{i-1,j} + f_{i+1,j} + f_{i,j-1} + f_{i,j+1} - 4f_{i,j}}{h^2}.$$

Two iterative procedures for solving Laplace's equation are Jacobi and Gauss–Seidel iteration. These can be accelerated with so-called overrelaxation.
- Numerical derivatives acting on complex exponentials $f(x) = \exp(jkx)$ are useful when analyzing finite difference schemes. The first-order derivative on a staggered grid gives

$$\frac{f'_{i+\frac{1}{2}}}{f_{i+\frac{1}{2}}} = \frac{1}{f_{i+\frac{1}{2}}} \frac{f_{i+1} - f_i}{h} = \frac{2j}{h} \sin\left(\frac{kh}{2}\right).$$

First-order derivatives across two cells with no staggering should be avoided, since

$$\frac{f'_i}{f_i} = \frac{1}{f_i} \frac{f_{i+1} - f_{i-1}}{2h} = \frac{j}{h} \sin(kh),$$

which is nonmonotonic and gives a zero derivative for solutions that vary on the scale of the grid, i.e., $kh \to \pi$.

The second-order derivative gives

$$\frac{f_i''}{f_i} = \frac{1}{f_i}\frac{f_{i+1} - 2f_i + f_{i-1}}{h^2} = -\frac{4}{h^2}\sin^2\left(\frac{kh}{2}\right).$$

- Higher-order finite-difference stencils can be derived by means of the Taylor expansion and two important fourth-order approximations are

$$f'_{i+\frac{1}{2}} \approx \frac{-f_{i+2} + 27f_{i+1} - 27f_i + f_{i-1}}{24h}$$

$$f_i'' \approx \frac{-f_{i+2} + 16f_{i+1} - 30f_i + 16f_{i-1} - f_{i-2}}{12h^2}$$

Problems

P.3-1 Use the technique in Sect. 3.1.1 to solve the Laplace's equation at the midpoint of a square 3×3 grid where the potential is known on the boundary. How does the solution depend on the cell size?

P.3-2 Show that if the grid is nonuniform, the finite difference approximation of the second-order derivative is

$$f''(x_i) \approx \frac{2}{x_{i+1} - x_{i-1}}\left(\frac{f_{i+1} - f_i}{x_{i+1} - x_i} - \frac{f_i - f_{i-1}}{x_i - x_{i-1}}\right).$$

Derive the leading error term for this finite difference approximation. A nonuniform grid implies that $x_{i+1} - x_i$ does not have to be equal to $x_i - x_{i-1}$. Discuss when nonuniform grids can be useful for computations.

P.3-3 Derive a finite difference expression for $f'(0)$ in terms of $f(0)$, $f(h)$, and $f(2h)$ that has an $O(h^2)$ error.

P.3-4 For a problem with the grid points $x_i = ih$, where $i = 0, 1, 2, \ldots$, derive a finite difference approximation of the Neumann boundary condition $f'(0) = 0$ by the use of a "ghost" grid point $x_{-1} = -h$ (outside the computational domain) such that the error is $O(h^2)$.

P.3-5 The capacitance can also be computed from $C = 2W/V^2$, where W is the electrostatic energy and V the potential difference between the two conductors of the capacitor. Write down an expression for W in terms of the electrostatic potential distribution and suggest a method for computing W given the finite difference solution to an electrostatic problem.

P.3-6 Discuss how the derivative operators in (3.9) and (3.10) can be related to, and useful in the context of, the one-way wave equation $\partial f/\partial x \pm (j\omega/c)f = 0$, where c is the speed of the wave.

P.3-7 Show that the Helmholtz equation, $\partial^2 f/\partial x^2 + (\omega/c)^2 f = 0$, can be factorized into

$$\left(\frac{\partial}{\partial x} + \frac{j\omega}{c}\right)\left(\frac{\partial}{\partial x} - \frac{j\omega}{c}\right) f = 0$$

and interpret the two factors of the Helmholtz operator. Discretize the above factorized operator by finite differences (on staggered grids) and multiply the two factors to derive the corresponding Helmholtz operator.

P.3-8 Discuss how the derivative operators in (3.9) and (3.10) can be related to, and useful in the context of, the wave equation $\partial^2 f/\partial x^2 - c^{-2}\partial^2 f/\partial t^2 = 0$, where c is the speed of the wave $f = f(x,t)$. Here, $f(x+ct)$ and $f(x-ct)$ solve the wave equation, and the lines where $x + ct$ and $x - ct$ are constant are referred to as characteristics.

P.3-9 Demonstrate that the Helmholtz equation is equivalent to the two coupled equations $\partial f/\partial x + (j\omega/c)g = 0$ and $\partial g/\partial x + (j\omega/c)f = 0$. What is the meaning of the new function g? How should the first-order system of coupled equations be discretized by finite differences?

P.3-10 Show that the analysis with complex exponentials applied to (3.16) gives $\lambda = (2/h)\tan(kh/2)$, so that $\lambda \to \infty$ as $kh \to \pi$.

P.3-11 The 1D Helmholtz equation for a transversal wave E_z in a homogeneous medium with losses reads

$$\left(-\frac{\partial^2}{\partial x^2} + j\omega\mu\sigma - \omega^2\mu\epsilon\right) E_z = 0.$$

Use the finite difference approximation to discretize this equation. Calculate and compare the dispersion relation of the continuous and the discretized problems. Does the discretized problem reproduce the physics for well-resolved solutions? What happens for poorly resolved solutions? How does the angular frequency ω and the material parameters μ, ϵ, and σ influence the accuracy of the dispersion relation of the discretized equation?

Computer Projects

C.3-1 Write down the system of linear equations that results from the discretization of the capacitance problem shown in Fig. 3.1. Let $c = d = 3a = 3b$ and use a square grid with one grid point between the inner and outer conductors. Let the potential be ϕ_1 on the inner conductor and ϕ_2 on the outer conductor. How are these boundary conditions incorporated into the system of linear equations? Is it possible to use symmetries in the solution of this problem?

Generalize the result so that it is possible to specify the number of points between the inner and outer conductors. Write a computer program that generates the system of linear equations $\mathbf{Af} = \mathbf{b}$ in terms of a matrix $\mathbf{A}$ and a vector $\mathbf{b}$, where the solution vector $\mathbf{f}$ stores the potential values at grid points between the inner and outer conductors.

C.3-2 Write a computer program that uses Jacobi and Gauss–Seidel iteration to solve for the electrostatic potential on a square domain of side a. Use the boundary conditions $\phi(x,0) = \phi(0,y) = 0$, $\phi(x,a) = \phi_0 \cdot (x/a)$ and $\phi(a,y) = \phi_0 \cdot (y/a)$, where ϕ_0 is a constant. Study and compare the convergence of the iterative methods in Sect. 3.1.1. Implement the overrelaxation method and investigate how the value of R influences the convergence. The analytical solution to this problem is $\phi(x,y) = \phi_0 \cdot (xy/a^2)$.

C.3-3 Use the finite difference scheme to compute the capacitance for a coaxial cable of two concentric circular cylinders with inner radius a and outer radius b. For this case, the capacitance per unit length can be calculated analytically, and it is $2\pi\epsilon/\ln(b/a)$. The circular boundaries do not fall on grid points in a natural way, and one way to proceed is to approximate these boundaries in some sense given the structured Cartesian grid. This type of approximation is often referred to as the staircase approximation. How does the error depend on the cell size h? Can you extrapolate the results to zero cell size?

C.3-4 Try to reformulate the previous problem using polar coordinates (it can be reduced to a 1D problem) to avoid the staircase approximation and use the finite difference scheme to solve for the capacitance. Determine the order of convergence. Is it possible to extrapolate the capacitance to zero cell size?

Chapter 4
Eigenvalue Problems

Maxwell's equations can be solved either in the time domain, by evolving an initial condition in time, or in the frequency domain, assuming harmonic $\exp(j\omega t)$ time dependence. In both cases, the application can be either

- a driven system, where one seeks the response to a source, for instance an antenna, or
- an eigenvalue calculation, where one seeks the natural oscillation frequencies of the system.

The field solution to the eigenvalue calculation is a nonzero field that satisfies the homogeneous problem, i.e. the field problem without sources. This type of situation occurs typically for an electromagnetic system with a source that is nonzero for a finite time. Once the source vanishes, the field solution can be expressed as a superposition of eigenmodes, where each eigenmode oscillates at a particular eigenfrequency. For an electromagnetic system that features losses of some type, the eigenfrequencies are in general complex and, as a consequence, the corresponding eigenmodes are damped such that their instantaneous amplitudes decrease with respect to time. In this chapter, we consider eigenvalue problems in both the frequency domain and the time domain. In addition, we analyze a popular explicit finite-difference time-stepping scheme by means of decomposing the time-domain field solution into its constituent eigenmodes. This analysis yields a stability condition for the time step, which is referred to as the Courant condition. The analysis itself is referred to as von Neumann stability analysis and it can be applied to a wide range of time-stepping schemes.

4.1 Maxwell's Equations

In a linear, dispersion-free medium (i.e., ϵ and μ depend only on the coordinate vector), Maxwell's equations can be written as the single second-order curl-curl equation (1.11) for the electric field

T. Rylander et al., *Computational Electromagnetics*, Texts in Applied Mathematics 51, DOI 10.1007/978-1-4614-5351-2_4,

$$-\epsilon \frac{\partial^2 \boldsymbol{E}}{\partial t^2} = \nabla \times \frac{1}{\mu} \nabla \times \boldsymbol{E} + \frac{\partial \boldsymbol{J}}{\partial t}. \tag{4.1}$$

In the absence of sources, $\boldsymbol{J} = 0$, and with harmonic time dependence $\exp(j\omega t)$, the curl-curl equation gives the following eigenvalue problem:

$$\omega_m^2 \epsilon \boldsymbol{E}_m = \nabla \times \frac{1}{\mu} \nabla \times \boldsymbol{E}_m. \tag{4.2}$$

For nontrivial solutions ($\boldsymbol{E}_m \neq \boldsymbol{0}$), ω_m^2 plays the role of an eigenvalue, and $\boldsymbol{E}_m$ is the corresponding eigenfunction, or eigenmode. (Sometimes the subindex m is omitted in order to simplify the notation.) If the region Ω, where (4.2) applies, is a closed cavity with a perfectly conducting boundary $\partial\Omega$ (i.e., $\hat{\boldsymbol{n}} \times \boldsymbol{E} = 0$), the operator on the right-hand side, $L \equiv \nabla \times \mu^{-1} \nabla \times$, is self-adjoint, that is,

$$\int_\Omega \boldsymbol{E}_1 \cdot L[\boldsymbol{E}_2] dV = \int_\Omega \boldsymbol{E}_2 \cdot L[\boldsymbol{E}_1] dV \tag{4.3}$$

for all vector fields $\boldsymbol{E}_1$ and $\boldsymbol{E}_2$ that satisfy the boundary conditions. This can be shown using the vector identity

$$\nabla \cdot \left[\boldsymbol{A} \times \frac{1}{\mu} (\nabla \times \boldsymbol{B}) \right] = \frac{1}{\mu} (\nabla \times \boldsymbol{A}) \cdot (\nabla \times \boldsymbol{B}) - \boldsymbol{A} \cdot \nabla \times \frac{1}{\mu} \nabla \times \boldsymbol{B}. \tag{4.4}$$

For all electric fields $\boldsymbol{E}_1$ and $\boldsymbol{E}_2$ satisfying the boundary condition, $\hat{\boldsymbol{n}} \times \boldsymbol{E} = 0$, (4.4) gives

$$\begin{aligned} \int_\Omega \boldsymbol{E}_1 \cdot \nabla \times \frac{1}{\mu} \nabla \times \boldsymbol{E}_2 dV &= \int_\Omega \frac{1}{\mu} \nabla \times \boldsymbol{E}_1 \cdot \nabla \times \boldsymbol{E}_2 dV \\ &= \int_\Omega \boldsymbol{E}_2 \cdot \nabla \times \frac{1}{\mu} \nabla \times \boldsymbol{E}_1 dV, \end{aligned} \tag{4.5}$$

where we have applied integration by parts twice. Integrating (4.2), multiplied by the complex conjugate of $\boldsymbol{E}$ over Ω, and integrating by parts once, we obtain $\omega^2 \int_\Omega \epsilon |\boldsymbol{E}|^2 dV = \int_\Omega \mu^{-1} |\nabla \times \boldsymbol{E}|^2 dV$. This gives the following expression for the eigenvalue:

$$\omega^2 = \frac{\int_\Omega \mu^{-1} |\nabla \times \boldsymbol{E}|^2 dV}{\int_\Omega \epsilon |\boldsymbol{E}|^2 dV}, \tag{4.6}$$

which is manifestly real and nonnegative. Thus, the eigenfrequencies ω are real for any lossless region bounded by perfect conductors. Damping can appear if there is dissipation of energy, for example from regions with finite electrical conductivity, or if the region is not completely enclosed by a perfect conductor.

Review Questions

4.1-1 What is an eigenvalue problem? What does the solution consist of and physically correspond to? To what extent is the solution uniquely defined?
4.1-2 What is required for an operator to be self-adjoint?
4.1-3 Show that (4.5) is valid.
4.1-4 Show that the eigenfrequencies ω are real for any lossless region bounded by perfect conductors. What are the physical implications of this result?

4.2 Model Problems

In the previous section we showed that Maxwell's equations are self-adjoint in the absence of losses, and that this leads to real eigenfrequencies. Self-adjoint equations occur in many branches of science and technology. One example is the Schrödinger equation, where real eigenvalues describe well-defined energy levels of states with infinite lifetime. Another example is provided by the equations of linear elasticity, which have many properties in common with Maxwell's equations. This similarity comes from the fact that both can be written as a vector equation with second-order derivatives in time and space. The only difference is that the curl-curl operator of the Maxwell equations is replaced by another second-order vector operator, involving the modulus of elasticity for bulk compression and shearing. Because of the many similarities between the two fields, it has been possible to carry over techniques originally developed in computational mechanics (see, e.g., [38]) to CEM.

The self-adjoint curl-curl equation (4.2) leads us to consider eigenvalue problems of the type

$$L[f] = -\omega^2 f \quad \text{in } \Omega \tag{4.7}$$

together with a suitable boundary condition on $\partial\Omega$. We will assume that L is a linear self-adjoint operator with nonpositive eigenvalues. As a simple example to illustrate general principles, we will study the 1D Helmholtz equation:

$$\frac{d^2 f}{dx^2} = -k^2 f, \quad 0 < x < a, \quad f(0) = f(a) = 0. \tag{4.8}$$

This equation models many 1D wave phenomena, not only in electromagnetics. We will use it to introduce both frequency- and time-domain techniques that will be used later to determine eigenfrequencies of more complex electromagnetic systems in two and three dimensions.

The eigenvalue problem (4.8) is easy to solve analytically. The solutions of the differential equation are of the form $f = A\cos kx + B\sin kx$. The boundary condition $f(0) = 0$ gives $A = 0$, and then $f(a) = 0$ gives $\sin ka = 0$. Therefore, the wavenumber k can take the following values:

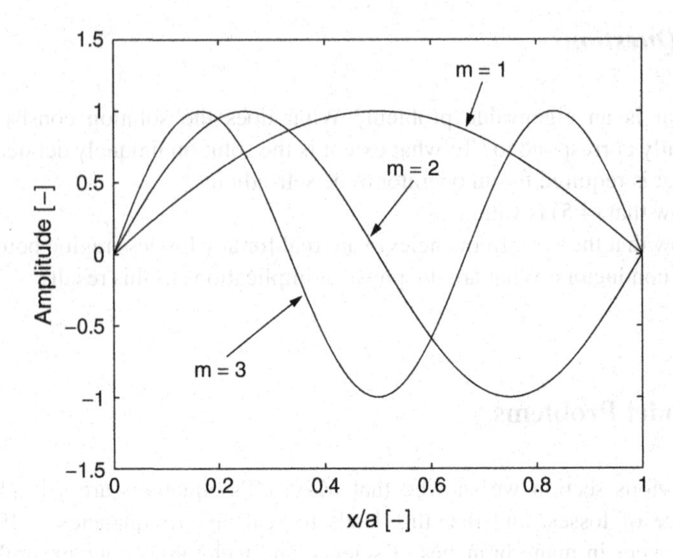

Fig. 4.1 The three lowest eigenmodes of the 1D Helmholtz equation with $f = 0$ on the boundaries

$$k_m = \frac{m\pi}{a}, \quad m \text{ an integer,}$$

so the eigenvalues $-k_m^2 = -m^2\pi^2/a^2$ are all real and negative. The three lowest eigenfunctions, or eigenmodes, are shown in Fig. 4.1.

Review Question

4.2-1 Calculate analytical eigenvalues and eigenfunctions to the eigenvalue problem $d^2f/dx^2 = -k^2 f$ with $f(0) = f(a) = 0$.

4.3 Frequency-Domain Eigenvalue Calculation

Frequency-domain eigenvalue problems of the form $L[f] = \lambda f$ are generally transformed into corresponding algebraic eigenvalue problems of the form $\mathbf{Af} = \lambda\mathbf{f}$ by, for example, a finite difference approximation. Therefore, the numerical solution of a frequency-domain eigenvalue problem involves the solution of an algebraic eigenvalue problem.

4.3.1 MATLAB: The 1D Helmholtz Equation

To discretize the 1D Helmholtz equation (4.8) by finite differences, we divide the interval $[0, a]$ into N subintervals of equal length $h = a/N$. The simplest finite difference approximation of (4.8) is

$$\frac{f_{i-1} - 2f_i + f_{i+1}}{h^2} = -k^2 f_i, \quad i = 1, 2, \ldots, N-1. \tag{4.9}$$

The boundary conditions are $f_0 = f_N = 0$, so there is no reason to include f_0 and f_N as unknowns. Equation (4.9) can be written as a linear system with an $(N-1) \times (N-1)$ matrix $\mathbf{A}$:

$$\mathbf{A}\mathbf{f} = \lambda \mathbf{f}.$$

Note that the matrix $\mathbf{A}$ is tridiagonal, with nonzero elements on the main diagonal and one lower and one upper subdiagonal; for six interior points, it is

$$\mathbf{A} = \frac{1}{h^2}\begin{pmatrix} -2 & 1 & 0 & 0 & 0 & 0 \\ 1 & -2 & 1 & 0 & 0 & 0 \\ 0 & 1 & -2 & 1 & 0 & 0 \\ 0 & 0 & 1 & -2 & 1 & 0 \\ 0 & 0 & 0 & 1 & -2 & 1 \\ 0 & 0 & 0 & 0 & 1 & -2 \end{pmatrix}.$$

When n is large, $\mathbf{A}$ consists mostly of zeros, and this can be exploited by saving the matrix in sparse form (see MATLAB example below). Note that when the right-hand side is as simple as in (4.9), the physical eigenvalues $-k^2$ are simply the eigenvalues of the matrix $\mathbf{A}$. These eigenvalues can be computed with the MATLAB routine `eig`, which computes all eigenvalues and corresponding eigenvectors of an algebraic eigenvalue problem. We will use this routine without discussing *how* it finds the eigenvalues. The following MATLAB program computes the eigenvalues, that correspond to wavenumbers, for the discretized 1D Helmholtz equation.

```
% ------------------------------------------------------------
% Compute eigenvalues of 1D Helmholtz equation using FD
% ------------------------------------------------------------
function k = HFD1D(a, N)

% Arguments:
%     a = length of interval
%     N = number of subintervals (equal length)
% Returns:
%     k = eigenvalues

h = a/N;                    % Grid size
A = spalloc(N-1, ...        % Allocate sparse matrix
            N-1, ...        % with 3*(N-1) nonzeros
            3*(N-1));
```

Table 4.1 The two lowest wavenumbers from FD discretizations with different resolutions

N [-]	h [m]	k_1 [1/m]	k_2 [1/m]
10	0.1000	0.99589 27352 4357	1.96726 32861 6693
20	0.0500	0.99897 22332 4854	1.99178 54704 8714
30	0.0333	0.99954 31365 0068	1.99634 65947 4160
40	0.0250	0.99974 29988 6918	1.99794 44664 9703

```
d = -2/h^2;                  % Value of diagonal entries
s = 1/h^2;                   % Value of upper and lower
                             % diagonal entries

% Initialize the diagonal entries
for i = 1:N-1
  A(i,i) = d;   % Diagonal entries
end

% Initialize the upper and lower diagonal entries
for i = 1:N-2
  A(i,i+1) = s; % Upper diagonal entries
  A(i+1,i) = s; % Lower diagonal entries
end

% Computing the eigenvalues
lambda = eig(A);
k = sqrt(sort(-lambda));
```

For this small example, we can rely on the MATLAB routine `eig`. It should be noted that `eig` is limited to systems with at most a few thousand unknowns. This means it is very useful in one dimension, and works for moderate-sized 2D problems. In three dimensions, more powerful routines, such as the MATLAB routine `eigs`, are generally needed.

We calculate the first two numerical wavenumbers k on the interval $[0, \pi]$ for four different resolutions. The analytical results are $k = 1, 2, 3, \ldots$, and the numerical results are shown in Table 4.1.

Plots of k_m versus h^p show a straight line when $p = 2$, which means that the convergence is quadratic. Extrapolation of the first eigenvalue to zero cell size using `polyfit` gives the following values for k_1: linear extrapolation 0.99999 93697 896, quadratic 0.99999 99999 437, and cubic 0.99999 99999 997, which is very close to the exact value 1. For the second eigenvalue, linear extrapolation gives 1.99997 98747 162, quadratic 1.99999 99928 090, and cubic 1.99999 99999 989. Thus, the two lowest eigenvalues could be computed with 12-digit accuracy using the cubic fit for extrapolation, even though the computations have only about 4-digit accuracy. The accuracy of the extrapolated values may at first be surprising, but it is typical for problems where the solution is completely regular, i.e., has bounded derivatives of arbitrarily high order. However, if the problem contains some singular behavior, caused for instance by a reentrant 270°-degree corner, as in Fig. 3.1, or a tip in three dimensions, the derivatives of the solution will diverge at the corner, the order of convergence decreases, and extrapolation becomes more difficult.

The error is larger for the second eigenmode. The second eigenmode oscillates twice as fast and needs twice the resolution to be computed with the same accuracy as the first, as is confirmed by Table 4.1.

Review Questions

4.3-1 Use finite differences to discretize the eigenvalue problem $d^2f/dx^2 = -k^2 f$ with $f(0) = f(a) = 0$. Write down the corresponding matrix eigenvalue problem.

4.3-2 What is the order of convergence for k in (4.9)?

4.3-3 Why is the error, in general, larger for higher eigenmodes? What situations could change this?

4.4 Time-Domain Eigenvalue Calculation

One common way of determining eigenfrequencies in CEM is to time-step a solution, using for example a finite difference program, record the field at some location, and then Fourier transform this signal to locate its main frequency components. This technique can be used for more general methods than the finite differences. It can be used to find the eigenvalues of any spatial operator L with real and negative eigenvalues,

$$L[f] = -\omega^2 f. \tag{4.10}$$

Equation (4.10) is written in such a form that it is the frequency-domain form of the time-domain equation

$$\frac{\partial^2 f}{\partial t^2} = L[f], \tag{4.11}$$

which is, most likely, what the eigenvalue problem (4.10) was derived from.

The simplest time-discretization of (4.11) is

$$\frac{f^{(n+1)} - 2f^{(n)} + f^{(n-1)}}{(\Delta t)^2} = L[f^{(n)}], \tag{4.12}$$

where Δt is the time step. An important advantage of this formulation is that the time-stepping is *explicit*, that is, no matrix inversion is needed to compute $f^{(n+1)}$:

$$f^{(n+1)} = 2f^{(n)} - f^{(n-1)} + (\Delta t)^2 L[f^{(n)}]. \tag{4.13}$$

Such time-stepping schemes, often referred to as "leap-frog," are very efficient, and allow determination of the complete eigenvalue spectrum of (4.10). An important issue for explicit time-stepping schemes is how to choose the time-step Δt. This is mainly determined by stability.

4.4.1 Stability Analysis

Before working out a specific example, we discuss how one can analyze the stability of a time-stepping algorithm such as (4.13). The following technique is known as von Neumann stability analysis.

The analysis is based on the fact that any *discrete* time equation, which has no explicit time dependence, has solutions of the form $f^{(n)} = f_\omega \rho^n$, that is, geometrical sequences in discrete time. This is true even if the equation involves space-dependent coefficients, as long as it has no explicit time-dependence. Here, ρ is called the amplification factor of the eigenmode f_ω, and stability requires $|\rho| \leq 1$ for *all* eigenmodes. Substituting $f^{(n)} = f_\omega \rho^n$ into (4.13), and using $L[f_\omega] = -\omega^2 f_\omega$, we obtain a quadratic equation for the amplification factor

$$\rho^2 - [2 - (\omega \Delta t)^2]\rho + 1 = 0 \tag{4.14}$$

with the solutions

$$\rho = 1 - \frac{1}{2}(\omega \Delta t)^2 \pm j\omega \Delta t \sqrt{1 - \frac{1}{4}(\omega \Delta t)^2}. \tag{4.15}$$

If $(\omega \Delta t)^2 \leq 4$, there are two complex conjugate solutions such that

$$|\rho|^2 = (\mathrm{Re}\rho)^2 + (\mathrm{Im}\rho)^2 = 1.$$

On the other hand, if $(\omega \Delta t)^2 > 4$, there are two real solutions, whose product is unity, so one of them has modulus larger than 1. Figure 4.2 shows how the roots move in the complex plane as $\omega \Delta t$ varies.

The roots stay on the unit circle $|\rho| = 1$ as long as $|\omega \Delta t| \leq 2$, but when $|\omega \Delta t| > 2$, one root has modulus larger than unity. Therefore, if $|\omega \Delta t| > 2$, the solution will grow exponentially in time, and the scheme for time-stepping is unstable. Thus, the explicit time-stepping scheme in (4.13) has a stability limit for the time-step: $\Delta t \leq 2/|\omega|$. Since this has to hold for *all* the eigenmodes of (4.10), the condition on the time-step for the explicit scheme is

$$\Delta t \leq \frac{2}{|\omega_{\max}|}. \tag{4.16}$$

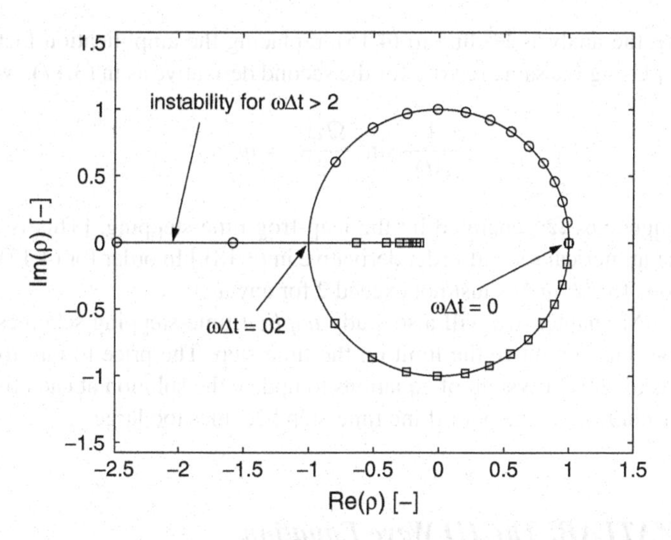

Fig. 4.2 Trajectories in the complex plane of the two roots for the amplification factor ρ in (4.15)

This means that the time-step times the highest eigenfrequency $f_{\max} = \omega_{\max}/2\pi$ should be at most $1/\pi$.

If we apply this stability limit to the operator $L = d^2/dx^2$ discretized on a uniform grid with cell size h, the largest numerical eigenvalue is $\omega^2_{\max} = 4/h^2$ [see (3.17)]. Thus, $\omega_{\max} = 2/h$, and stability requires $\Delta t \leq 2/\omega_{\max} = h$. We conclude that the time-step for our simple explicit scheme for the wave equation $\partial^2 f/\partial t^2 = \partial^2 f/\partial x^2$ should not be larger than the space step, for stability reasons.

We can also see how well the time-stepping reproduces the true oscillation frequency. The amplification factor per time-step ought to be

$$\exp(\pm j\omega t) = 1 \pm j\omega t - \frac{1}{2}(\omega \Delta t)^2 \mp \frac{j}{6}(\omega \Delta t)^3 + \cdots,$$

whereas (4.15) gives

$$\rho = 1 \pm j\omega t - \frac{1}{2}(\omega \Delta t)^2 \mp \frac{j}{8}(\omega \Delta t)^3 + \cdots.$$

The difference between ρ and $\exp(j\omega t)$ is $\pm j(\omega \Delta t)^3/24$, which corresponds to a relative frequency error of $(\omega \Delta t)^2/24$.

The von Neumann stability analysis is closely related to the analysis in Sec. 3.2.3. To see the connection, assume that the solution f of the time-discretized problem varies harmonically in time, $f \propto \exp(j\Omega t)$, i.e., $f^{(n)} \propto \exp(jn\Omega \Delta t)$. We will examine how the frequency Ω of the time-discretized solution is related to ω of the frequency-domain eigenvalue problem $L[f] = -\omega^2 f$. [Of course, this is just redoing the analysis leading to (4.15), replacing the amplification factor ρ by $\exp(j\Omega t)$.] Using the same rewrite for the second derivative as in (3.17), we obtain

$$\frac{4}{(\Delta t)^2}\sin^2\frac{\Omega\Delta t}{2}=\omega^2 \tag{4.17}$$

for the frequencies Ω generated by the leap-frog time-stepping. [This is also the same as the numerical second-order derivative in (3.18).] In order for (4.17) to have real solutions for Ω, $\omega\Delta t$ must not exceed 2 for any ω.

In the FEM chapter, we will also study *implicit* time-stepping schemes, which make it possible to remove the limit on the time-step. The price to pay for this is that one has to solve a system of equations to update the solution at each time step. Also, the accuracy may be poor if the time-step becomes too large.

4.4.2 MATLAB: The 1D Wave Equation

As a simple illustration of how to extract spectral information by explicit time-stepping, we seek the spectrum $-\omega^2$ of the operator $L = \partial^2/\partial x^2$ on the interval $0 < x < a$ with the boundary conditions $f(0,t) = f(a,t) = 0$. The true eigenfrequencies are

$$\omega_m = \frac{m\pi}{a}, \quad m = 1,2,\ldots.$$

The spectrum of L can be found by solving the wave equation

$$\frac{\partial^2 f}{\partial t^2}=\frac{\partial^2 f}{\partial x^2}, \quad 0<x<a, \quad f(0,t)=f(a,t)=0. \tag{4.18}$$

We use the simplest finite difference scheme:

$$f_i^{(n+1)}=2f_i^{(n)}-f_i^{(n-1)}+\left(\frac{\Delta t}{\Delta x}\right)^2\left(f_{i+1}^{(n)}+f_{i-1}^{(n)}-2f_i^{(n)}\right). \tag{4.19}$$

We will write this as a MATLAB function that records two signals [$f(t)$ at two locations, the midpoint and a point close to the left boundary] and stores them in arrays to be analyzed afterwards. More than one signal is recorded because some eigenmodes can be undetected if the eigenfunction f has a node (i.e., zero amplitude) at the "detector" location. An eigenmode may also be undetected if the initial condition does not excite it at sufficient amplitude.

```
% ---------------------------------------------------------------
% Time step 1D wave equation using two time-levels f0 & f1
% ---------------------------------------------------------------
function [omega, s1, s2] = Wave1D(a, time, nx)
```

```
% Arguments:
%    a     = the length of the interval
%    time  = the total time interval for the simulation
%    nx    = the number of subintervals in the domain (0,a)
% Returns:
%    omega = the angular frequencies
%    s1    = the complex Fourier transform of data at x = a/5
%    s2    = the complex Fourier transform of data at x = a/2

f0          = randn(nx+1, 1); % Initialize with random numbers
f0(1,1)     = 0;              % Boundary condition at x = 0
f0(nx+1,1)  = 0;              % Boundary condition at x = a

f1          = randn(nx+1, 1); % Initialize with random numbers
f1(1,1)     = 0;              % Boundary condition at x = 0
f1(nx+1,1)  = 0;              % Boundary condition at x = a

dx          = a/nx;           % The cell size
d2tmax      = 1.9*dx;         % The time step must satisfy
                              % 2*dt < 2*dx for stability

ntime = round(time/d2tmax + 1);  % The number of time steps
dt = time/(2*ntime);             % The time step

% Initialize the coefficient matrix for updating the solution f
A = spalloc(nx+1,nx+1,3*(nx+1)); % Sparse empty matrix with
                                 % 3*(nx+1) nonzero entries
for i = 2:nx
  A(i,i)    = 2*(1-(dt/dx)^2);   % Diagonal entries
  A(i,i+1)  = (dt/dx)^2;         % Upper diagonal entries
  A(i,i-1)  = (dt/dx)^2;         % Lower diagonal entries
end

% Time step and sample the solution
% Sample location #1 is close to the left boundary
% Sample location #2 is at the midpoint of the domain
for itime = 1:ntime % Every 'itime' means two time steps 'dt'

  f0                  = A*f1 - f0;            % Update
  sign1(2*itime-1)    = f0(round(1+nx/5));    % Sample at location #1
  sign2(2*itime-1)    = f0(round(1+nx/2));    % Sample at location #2

  f1                  = A*f0 - f1;            % Update
  sign1(2*itime)      = f1(round(1+nx/5));    % Sample at location #1
  sign2(2*itime)      = f1(round(1+nx/2));    % Sample at location #2

end

% Compute the discrete Fourier transform of
% the time-domain signals
spectr1      = fft(sign1);
spectr2      = fft(sign2);

% In the MATLAB implementation of the function fft(),
```

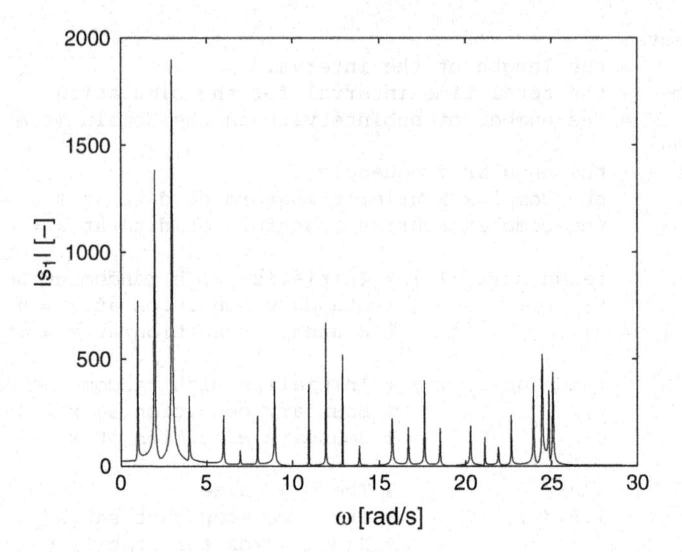

Fig. 4.3 Amplitude of Fourier coefficient s_1 (measured at one-fifth from the left boundary) versus angular frequency for the 1D wave equation. Every fifth mode is undetected because the detector is located at a node for the eigenfunction

```
% the first half of the output corresponds to positive frequency
s1(1:ntime)  = spectr1(1:ntime);
s2(1:ntime)  = spectr2(1:ntime);

% Frequency vector for use with 's1' and 's2'
omega        = (2*pi/time)*linspace(0, ntime-1, ntime);
```

We call the routine by

```
[omega,s1,s2] = Wave1D(pi,200,30);
```

to compute the spectrum of the second derivative on the interval $[0, \pi]$. Figures 4.3 and 4.4 show the absolute values of s_1 and s_2 versus angular frequency. The spectral peaks fall very close to integers, as they should. Because of the spatial locations of the observation points, the even peaks are absent in s_2 and those divisible by 5 in s_1. These are the eigenmodes that have zero amplitudes (nodes) at the respective observation points.

A significant advantage of such a time-domain calculation is that we can find the whole spectrum (except the few peaks that are accidentally missed) from a single simulation.

4.4.3 *Extracting the Eigenfrequencies*

Let us briefly consider how to extract the eigenfrequencies from a time-domain simulation. We first run the simulation and record the signals. The longer the simulation

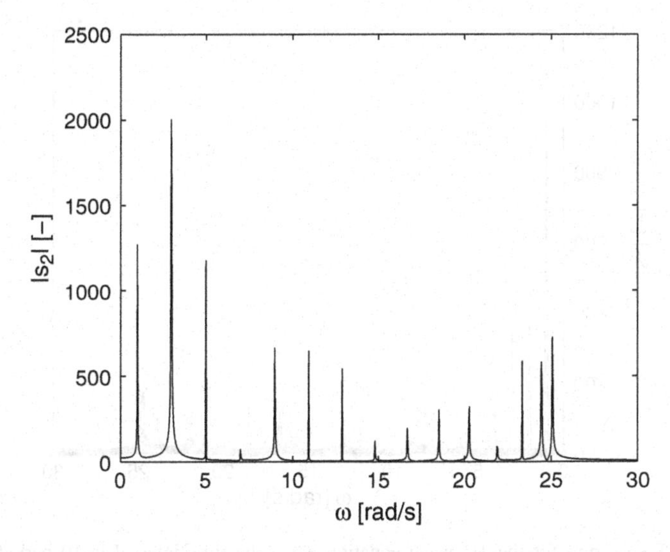

Fig. 4.4 Amplitude of Fourier coefficient s_2 (measured at the midpoint of the interval) versus angular frequency for the 1D wave equation. All modes with even number are undetected because the detector is at a node for those modes

is run, the sharper the spectral peaks become, and the better the eigenfrequencies are determined, but the convergence of the estimated frequencies is slow. One can see that when there is no damping, the estimates are sensitive to how close the various frequency components are to making an integer number of oscillations during the simulation. This is because the fast Fourier transform (FFT), which is used to transform the recorded signal into the frequency domain, treats the signal as if it were periodic with a period equal to the simulated time. If the time interval is not an integer number of wave periods, either the signal or its time derivative will have a jump at the end of the time window, and this broadens the Fourier spectrum of a sinusoidal signal. As an example, compare the spectrum obtained by calling the time-stepping routine by `Wave1D(pi,20*pi,30)`, which gives 10 (analytical) oscillation periods for the first mode, and where all the low-order modes make approximately an integer number of oscillations, with that obtained from `Wave1D(pi,21*pi,30)`, where the first mode has 10.5 oscillation periods and all the odd modes will be strongly broadened by the FFT. In the first case, where the low-order modes make an integer number of oscillations, the FFT finds very sharp peaks for these modes, despite the rather short time interval; see Fig. 4.5. In the second case, see Fig. 4.6, the odd modes, with half-integer number of periods, are broad.

One way to avoid the dependence on how the time sequence is terminated is to multiply the time signal by an exponential damping factor $\exp(-\gamma t)$, and choose γ such that $\gamma t_{\max}$ is large enough, say in the range of 3 to 5. (This makes the FFT an approximation of the Laplace transform.) Now the FFT produces a cleaner spectrum. The frequencies can be extracted almost automatically by fitting the

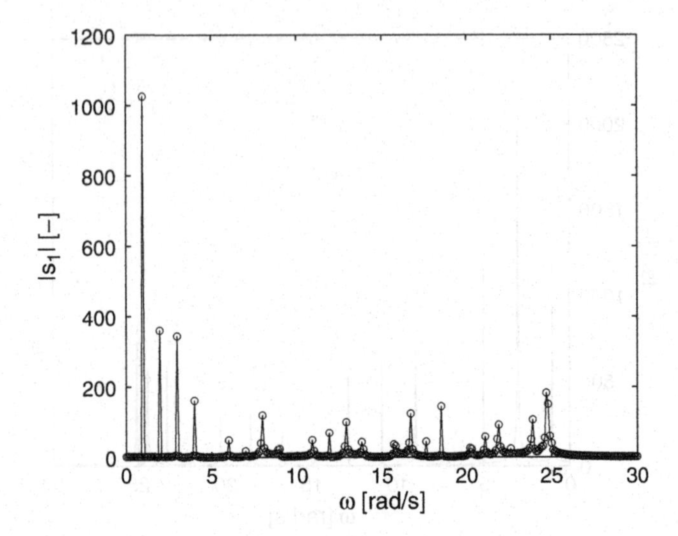

Fig. 4.5 FFT spectrum for the 1D wave equation when the time interval is 10 periods for the lowest mode (and an integer number of modes for all the lowest modes)

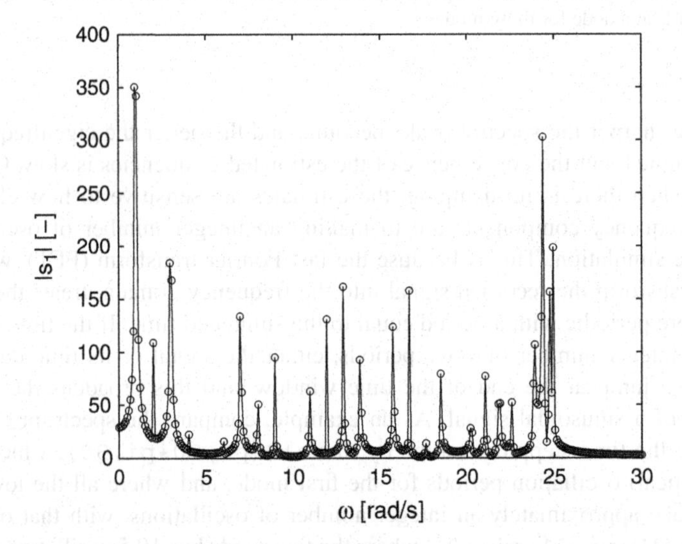

Fig. 4.6 FFT spectrum for the 1D wave equation when the time interval is 10.5 periods for the lowest mode (and a half-integer number of modes for all the odd modes)

output from the FFT (Laplace transform) to a so-called Padé approximation. This consists in fitting the frequency response by a ratio of polynomials

$$s(\omega) = \frac{P(\omega)}{Q(\omega)}. \tag{4.20}$$

The idea behind this (which is correct only when the signal decays to zero at the end of the recorded interval) is that we expect the Laplace transform to consist of simple poles

$$s(\omega) \approx \sum_n \frac{c_n}{\omega - \omega_n}, \tag{4.21}$$

and this pole expansion is a rational function of the same type as the Padé approximation (4.20).

4.4.4 MATLAB: Padé Approximation

The following MATLAB function computes the coefficients of P and Q and then uses MATLAB's `residue` function to find the poles ω_n and residues c_n in (4.21).

```
% ---------------------------------------------------------------
% Pade approximation for s(omega)
% ---------------------------------------------------------------
function [poles, res] = Pade(omega, s, l, n)

% Arguments:
%     omega = the array of the independent frequency
%     s     = the function of omega to be Pade approximated as
%             the ratio of polynomials P(omega)/Q(omega)
%     l     = discrete index of center frequency
%     n     = degree of polynomials P and Q
% Returns:
%     poles = the poles of the Pade approximation
%     res   = the residues of the Pade approximation

% Setup the matrix for computing coefficients of P and Q
A = zeros(2*n+1);
for i = 1:2*n+1
  % Shift frequencies
  oshift(i) = omega(l-1-n+i)-omega(l);

  % P entries
  for k = 1:n+1
    A(i,k) = oshift(i)^(k-1);
  end

  % Q entries
  for k = 1:n
    A(i,n+1+k) = -s(l-1-n+i)*oshift(i)^k;
  end

  % Q_0 set to 1
  x(i) = s(l-1-n+i);
end

% Compute the coefficients
```

```
coef = (A\(x.')).';

for k = 1:n+1
  P(k) = coef(n+2-k);
end

for k = 1:n
  Q(k) = coef(2*n+2-k);
end
Q(n+1) = 1;

% Find the poles and the residues
[res, poles] = residue(P, Q);
poles        = poles + omega(1);  % Restore the frequency shift
```

Applying this routine to an approximate Laplace transform, one can make the frequencies converge very well with about 10 periods of oscillation. A standard method used for frequency determination in the literature is Prony's method; see for instance [80]. However, more modern techniques of signal processing can be used to give much more efficient extraction of frequencies, in particular when the frequency spectrum is dense [65].

Review Questions

4.4-1 How are the eigenvalues extracted from a time-domain eigenvalue calculation? Can the corresponding eigenmodes be extracted in a simple way?
4.4-2 What considerations should be taken into account in selecting the time-step Δt?
4.4-3 What is an explicit time-stepping method?
4.4-4 Describe the meaning and the use of the amplification factor in words.
4.4-5 How does the highest eigenfrequency relate to the maximal stable time-step for (4.13)?
4.4-6 How well is the true oscillation frequency reproduced by (4.15)? Quantify your answer.
4.4-7 How do the excitation and detector positions influence the frequency spectrum computed from a time-domain method?
4.4-8 Why are the frequency estimates of the FFT sensitive to how close the various undamped resonances are to making an integer number of oscillations during the simulation?

Summary

- The solution of the eigenvalue problem $L[f_m] = \lambda_m f_m$ consists of pairs of eigenvalues λ_m and eigenvectors f_m, where the pairs typically are indexed by an integer m. (Sometimes the subindex m is omitted in order to simplify the notation.) Here, the operator L and boundary conditions are given. For Maxwell's equations, we have
$$\nabla \times \mu^{-1} \nabla \times \boldsymbol{E}_m = \omega_m^2 \epsilon \boldsymbol{E}_m,$$
where the eigenfunction is $\boldsymbol{E}_m$ and the eigenvalue is ω_m^2.
- For the 1D Helmholtz equation $d^2 f/dx^2 = -k^2 f$ on the interval $0 < x < a$ with the boundary conditions $f(0) = f(a) = 0$, the eigenvalues are $k^2 = (\pi m/a)^2$ with integer $m = 1, 2, \ldots$ for the continuous problem, and the discretized problem has
$$k^2 = \frac{4}{h^2} \sin^2 \left(\frac{\pi m h}{2a} \right)$$
for the cell size h and $m = 1, 2, \ldots, N$, where N is the number of internal nodes in the grid.
- A time-domain computation of eigenvalues is based on the inverse Fourier transform of $L[f] = -\omega^2 f$, i.e., $L[f] = \partial^2 f/\partial t^2$, and a finite difference discretization with respect to time gives
$$L[f^{(n)}] = \frac{f^{(n+1)} - 2f^{(n)} + f^{(n-1)}}{(\Delta t)^2}.$$
The substitution $f^{(n)} = f_\omega \rho^n$, where ρ is an amplification factor and $L[f_\omega] = -\omega^2 f_\omega$, gives
$$\rho = 1 - \frac{1}{2}(\omega \Delta t)^2 \pm j\omega \Delta t \sqrt{1 - \frac{1}{4}(\omega \Delta t)^2}.$$
We have $|\rho| = 1$ for $\omega \Delta t \leq 2$. If $\Delta t < 2/|\omega_{\max}|$, no mode will grow, and every mode is multiplied by a phase-factor in each time-step. Thus, stable time-stepping is achieved for $\Delta t < 2/|\omega_{\max}|$, where $\omega_{\max}$ is the highest eigenfrequency.
- The output $s(t)$ from a time-domain simulation can be represented by its Fourier transform:
$$s(\omega) \approx \frac{P(\omega)}{Q(\omega)} = \sum_n \frac{c_n}{\omega - \omega_n}.$$
Peaks in the spectrum of $s(\omega)$ fall close to the resonance frequencies ω_n.

Problems

P.4-1 Calculate the eigenvalues k^2 of the vector wave equation $\nabla \times \nabla \times \boldsymbol{E} = k^2\boldsymbol{E}$ for a 2D rectangular cavity with PEC boundaries. Consider the two cases with $\boldsymbol{E} = \hat{z}E_z(x, y)$ and $\boldsymbol{E} = \hat{x}E_x(x, y) + \hat{y}E_y(x, y)$, where the second case is easier to treat if it is reformulated in terms of the magnetic field.

P.4-2 Show that the eigenvalues of the *discretized* 1D Helmholtz equation (4.9), for $a = \pi$, are

$$-k^2 = -\frac{4}{h^2}\sin^2\frac{mh}{2}, \quad m = 1, 2, 3, \ldots,$$

and find how the error in k depends on the mode number and resolution.

P.4-3 Let the electric field be $\boldsymbol{E} = \hat{z}E_z(x)$ for a 1D cavity with PEC walls and constant μ and ϵ. Use the finite difference scheme and show that (4.6) can be rewritten as

$$\omega^2 = \frac{1}{\mu\epsilon}\frac{\mathbf{e}^{\mathrm{T}}\mathbf{A}\mathbf{e}}{\mathbf{e}^{\mathrm{T}}\mathbf{e}},$$

where $\mathbf{e}$ is a vector with the electric field at the interior grid points. Determine $\mathbf{A}$ and interpret the products $\mathbf{e}^{\mathrm{T}}\mathbf{A}\mathbf{e}$ and $\mathbf{e}^{\mathrm{T}}\mathbf{e}$ in terms of a numerical integration scheme.

P.4-4 In one dimension, Helmholtz equation gives $L = d^2/dx^2$. Find a nonzero solution f that yields $L[f] = 0$ and solve (4.10) and (4.11) for that particular solution. Can this solution exist in a region of finite size, and if so, what boundary conditions are satisfied by this solution?

P.4-5 Consider the questions in the previous exercise when the operator $L = d^2/dx^2$ is discretized by finite differences. How do you treat the boundary conditions so that the order of convergence associated with the finite difference stencils of the interior grid points is preserved? How does the discretized problem compare to its continuous counterpart? Does the discretized problem have a nonzero solution f with $L[f] = 0$?

P.4-6 Discretize $L = \partial^2/\partial x^2$ with finite differences so that the dominant term in the error is $O(h^4)$ (more than three points are needed) and derive the stability limit on Δt for (4.13). Compare the stability limit with the case in which the error is $O(h^2)$.

P.4-7 Compute the discrete Fourier transform of the signal $\sin(\omega t)$ sampled at $t = n\Delta t$, where $n = 0, 1, \ldots, N-1$. Compare some arbitrarily chosen value of ω with the special case $\omega = 2\pi q/(N\Delta t)$ for some integer $q = 0, 1, \ldots, N-1$. How and why do these cases differ?

P.4-8 For three resonances, rewrite (4.21) as a ratio of polynomials $s(\omega) = P(\omega)/Q(\omega)$. Consider the output signal $y(\omega) = s(\omega)x(\omega)$, where $x(\omega)$ is the input signal to the system. Use the inverse Fourier transform to derive the time-domain expression for $y(\omega)$. Interpret your findings.

Computer Projects

C.4-1 The transverse electric (TE) modes and the corresponding eigenvalues k_t^2 for a closed metal waveguide satisfy

$$-\nabla^2 H_z = k_t^2 H_z \ \text{in } S,$$
$$\hat{\boldsymbol{n}} \cdot \nabla H_z = 0 \ \text{on } L.$$

Similarly, the transverse magnetic (TM) modes and their eigenvalues k_t^2 fulfill

$$-\nabla^2 E_z = k_t^2 E_z \ \text{in } S,$$
$$E_z = 0 \ \text{on } L.$$

Here, the metal boundary of the waveguide cross section is denoted by L, and it encloses the interior S of the waveguide. Write a program that solves for the eigenmodes and the eigenvalues based on a finite difference discretization of the TE and TM problem for a waveguide with rectangular cross section. The analytical eigenvalues are $k_t^2 = (m\pi/a)^2 + (n\pi/b)^2$ for integers m and n excluding the combination $m = n = 0$ for the TE case and $mn = 0$ for the TM case. Here, the rectangular cross section has width a and height b.

C.4-2 Equation (4.2) with losses and constant permeability is given by $\nabla\times\nabla\times\boldsymbol{E} = \mu(\omega^2\epsilon - j\omega\sigma)\boldsymbol{E}$, and for a problem with $\boldsymbol{E} = \hat{\boldsymbol{z}}E_z(x, y)$, we get

$$-\nabla^2 E_z = \mu(\omega^2\epsilon - j\omega\sigma)E_z,$$

which is a nonlinear eigenvalue problem in ω. Rewrite this problem to a linear eigenvalue problem in terms of E_z and ωE_z. Implement a finite-difference algorithm and solve for the resonance frequencies and quality factors of a square cavity with a boundary of a PEC. For constant material parameters, derive the analytical eigenfrequencies and compare the numerical and analytical results. How is the spectrum influenced by losses? Explore the case in which $\sigma > 0$ in a part of the domain and study the dependence of the lowest eigenmodes as a function of σ. Try to explain your findings.

Chapter 5
The Finite-Difference Time-Domain Method

The finite-difference time-domain (FDTD) scheme is one of the most popular computational methods for microwave problems; it is simple to program, highly efficient, and easily adapted to deal with a variety of problems. The FDTD scheme is typically formulated on a structured Cartesian grid and it discretizes Maxwell's equations formulated in the time domain. The derivatives with respect to space and time are approximated by finite-differences, where the field components of the electric and magnetic field are staggered in space with respect to each other in a particular manner that is tailored for Maxwell's equations.

A major weakness of the method lies in the way it deals with boundaries that are not aligned with the Cartesian grid: for oblique boundaries, FDTD programs typically resort to the "staircase approximation." The error due to the staircase approximation can be difficult to assess, but some examples can be found in the literature [14, 63]. The finite element method (FEM), which will be discussed in Chap. 6, is better suited for problems with oblique and curved boundaries and fine structures that may need higher resolution locally.

However, the FDTD allows for explicit time-stepping, and this makes it much more efficient than time-domain FEM, which in general is implicit (i.e., a system of equations must be solved at each time step). Another advantage of the FDTD is that no matrix has to be stored. This reduces memory consumption and makes it possible to solve problems with a very large number of unknowns.

The FDTD has a time-step limit $\Delta t < h/c\sqrt{3}$ in three dimensions, where Δt is the time-step, h is the cell size, and c is the speed of light (in vacuum, the speed of light is $c_0 = 299\,792\,458$ m/s). This is a serious limitation in problems involving time scales much longer than it takes a light wave to cross the simulation region. An important example of this is eddy current problems, in which the FDTD cannot be used because of its short limit to the time-step.

The type of problems for which the FDTD is particularly suited involves the propagation of electromagnetic waves and geometries where characteristic lengths are comparable to a wavelength. This typically includes microwave problems. Similar conditions also apply for optical devices whose dimensions are comparable to the wavelength.

T. Rylander et al., *Computational Electromagnetics*, Texts in Applied Mathematics 51, DOI 10.1007/978-1-4614-5351-2_5,

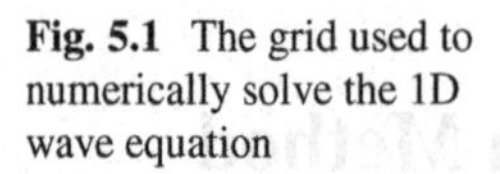

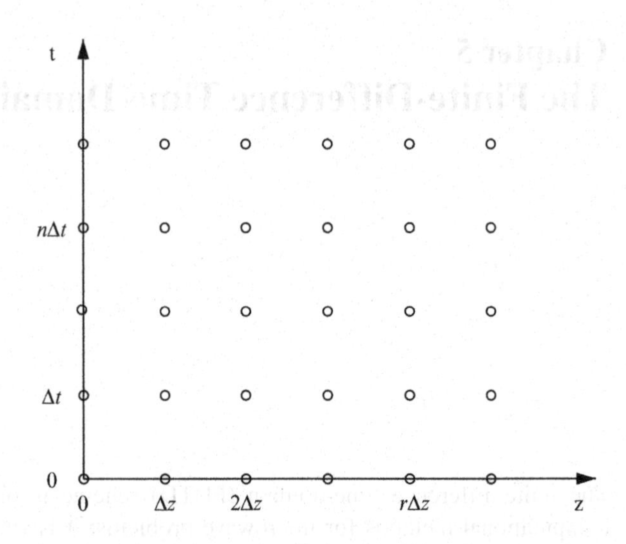

Fig. 5.1 The grid used to numerically solve the 1D wave equation

A powerful way to find several resonant frequencies of a microwave cavity is to perform an FDTD simulation and then Fourier transform selected signals in time. This is the same procedure that we discussed for finding the eigenvalues of the 1D Helmholtz equation in Chap. 4. For many applications, for instance scattering problems, selected time signals from an FDTD simulation can be Fourier transformed while the simulation proceeds, and a single FDTD run can produce frequency-domain results at any desired number of frequencies. This is a major advantage of time-domain methods.

The FDTD algorithm was originally proposed by K.S. Yee in 1966 [93]. Since then, it has been used for a variety of applications, and many extensions of the basic algorithm have been developed. The literature on the FDTD is vast, and over the period 1975–1995 the number of research papers in which the FDTD method was used grew exponentially in time. By now, the FDTD is considered a basic tool in CEM, and research articles now tend to be on more complicated methods. The books by Taflove et al. [80–82] give a good overview and describe many important extensions of the FDTD.

5.1 The 1D Wave Equation

To solve the wave equation (1.12) numerically, we divide the z-axis into intervals of length Δz and the time axis into intervals of length Δt (see Fig. 5.1).

Let $|_r$ be an index that refers to the z-coordinate and let $|^n$ refer to the time coordinate such that $E|_r^n = E(r\Delta z, n\Delta t)$. We get the discrete equation by using standard difference approximations for the derivatives:

$$\frac{E|_r^{n+1} - 2\,E|_r^n + E|_r^{n-1}}{(\Delta t)^2} = c^2\,\frac{E|_{r+1}^n - 2\,E|_r^n + E|_{r-1}^n}{(\Delta z)^2}. \tag{5.1}$$

Equation (5.1) gives an *explicit* expression for E at the next time level $n+1$ in terms of E at the previous levels:

$$E|_r^{n+1} = 2\,E|_r^n - E|_r^{n-1} + \left(\frac{c\Delta t}{\Delta z}\right)^2 (E|_{r+1}^n - 2\,E|_r^n + E|_{r-1}^n), \tag{5.2}$$

which is identical to (4.19) when the speed of the wave c is set to unity. Two time levels of E must be given as initial conditions. For the analytical wave equation one needs E and $\partial E/\partial t$ as functions of z at $t = 0$.

The dispersion relation for the finite difference approximation in (5.1) can be found by substituting $E|_r^n$ with $\exp(j\omega\, n\Delta t - jk\, r\Delta z)$ and dividing the equation by $\exp(j\omega\, n\Delta t - jk\, r\Delta z)$:

$$\frac{e^{j\omega\Delta t} - 2 + e^{-j\omega\Delta t}}{(\Delta t)^2} = c^2\,\frac{e^{-jk\Delta z} - 2 + e^{jk\Delta z}}{(\Delta z)^2}.$$

This can be rewritten as

$$\left(\frac{e^{j\omega\Delta t/2} - e^{-j\omega\Delta t/2}}{2j}\right)^2 = \left(\frac{c\Delta t}{\Delta z}\right)^2 \left(\frac{e^{jk\Delta z/2} - e^{-jk\Delta z/2}}{2j}\right)^2.$$

Taking the square root, we get the dispersion relation for the numerical scheme:

$$\sin\frac{\omega\Delta t}{2} = \pm\frac{c\Delta t}{\Delta z}\,\sin\frac{k\Delta z}{2}. \tag{5.3}$$

For the numerical solutions, the angular frequency ω is only *approximately* a linear function of the wavenumber k, unless $\Delta z = c\Delta t$. Consequently, waves with different wavenumbers will propagate with different velocities. This means that a wave package containing several different spatial frequencies will change shapes as it propagates. This is referred to as the dispersion of the numerical scheme, or numerical dispersion for short.

5.1.1 Dispersion and Stability

How does the choice of Δt and Δz affect the dispersion? Equation (5.3) shows that the important parameter is $R = c\Delta t/\Delta z$, that is, how many grid cells the exact solution propagates in one time-step. Dispersion relations for different values of $R \le 1$ are shown in Fig. 5.2.

We have the following distinct situations:

$R = 1$: If $\Delta t = \Delta z/c$, then $R = 1$ and (5.3) simplifies to $\omega = \pm ck$, which is exactly the analytical dispersion relation (1.14). This choice of Δt is called *the magic time step*. The errors of the spatial and temporal difference approximations

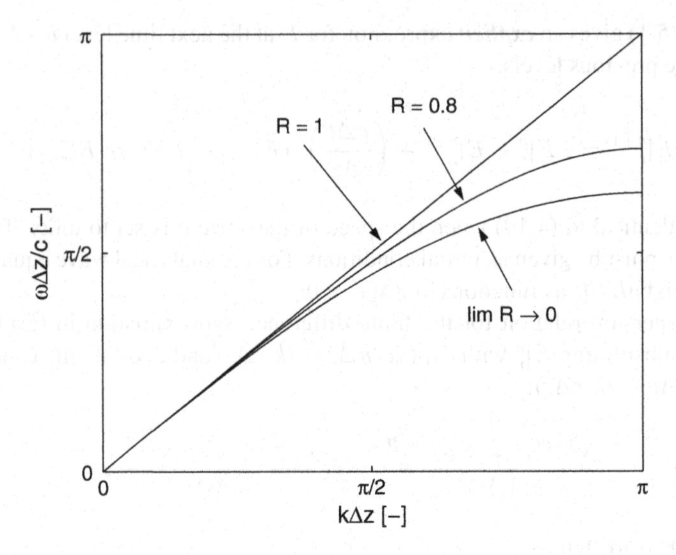

Fig. 5.2 Numerical dispersion relations for different values of $R = c\Delta t/\Delta z$

cancel, and the signals propagate exactly one cell per time-step, in either direction.

$R < 1$: If $\Delta t < \Delta z/c$, the numerical dispersion relation differs from the analytical. The smaller R is, the stronger is the numerical dispersion (see Fig. 5.2). The dispersion properties improve as Δt approaches the magic time-step.

$R > 1$: If $\Delta t > \Delta z/c$, then $R > 1$ and (5.3) yields complex angular frequencies for wavenumbers such that $|\sin k\Delta z/2| > \Delta z/c\Delta t = 1/R$. As a consequence, some waves will be exponentially growing in time, i.e., the algorithm is *unstable*. This exemplifies the type of instability discussed in Sect. 4.4.1. When $c\Delta t > \Delta z$, the signal of the true solution propagates more than one cell per time-step, and that is not possible with the explicit scheme in (5.2), which involves only nearest neighbors. The stability condition $c\Delta t \leq \Delta z$ is often called the Courant (or Courant–Friedrichs–Levy, CFL) condition. Similar conditions, implying that the signal can propagate at most one grid cell per time-step, hold for practically all explicit schemes for any differential equation.

5.1.1.1 Example: A Square Wave

A square wave can be represented as an infinite sum of harmonic components with different frequencies, and it is rich in high-frequency components. When such a wave propagates in a dispersive medium, the different sine waves propagate with different velocities, and the shape of the wave will change as it propagates. The 1D wave equation can be time-stepped using selected parts of the MATLAB function

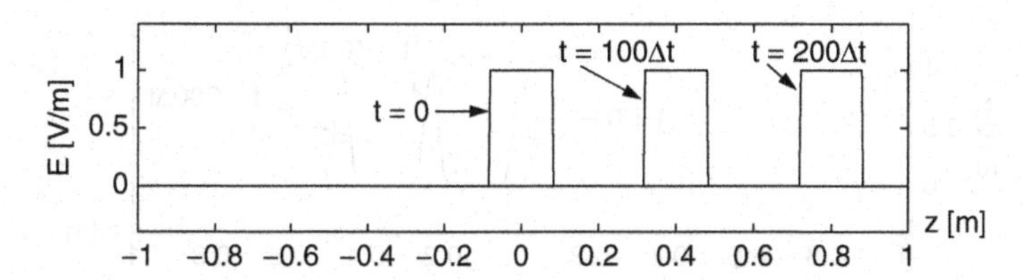

Fig. 5.3 Propagation of a square wave when Δt is equal to the magic time-step, $R = c\Delta t/\Delta z = 1$. There is no dispersion: the shape of the pulse stays the same as it propagates

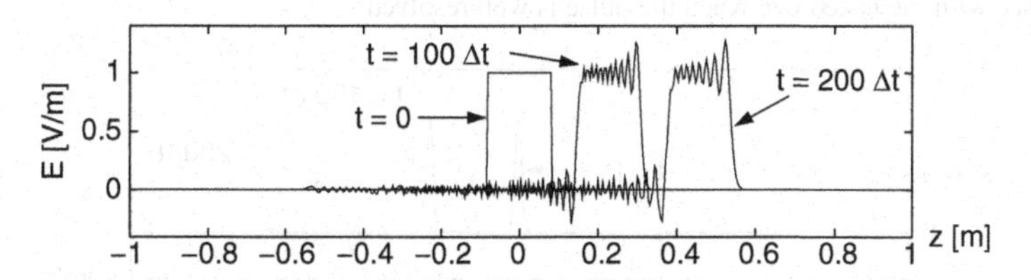

Fig. 5.4 Propagation of a square wave when Δt is smaller than the magic time-step, $R = c\Delta t/\Delta z = 1/\sqrt{3} \approx 0.58$. In this case, there is significant numerical dispersion: the shape of the pulse changes as it propagates

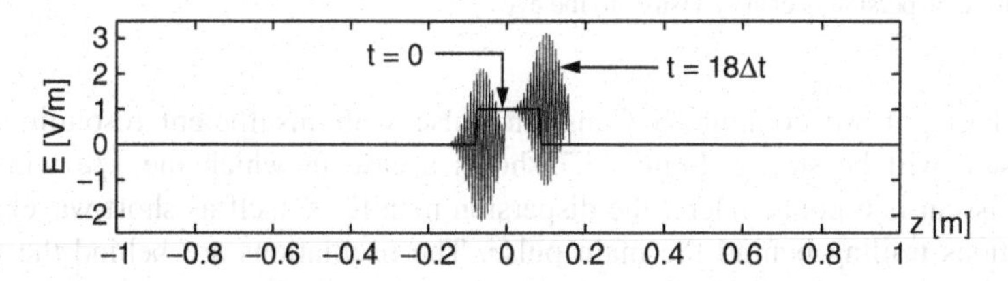

Fig. 5.5 Propagation of a square wave when Δt is slightly greater than the magic time-step, $R = c\Delta t/\Delta z = 1.01$. The scheme is unstable and the wave amplitude increases rapidly in an unphysical way

`Wave1D` given in Sect. 4.4.2. Figures 5.3–5.5 show the propagation of a square wave for three different values of R.

5.1.1.2 Example: A Smooth Wave

An initial condition in the form of a square wave highlights the dispersion of the numerical scheme. As a second example, we take as initial condition a Gaussian pulse that is well resolved on the grid, with 12 points across the $1/e$ width; see Fig. 5.6. This pulse can propagate many pulse widths before the dispersion becomes apparent to the eye, even when $R = 1/\sqrt{3}$. This illustrates an important point: numerical results are accurate only when the solution is well resolved by the grid. Of course, a square wave is not well resolved on *any* grid.

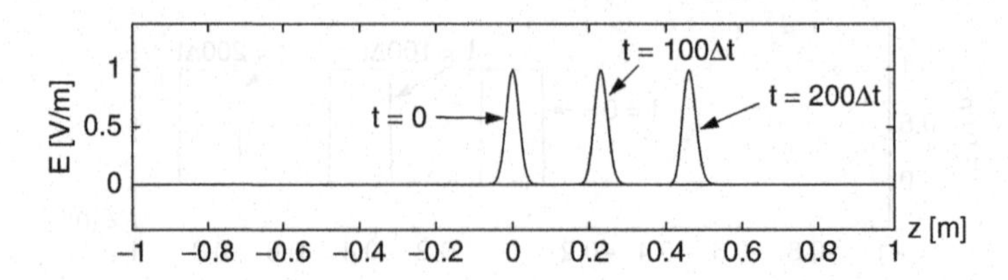

Fig. 5.6 Propagation of a Gaussian pulse with 12 points across the $1/e$ width when Δt is smaller than the magic time-step, $R = c\Delta t/\Delta z = 1/\sqrt{3}$. Although the scheme has some dispersion, it is hard to see with the naked eye when the pulse is well resolved

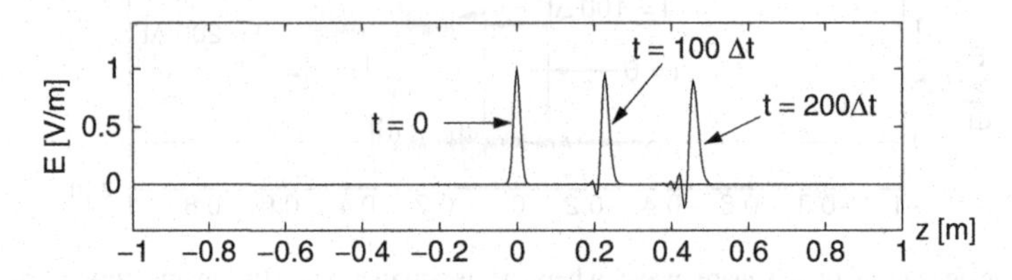

Fig. 5.7 Propagation of a Gaussian pulse with 6 points across the $1/e$ width when Δt is smaller than the magic time-step, $R = c\Delta t/\Delta z = 1/\sqrt{3}$. Here the resolution is not very good, and the effect of the dispersion is clearly visible to the eye

Similarly, if we compute a Gaussian pulse with insufficient resolution, the dispersion will be strong. Figure 5.7 shows a case in which the $1/e$ width of the Gaussian is 6 points. Here, the dispersion manifests itself as short-wavelength oscillations trailing behind the main pulse. The oscillations are behind the main pulse because the phase velocity is smaller for short wavelengths.

Review Questions

5.1-1 List some pros and cons of the FDTD scheme.

5.1-2 What is a dispersion relation? Derive the dispersion relation for the 1D wave equation discretized by the standard finite difference approximation. Compare the numerical dispersion relation with its analytical counterpart.

5.1-3 Under what conditions will $E(z,t) = E^+(z - c\,t) + E^-(z + c\,t)$ satisfy the *discretized* 1D wave equation?

5.1-4 Generally, higher resolutions lead to more accurate results, but in some cases this is not true. Give an example of this and explain why.

5.1-5 Explain how and why a pulse is distorted when propagated by the wave equation discretized by finite differences.

5.2 The FDTD Method: Staggered Grids

The wave equation is a second-order differential equation for the electric field only. It can also be stated as a system of coupled first-order differential equations for both $\boldsymbol{E}$ and $\boldsymbol{H}$. In three dimensions, Maxwell's equations (1.9)–(1.10) in a source-free region give six scalar equations, three for Ampère's law,

$$\epsilon \frac{\partial E_x}{\partial t} = \frac{\partial H_z}{\partial y} - \frac{\partial H_y}{\partial z}, \tag{5.4}$$

$$\epsilon \frac{\partial E_y}{\partial t} = \frac{\partial H_x}{\partial z} - \frac{\partial H_z}{\partial x}, \tag{5.5}$$

$$\epsilon \frac{\partial E_z}{\partial t} = \frac{\partial H_y}{\partial x} - \frac{\partial H_x}{\partial y}, \tag{5.6}$$

and three for Faraday's law,

$$\mu \frac{\partial H_x}{\partial t} = \frac{\partial E_y}{\partial z} - \frac{\partial E_z}{\partial y}, \tag{5.7}$$

$$\mu \frac{\partial H_y}{\partial t} = \frac{\partial E_z}{\partial x} - \frac{\partial E_x}{\partial z}, \tag{5.8}$$

$$\mu \frac{\partial H_z}{\partial t} = \frac{\partial E_x}{\partial y} - \frac{\partial E_y}{\partial x}. \tag{5.9}$$

The FDTD is a finite difference scheme particularly suited to the structure of these six first-order equations. In particular, it uses difference formulas that are as *local* as possible and *centered.*

5.2.1 *One Space Dimension*

To illustrate the use of staggered grids, which is central to the FDTD, we will start with a 1D problem. Consider a plane wave propagating in the z-direction through a medium such that all quantities are constant in planes perpendicular to the z-axis. We assume that the electric field is oriented in the x-direction, and the magnetic field in the y-direction. Then, (5.4)–(5.9) reduce to

$$\epsilon \frac{\partial E_x}{\partial t} = -\frac{\partial H_y}{\partial z}, \tag{5.10}$$

$$\mu \frac{\partial H_y}{\partial t} = -\frac{\partial E_x}{\partial z}. \tag{5.11}$$

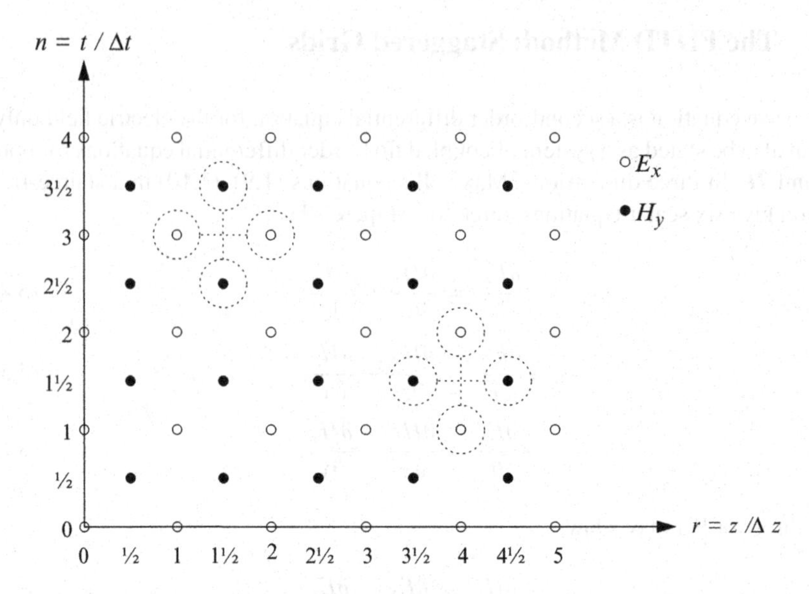

Fig. 5.8 Staggered grid used in the 1D leap-frog algorithm. The two "stencils" show which values of E_x and H_y are used in solving (5.13) with ($r = 1, n = 3$) and in solving (5.12) with ($r = 4$, $n = 1$)

The "trick" used to get a good algorithm is to put the different E- and H-components at different positions on the grid, and also to evaluate the equations at different positions. As we saw in Sect. 3.2, *first-order* derivatives are much more accurately evaluated on staggered grids, such that if a variable is located on the integer grid, its first derivative is best evaluated on the half-grid, and vice versa. This holds with respect to both space and time. Therefore, if we choose to place E_x on the integer points both in space and in time, H_y should be on the half-grids in both variables, as illustrated in Fig. 5.8. This arrangement is called "staggered grids."

Let $|_r$ be an index that refers to the z-coordinate and let $|^n$ refer to the time coordinate such that $f|_r^n \equiv f(r\Delta z, n\Delta t)$. Then, (5.10) is applied at integer space points (indexed by r) and half-integer time points (indexed by $n + 1/2$) using centered and local finite differences in both z and t. Similarly, (5.11) is applied at half-integer space points (indexed by $r + 1/2$) and integer time points (indexed by n) points, also using centered and local finite differences in both z and t. The finite difference approximation of (5.10)–(5.11) on the staggered grids reads

$$\frac{E_x|_r^{n+1} - E_x|_r^n}{\Delta t} = -\frac{1}{\epsilon}\frac{H_y|_{r+\frac{1}{2}}^{n+\frac{1}{2}} - H_y|_{r-\frac{1}{2}}^{n+\frac{1}{2}}}{\Delta z}, \tag{5.12}$$

$$\frac{H_y\big|_{r+\frac{1}{2}}^{n+\frac{1}{2}} - H_y\big|_{r+\frac{1}{2}}^{n-\frac{1}{2}}}{\Delta t} = -\frac{1}{\mu}\frac{E_x\big|_{r+1}^{n} - E_x\big|_{r}^{n}}{\Delta z}. \tag{5.13}$$

As initial conditions we need one time level for E_x and one for H_y.

For problems with variable permittivity and permeability, it is important to keep in mind that (5.12) is evaluated on the integer grid and (5.13) is evaluated on the half-grid. Consequently, it is natural to sample the permittivity on the integer grid that gives $\epsilon = \epsilon(z_r)$ with $zr = r\Delta z$. Similarly, the permeability is evaluated on the half-grid which gives that $\mu = \mu(z_{r+\frac{1}{2}})$.

Interfaces between regions with homogeneous but different material parameters can be treated in the following way: we place a grid point z_r (where the electric field is defined) at the interface and choose the permittivity at this grid point to be the average of the permittivities in the two media sharing the interface, i.e., $\epsilon = (\epsilon_A + \epsilon_B)/2$ at z_r, where ϵ_A and ϵ_B denote the permittivities in the two media. The permeability is then unproblematic, since it is evaluated at least half a cell from the interface. This approach maintains the order of convergence for the FDTD scheme, whereas other approaches may yield deteriorated convergence properties.

It is instructive to eliminate H_y from (5.12)–(5.13):

$$
\begin{aligned}
&\frac{E_x\big|_r^{n+1} - 2E_x\big|_r^{n} + E_x\big|_r^{n-1}}{(\Delta t)^2} \\
\{\text{rearrange}\} &= \frac{1}{\Delta t}\left(\frac{E_x\big|_r^{n+1} - E_x\big|_r^{n}}{\Delta t} - \frac{E_x\big|_r^{n} - E_x\big|_r^{n-1}}{\Delta t}\right) \\
\{(5.12)\} &= -\frac{1}{\epsilon\Delta t}\left(\frac{H_y\big|_{r+\frac{1}{2}}^{n+\frac{1}{2}} - H_y\big|_{r-\frac{1}{2}}^{n+\frac{1}{2}}}{\Delta z} - \frac{H_y\big|_{r+\frac{1}{2}}^{n-\frac{1}{2}} - H_y\big|_{r-\frac{1}{2}}^{n-\frac{1}{2}}}{\Delta z}\right) \\
\{\text{rearrange}\} &= -\frac{1}{\epsilon\Delta z}\left(\frac{H_y\big|_{r+\frac{1}{2}}^{n+\frac{1}{2}} - H_y\big|_{r+\frac{1}{2}}^{n-\frac{1}{2}}}{\Delta t} - \frac{H_y\big|_{r-\frac{1}{2}}^{n+\frac{1}{2}} - H_y\big|_{r-\frac{1}{2}}^{n-\frac{1}{2}}}{\Delta t}\right) \\
\{(5.13)\} &= \frac{1}{\epsilon\mu\Delta z}\left(\frac{E_x\big|_{r+1}^{n} - E_x\big|_r^{n}}{\Delta z} - \frac{E_x\big|_r^{n} - E_x\big|_{r-1}^{n}}{\Delta z}\right) \\
\{\text{rearrange}\} &= c^2\frac{E_x\big|_{r+1}^{n} - 2E_x\big|_r^{n} + E_x\big|_{r-1}^{n}}{(\Delta z)^2}.
\end{aligned} \tag{5.14}
$$

Thus, E_x evolved according to the coupled first-order equations (5.12)–(5.13) on the staggered grid satisfies the 1D wave equation on standard integer grids, which we studied in Sect. 5.1. Therefore, the dispersion properties and the stability condition of the coupled first-order system are the same as for the wave equation; for instance, $\Delta t \leq \Delta z/c$ is necessary for stability.

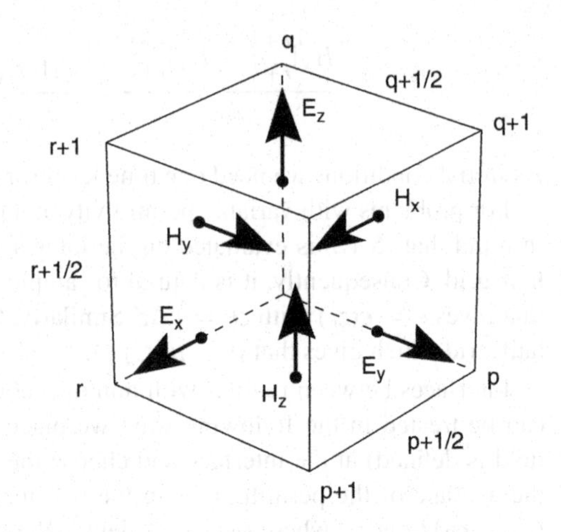

Fig. 5.9 Unit cell in the 3D FDTD algorithm

If we had not used staggered grids for E_x and H_y, but taken the first derivative in z across two cells, then the resulting difference approximation for the second-order z-derivative in (5.14) would involve $E_x|_{r+2}^n$ and $E_x|_{r-2}^n$. This is less accurate and makes the grids with r even and odd decouple. E_x components with r odd would evolve completely independently of those with r even. We conclude that in order to get the same accuracy and robustness as the 1D wave equation for E_x, it is necessary to place one of E_x and H_y on a half-grid; that is, we must use a staggered grid for the coupled first-order equations.

5.2.2 *Three Space Dimensions*

The Yee scheme extends the staggering to three dimensions with a special arrangement of all the components of $\boldsymbol{E}$ and $\boldsymbol{H}$. The electric field components are computed at "integer" time-steps and the magnetic field at "half-integer" time-steps. Space is divided into bricks with sides Δx, Δy, and Δz (usually one uses cubes with $\Delta x = \Delta y = \Delta z = h$). The different field components are placed in the grid according to the unit cell shown in Fig. 5.9.

The electric field components are placed at the midpoints of the corresponding edges; E_x is placed at the midpoints of edges oriented in the x-direction, E_y at the midpoints of edges oriented in the y-direction, and E_z at the midpoints of edges oriented in the z-direction. Thus, E_x is on the half-grid in x and on the integer grids in y and z, etc. The magnetic field components are placed at the centers of the faces of the cubes and oriented normal to the faces. H_x components are placed at the centers of faces in the yz-plane, H_y components are centered on faces in the xz-plane, and H_z components are centered on faces in the xy-plane. Thus, H_x is on the integer grid in x and on the half-grids in y and z, etc. This arrangement was introduced by Yee [93], and the unit cell in Fig. 5.9 is also known as the Yee cell.

Let $|_{p,q,r}$ be indices that refer to the x, y, and z coordinates and let $|^n$ refer to the time coordinate such that $f|^n_{p,q,r} \equiv f(p\Delta x, q\Delta y, r\Delta z, n\Delta t)$. With the Yee arrangement for the field components, the finite difference approximation of Maxwell's equations (5.4)–(5.9) reads

$$\epsilon \frac{E_x|^{n+1}_{p+\frac{1}{2},q,r} - E_x|^{n}_{p+\frac{1}{2},q,r}}{\Delta t} = \frac{H_z|^{n+\frac{1}{2}}_{p+\frac{1}{2},q+\frac{1}{2},r} - H_z|^{n+\frac{1}{2}}_{p+\frac{1}{2},q-\frac{1}{2},r}}{\Delta y} - \frac{H_y|^{n+\frac{1}{2}}_{p+\frac{1}{2},q,r+\frac{1}{2}} - H_y|^{n+\frac{1}{2}}_{p+\frac{1}{2},q,r-\frac{1}{2}}}{\Delta z}, \tag{5.15}$$

$$\epsilon \frac{E_y|^{n+1}_{p,q+\frac{1}{2},r} - E_y|^{n}_{p,q+\frac{1}{2},r}}{\Delta t} = \frac{H_x|^{n+\frac{1}{2}}_{p,q+\frac{1}{2},r+\frac{1}{2}} - H_x|^{n+\frac{1}{2}}_{p,q+\frac{1}{2},r-\frac{1}{2}}}{\Delta z} - \frac{H_z|^{n+\frac{1}{2}}_{p+\frac{1}{2},q+\frac{1}{2},r} - H_z|^{n+\frac{1}{2}}_{p-\frac{1}{2},q+\frac{1}{2},r}}{\Delta x}, \tag{5.16}$$

$$\epsilon \frac{E_z|^{n+1}_{p,q,r+\frac{1}{2}} - E_z|^{n}_{p,q,r+\frac{1}{2}}}{\Delta t} = \frac{H_y|^{n+\frac{1}{2}}_{p+\frac{1}{2},q,r+\frac{1}{2}} - H_y|^{n+\frac{1}{2}}_{p-\frac{1}{2},q,r+\frac{1}{2}}}{\Delta x} - \frac{H_x|^{n+\frac{1}{2}}_{p,q+\frac{1}{2},r+\frac{1}{2}} - H_x|^{n+\frac{1}{2}}_{p,q-\frac{1}{2},r+\frac{1}{2}}}{\Delta y}, \tag{5.17}$$

and

$$\mu \frac{H_x|^{n+\frac{1}{2}}_{p,q+\frac{1}{2},r+\frac{1}{2}} - H_x|^{n-\frac{1}{2}}_{p,q+\frac{1}{2},r+\frac{1}{2}}}{\Delta t} = \frac{E_y|^{n}_{p,q+\frac{1}{2},r+1} - E_y|^{n}_{p,q+\frac{1}{2},r}}{\Delta z} - \frac{E_z|^{n}_{p,q+1,r+\frac{1}{2}} - E_z|^{n}_{p,q,r+\frac{1}{2}}}{\Delta y}, \tag{5.18}$$

$$\mu \frac{H_y|^{n+\frac{1}{2}}_{p+\frac{1}{2},q,r+\frac{1}{2}} - H_y|^{n-\frac{1}{2}}_{p+\frac{1}{2},q,r+\frac{1}{2}}}{\Delta t} = \frac{E_z|^{n}_{p+1,q,r+\frac{1}{2}} - E_z|^{n}_{p,q,r+\frac{1}{2}}}{\Delta x} - \frac{E_x|^{n}_{p+\frac{1}{2},q,r+1} - E_x|^{n}_{p+\frac{1}{2},q,r}}{\Delta z}, \tag{5.19}$$

$$\mu \frac{H_z|^{n+\frac{1}{2}}_{p+\frac{1}{2},q+\frac{1}{2},r} - H_z|^{n-\frac{1}{2}}_{p+\frac{1}{2},q+\frac{1}{2},r}}{\Delta t} = \frac{E_x|^{n}_{p+\frac{1}{2},q+1,r} - E_x|^{n}_{p+\frac{1}{2},q,r}}{\Delta y} - \frac{E_y|^{n}_{p+1,q+\frac{1}{2},r} - E_y|^{n}_{p,q+\frac{1}{2},r}}{\Delta x}. \tag{5.20}$$

The Yee scheme, or FDTD scheme, has proven very successful for microwave problems. All derivatives are centered and as compact as possible, that is, they are taken across a single cell.

Chapter 6 treats the FEM and, there, we return to this type of approximation and demonstrate how it can be deduced by means of the FEM formulated for a discretization with brick shaped finite elements. For such brick shaped elements, we use the so-called edge elements (also known as curl-conforming elements) to approximate the electric field, which is associated with the edges of the Cartesian grid. In addition, we exploit so-called face elements (also known as divergence-conforming elements) to approximate the magnetic field, which is associated with the faces of the Cartesian grid. The mathematical expressions for these field approximations are given in Appendix B.2.3 and their usage is described in Chap. 6. However, it can be useful to know already at this point that the FDTD scheme can be derived in a different manner. In fact, it is possible to view the FDTD scheme as a special case of the finite element method formulated for brick shaped elements. Such FEM arrangements for the spatial discretization are also used for frequency-domain microwave calculations and eddy current calculations.

5.2.3 *MATLAB: Cubical Cavity*

In this example we will use the FDTD to compute the resonant frequencies of an air-filled, cubical cavity with metal walls. By evolving the electric field in time and sampling it at some locations in the cavity, we get the electric fields at these locations as functions of time. We then use a discrete Fourier transform to find the resonant frequencies of the cavity.

5.2.3.1 Discretization

First the cavity must be discretized. Let us divide the cavity into $N_x \times N_y \times N_z$ cells. A cavity divided into $3 \times 4 \times 2$ cells is shown in Fig. 5.10. To store the fields both inside the cavity and on the cavity wall, we need to store the values of

$$\begin{aligned}
E_x \text{ at } 3 \times 5 \times 3 &= N_x \times (N_y+1) \times (N_z+1) \text{ positions,}\\
E_y \text{ at } 4 \times 4 \times 3 &= (N_x+1) \times N_y \times (N_z+1) \text{ positions,}\\
E_z \text{ at } 4 \times 5 \times 2 &= (N_x+1) \times (N_y+1) \times N_z \text{ positions,}
\end{aligned}$$

$$\begin{aligned}
H_x \text{ at } 4 \times 4 \times 2 &= (N_x+1) \times N_y \times N_z \text{ positions,}\\
H_y \text{ at } 3 \times 5 \times 2 &= N_x \times (N_y+1) \times N_z \text{ positions, and}\\
H_z \text{ at } 3 \times 4 \times 3 &= N_x \times N_y \times (N_z+1) \text{ positions.}
\end{aligned}$$

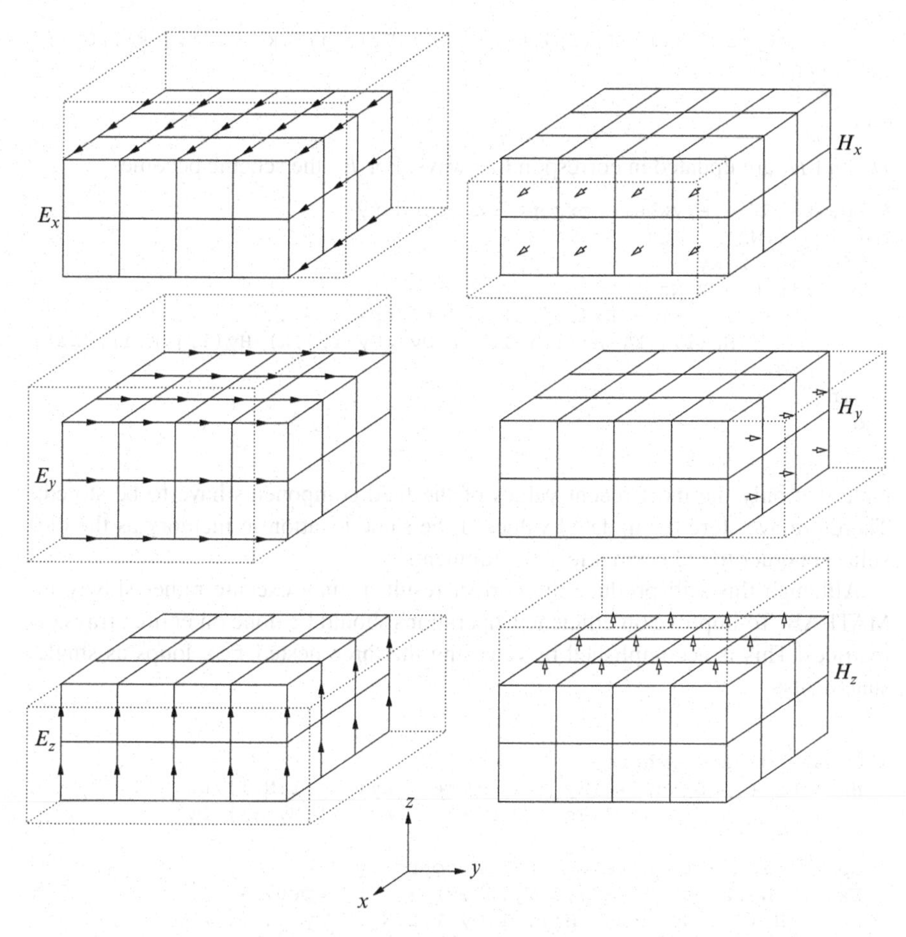

Fig. 5.10 An illustration of how the different field components are placed on a grid with $3 \times 4 \times 2$ cells. The dotted lines indicate the number of unknowns (cells) that have to be stored

5.2.3.2 Boundary Conditions

At microwave frequencies, metal surfaces behave, to a good approximation, as perfect electric conductors (PEC). Therefore, we set the tangential component of the electric field to zero on the metal boundaries.

Taking into account the arrangement of $\boldsymbol{E}$ and $\boldsymbol{H}$, with the PEC boundary condition, we can write FORTRAN-styled loops over indices, for updating H_x as follows:

```
% Update Hx
for i = 1:Nx+1
   for j = 1:Ny
      for k = 1:Nz
         Hx(i,j,k) = Hx(i,j,k) + (Dt/mu0)* ...
```

```
        ((Ey(i,j,k+1)-Ey(i,j,k))/Dz - (Ez(i,j+1,k)-Ez(i,j,k))/Dy);
      end
   end
end
```

H_y and H_z are updated in corresponding ways. For E_x the scheme becomes

```
% Update Ex everywhere except on boundary
for i = 1:Nx
   for j = 2:Ny
      for k = 2:Nz
         Ex(i,j,k) = Ex(i,j,k) + (Dt /eps0) * ...
           ((Hz(i,j,k)-Hz(i,j-1,k))/Dy-(Hy(i,j,k)-Hy(i,j,k-1))/Dz);
      end
   end
end
```

Note that only the most recent values of the field components have to be stored. Therefore, we store the updated values at the same location in memory as the old values in order to reduce memory requirements.

Although this will produce the correct result, it may execute rather slowly in MATLAB. To improve on efficiency, operations should be done on entire arrays or matrices. This is accomplished by rewriting the three nested `for` loops as single statements:

```
% Update Hx everywhere
  Hx = Hx + (Dt/mu0)*((Ey(:,:,2:Nz+1)-Ey(:,:,1:Nz))/Dz ...
                    - (Ez(:,2:Ny+1,:)-Ez(:,1:Ny,:))/Dy);
...
% Update Ex everywhere except on boundary
  Ex(:,2:Ny,2:Nz) = Ex(:,2:Ny,2:Nz) + (Dt /eps0) * ...
      ((Hz(:,2:Ny,2:Nz)-Hz(:,1:Ny-1,2:Nz))/Dy ...
     - (Hy(:,2:Ny,2:Nz)-Hy(:,2:Ny,1:Nz-1))/Dz);
```

Finally, the differences in the discretized curl operator can be written even more compactly by using the `diff` function, as will be shown in the complete program that follows.

5.2.3.3 Initial Conditions

In order to observe an eigenfrequency in the resulting frequency spectrum, the corresponding eigenmode must be excited. An initial condition for $\boldsymbol{E}$ in the form of a random field ensures that most modes are excited. [This leads to $\nabla \cdot \boldsymbol{E} \neq 0$ in the initial condition. Since there is no electric current, the resulting electrical charge density $\rho = \epsilon_0 \nabla \cdot \boldsymbol{E}$ should be time-independent. Fortunately, one of the good properties of the FDTD scheme is that it preserves this property of Maxwell's equations exactly.]

5.2.3.4 Sampling

It is important to sample the fields in such a way that all desired frequencies (modes) are detected. With only a bit of bad luck, some modes will have a node (zero) at the chosen detector location. To avoid this problem, it is a good idea to record several field components at several detector locations.

5.2.3.5 Choice of Time Step

The larger the time step, the smaller the dispersion and the faster the simulation. Therefore, we choose Δt as big as possible, i.e., at the stability limit (5.33).

A MATLAB program that simulates the field inside a brick-shaped cavity with PEC walls is listed below. In the time-stepping part, (5.15)–(5.20) are evaluated using the MATLAB function `diff`. For a vector `X`, of length N, `diff(X)` is the vector `[X(2)-X(1) X(3)-X(2) ... X(N)-X(N-1)]` of length $N-1$. The second argument of `diff` is the order of the difference, in this case 1, for the first derivative. The third argument specifies the dimension in which differences are taken ($x \to 1,\ y \to 2,\ z \to 3$).

The frequency spectrum we get from the Fourier transform of the columns of `Et` is plotted in Fig. 5.11 together with the analytical resonant frequencies:

$$f_{mnp} = \frac{c}{2}\left[(m/L_x)^2 + (n/L_y)^2 + (p/L_z)^2\right]^{1/2}. \tag{5.21}$$

In this case there are two kinds of modes, referred to as TM$_{mnp}$ and TE$_{mnp}$ modes (see, e.g., [19]). For TM$_{mnp}$ modes, $m \neq 0, n \neq 0$. For TE$_{mnp}$ modes, $p \neq 0$, m or n is nonzero.

5.2.4 *Integral Interpretation of the FDTD Method*

The Yee-scheme, (5.15)–(5.20), can also be derived using the integral representation of Maxwell's equations:

$$\int_S \frac{\partial(\epsilon \boldsymbol{E})}{\partial t} \cdot d\boldsymbol{S} = \oint_{\partial S} \boldsymbol{H} \cdot d\boldsymbol{l}, \tag{5.22}$$

$$\int_S \frac{\partial(\mu \boldsymbol{H})}{\partial t} \cdot d\boldsymbol{S} = -\oint_{\partial S} \boldsymbol{E} \cdot d\boldsymbol{l}. \tag{5.23}$$

To obtain the equation for $\partial H_z/\partial t$ we first compute the surface integral over a face on the grid cells $z = r\Delta z$, $p\Delta x < x < (p+1)\Delta x$, $q\Delta y < y < (q+1)\Delta y$:

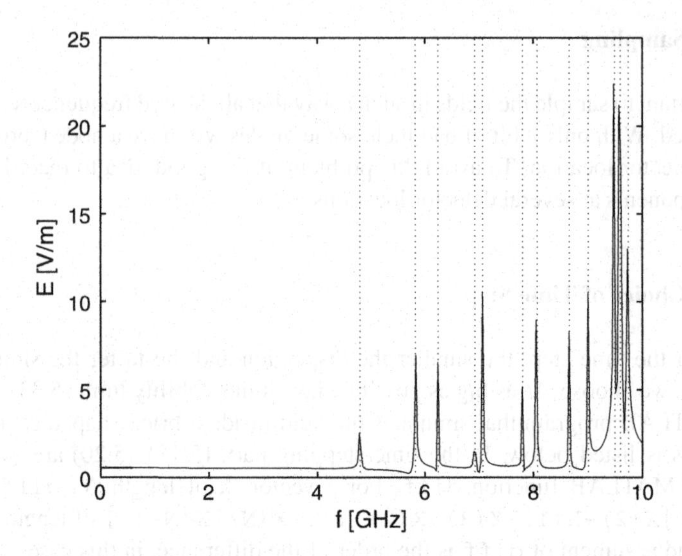

Fig. 5.11 Frequency spectrum obtained from an FDTD simulation of an air-filled brick-shaped cavity. The solid curve shows the frequency spectrum of the sum of the sampled E_x, E_y, and E_z components. The dotted lines show the exact eigenfrequencies

$$\int_S \frac{\partial(\mu \boldsymbol{H})}{\partial t} \cdot d\boldsymbol{S} \approx \mu \frac{H_z|^{n+\frac{1}{2}}_{p+\frac{1}{2},q+\frac{1}{2},r} - H_z|^{n-\frac{1}{2}}_{p+\frac{1}{2},q+\frac{1}{2},r}}{\Delta t} \Delta x \Delta y. \tag{5.24}$$

The corresponding line integral of $\boldsymbol{E}$ along the line circulating $H_z|_{p+\frac{1}{2},q+\frac{1}{2},r}$ according to the right-hand rule, shown in Fig. 5.12, is calculated as

$$\begin{aligned} \oint_{\partial S} \boldsymbol{E} \cdot d\boldsymbol{l} \approx\; & E_x|^n_{p+\frac{1}{2},q,r} \Delta x + E_y|^n_{p+1,q+\frac{1}{2},r} \Delta y \\ & - E_x|^n_{p+\frac{1}{2},q+1,r} \Delta x - E_y|^n_{p,q+\frac{1}{2},r} \Delta y. \end{aligned} \tag{5.25}$$

Here, the Yee arrangement has the nice property that the components of $\boldsymbol{E}$ that are needed for this integral appear exactly at the midpoint of the edges along which they are to be integrated.

Combining (5.23)–(5.25) we obtain

$$\begin{aligned} & \mu \frac{H_z|^{n+\frac{1}{2}}_{p+\frac{1}{2},q+\frac{1}{2},r} - H_z|^{n-\frac{1}{2}}_{p+\frac{1}{2},q+\frac{1}{2},r}}{\Delta t} \\ & = -\frac{E_y|^n_{p+1,q+\frac{1}{2},r} - E_y|^n_{p,q+\frac{1}{2},r}}{\Delta x} + \frac{E_x|^n_{p+\frac{1}{2},q+1,r} - E_x|^n_{p+\frac{1}{2},q,r}}{\Delta y} \end{aligned} \tag{5.26}$$

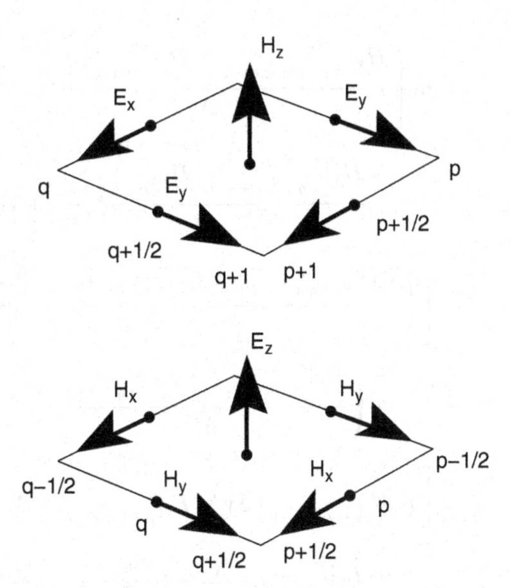

Fig. 5.12 An illustration showing how H_z and E_z are "circulated" by four electric and magnetic components respectively in the Yee grid

which is exactly the same as the previously derived (5.20).

Another interesting property of the Yee scheme is that the condition of solenoidal magnetic flux density (1.4) is implicitly enforced for all times, provided that the initial conditions are correct. To demonstrate this, we apply Gauss's theorem to (1.4), and this gives $\oint_S \boldsymbol{B} \cdot \hat{\boldsymbol{n}} dS = 0$, where the closed surface S is taken as the surface of the unit cell shown in Fig. 5.9. This integral is divided into three Cartesian components: $\oint_S \boldsymbol{B} \cdot \hat{\boldsymbol{n}}\, dS = \int_{S_p} \boldsymbol{B} \cdot \hat{\boldsymbol{n}}\, dS + \int_{S_q} \boldsymbol{B} \cdot \hat{\boldsymbol{n}}\, dS + \int_{S_r} \boldsymbol{B} \cdot \hat{\boldsymbol{n}}\, dS$. For example, S_p is the two surfaces in the yz-plane that are defined by constant index p and $p+1$. It is instructive to study the time derivative of $\int_{S_p} \boldsymbol{B} \cdot \hat{\boldsymbol{n}} dS$ in the discrete setting. The integrals over the surfaces p and $p+1$ are evaluated as in (5.24), and given this result, we form the time derivative (in the leap-frog sense) centered at n. Next, we change the order of the (numerical) time derivative and surface integral, which yields an expression that features the time derivative of the normal component of the magnetic field for the two surfaces. These are shown in (5.18), which is the x-component of Faraday's law, and we use this relation to replace the time derivative of the magnetic field with the curl of the electric field, still working only with the x-component. The last step is to rewrite the x-components of the curl into the circulation of the electric field along the contour of the surfaces p and $p+1$. Here are the detailed calculations:

$$
\frac{\partial}{\partial t} \int_{S_p} \boldsymbol{B} \cdot \hat{\boldsymbol{n}}\, dS \approx \frac{\mu_0}{\Delta t} \Bigg[\left(H_x\Big|^{n+\frac{1}{2}}_{p+1,q+\frac{1}{2},r+\frac{1}{2}} - H_x\Big|^{n+\frac{1}{2}}_{p,q+\frac{1}{2},r+\frac{1}{2}} \right) \Delta y \Delta z
$$

$$
- \left(H_x\Big|^{n-\frac{1}{2}}_{p+1,q+\frac{1}{2},r+\frac{1}{2}} - H_x\Big|^{n-\frac{1}{2}}_{p,q+\frac{1}{2},r+\frac{1}{2}} \right) \Delta y \Delta z \Bigg]
$$

$$= \mu_0 \left[\frac{H_x|^{n+\frac{1}{2}}_{p+1,q+\frac{1}{2},r+\frac{1}{2}} - H_x|^{n-\frac{1}{2}}_{p+1,q+\frac{1}{2},r+\frac{1}{2}}}{\Delta t} - \frac{H_x|^{n+\frac{1}{2}}_{p,q+\frac{1}{2},r+\frac{1}{2}} - H_x|^{n-\frac{1}{2}}_{p,q+\frac{1}{2},r+\frac{1}{2}}}{\Delta t} \right] \Delta y \Delta z$$

$$= \left[\left(\frac{E_y|^n_{p+1,q+\frac{1}{2},r+1} - E_y|^n_{p+1,q+\frac{1}{2},r}}{\Delta z} - \frac{E_z|^n_{p+1,q+1,r+\frac{1}{2}} - E_z|^n_{p+1,q,r+\frac{1}{2}}}{\Delta y} \right) - \left(\frac{E_y|^n_{p,q+\frac{1}{2},r+1} - E_y|^n_{p,q+\frac{1}{2},r}}{\Delta z} - \frac{E_z|^n_{p,q+1,r+\frac{1}{2}} - E_z|^n_{p,q,r+\frac{1}{2}}}{\Delta y} \right) \right] \Delta y \Delta z$$

$$= \left[E_y|^n_{p+1,q+\frac{1}{2},r+1} \Delta y - E_z|^n_{p+1,q+1,r+\frac{1}{2}} \Delta z - E_y|^n_{p+1,q+\frac{1}{2},r} \Delta y + E_z|^n_{p+1,q,r+\frac{1}{2}} \Delta z - E_y|^n_{p,q+\frac{1}{2},r+1} \Delta y - E_z|^n_{p,q,r+\frac{1}{2}} \Delta z + E_y|^n_{p,q+\frac{1}{2},r} \Delta y + E_z|^n_{p,q+1,r+\frac{1}{2}} \Delta z \right].$$

The corresponding results for the other two surface integrals, evaluated over S_q and S_r, are given by cyclic permutations of the final result for S_p. When these three expressions are added, we find that the circulations on the six faces of the cube give, in total, two contributions to each edge of the unit cell that cancel each other. Consequently, the condition of solenoidal magnetic flux density (1.4) is preserved numerically at all times, given appropriate initial conditions. A similar analysis can be applied to Gauss's law (1.3).

5.2.5 Dispersion Analysis in Three Dimensions

To simplify the dispersion analysis (and also to allow later comparison with the finite element approach in Chap. 6), we note that one can eliminate $\boldsymbol{H}$ by forming the second-order time derivative for $\boldsymbol{E}$, in the same way as we did for the 1D case in (5.14). Starting from (5.15)–(5.20), a somewhat lengthy calculation (assuming that ϵ and μ are constant) gives

$$\frac{1}{c^2} \frac{E_x|^{n+1}_{p+\frac{1}{2},q,r} - 2E_x|^n_{p+\frac{1}{2},q,r} + E_x|^{n-1}_{p+\frac{1}{2},q,r}}{(\Delta t)^2} \tag{5.27}$$

$$
\begin{aligned}
&= \frac{E_x|^n_{p+\frac{1}{2},q+1,r} - 2E_x|^n_{p+\frac{1}{2},q,r} + E_x|^n_{p+\frac{1}{2},q-1,r}}{(\Delta y)^2} \\
&+ \frac{E_x|^n_{p+\frac{1}{2},q,r+1} - 2E_x|^n_{p+\frac{1}{2},q,r} + E_x|^n_{p+\frac{1}{2},q,r-1}}{(\Delta z)^2} \\
&- \frac{E_y|^n_{p+1,q+\frac{1}{2},r} - E_y|^n_{p,q+\frac{1}{2},r} - E_y|^n_{p+1,q-\frac{1}{2},r} + E_y|^n_{p,q-\frac{1}{2},r}}{\Delta x \Delta y} \\
&- \frac{E_z|^n_{p+1,q,r+\frac{1}{2}} - E_z|^n_{p,q,r+\frac{1}{2}} - E_z|^n_{p+1,q,r-\frac{1}{2}} + E_z|^n_{p,q,r-\frac{1}{2}}}{\Delta x \Delta z}.
\end{aligned}
$$

This is the finite difference form of

$$
\frac{1}{c^2}\frac{\partial^2 E_x}{\partial t^2} = \left(\frac{\partial^2}{\partial y^2} + \frac{\partial^2}{\partial z^2}\right) E_x - \frac{\partial}{\partial x}\left(\frac{\partial E_y}{\partial y} + \frac{\partial E_z}{\partial z}\right),
$$

which, in turn, is the x-component of the curl-curl equation for $\boldsymbol{E}$:

$$
\frac{1}{c^2}\frac{\partial^2 \boldsymbol{E}}{\partial t^2} = \nabla^2 \boldsymbol{E} - \nabla(\nabla \cdot \boldsymbol{E}) = -\nabla \times \nabla \times \boldsymbol{E}. \tag{5.28}
$$

The dispersion relation for FDTD in three dimensions can be found in several different ways. For instance, one can start from the electric field formulation (5.28) for all three components and plug in a plane wave solution $\boldsymbol{E} = (e_x, e_y, e_z)\exp[j(\omega t - k_x x - k_y y - k_z z)]$. From the analysis in Sect. 3.2, we know that on staggered grids, where first-order derivatives are taken across one cell and second-order derivatives across two cells, numerical derivatives acting on such exponentials simply multiply the function by the following imaginary factors:

$$
\begin{aligned}
\frac{\partial}{\partial t} &\to D_t = \frac{2j}{\Delta t}\sin\frac{\omega \Delta t}{2}, \\
\frac{\partial}{\partial x} &\to D_x = \frac{-2j}{\Delta x}\sin\frac{k_x \Delta x}{2}, \\
\frac{\partial}{\partial y} &\to D_y = \frac{-2j}{\Delta y}\sin\frac{k_y \Delta y}{2}, \\
\frac{\partial}{\partial z} &\to D_z = \frac{-2j}{\Delta x}\sin\frac{k_z \Delta z}{2}.
\end{aligned} \tag{5.29}
$$

Thus, for complex exponentials, the matrix equation corresponding to the three vector components of (5.28) is

$$\begin{pmatrix} D_y^2 + D_z^2 - D_t^2/c^2 & -D_x D_y & -D_x D_z \\ -D_x D_y & D_x^2 + D_z^2 - D_t^2/c^2 & -D_y D_z \\ -D_x D_z & -D_y D_z & D_x^2 + D_y^2 - D_t^2/c^2 \end{pmatrix} \begin{pmatrix} e_x \\ e_y \\ e_z \end{pmatrix} = \begin{pmatrix} 0 \\ 0 \\ 0 \end{pmatrix}, \tag{5.30}$$

where $D_t = j\omega$ for the continuous case and $D_t = (2j/\Delta t)\sin(\omega\Delta t/2)$ for the discretized system, etc. By setting the determinant of the matrix to zero, we find two roots,

$$D_t^2 = c^2(D_x^2 + D_y^2 + D_z^2), \tag{5.31}$$

representing transverse electromagnetic waves with two polarizations $\boldsymbol{e} \perp \boldsymbol{k}$. We get the usual (and exact) dispersion relation for light waves $\omega^2 = c^2(k_x^2 + k_y^2 + k_z^2)$ by replacing $D_t \to j\omega$ and $D_{x,y,z} \to -jk_{x,y,z}$, where the polarizations of the two solutions are completely orthogonal as expected. In addition, there is one root $D_t^2 = 0$ of (5.30), which translates into $\omega = 0$. This represents an "electrostatic" solution with $\boldsymbol{e} \parallel \boldsymbol{k}$, i.e., a longitudinal, time-independent solution. Note that this solution does not propagate. It gives a purely static response of the electric field to space charge.

It is interesting to see how the electrostatic solutions are treated by the FDTD. Clearly, any electrostatic field $\boldsymbol{E} = -\nabla\phi$, with ϕ constant in time, and an arbitrary function of space, is a solution of the curl-curl equation (5.28). One can verify that a solution $\boldsymbol{E} = -\nabla\phi$ does not evolve in time with the FDTD algorithm. This time-independent solution corresponds to the root $D_t^2 = 0$ of (5.30). Thus, the Yee scheme preserves the null-space of the curl-curl operator, and this is one of its many good properties.

The numerical dispersion relation for the electromagnetic waves is obtained by substituting the discrete derivative operators (5.29) into the general dispersion relation (5.31):

$$\frac{\sin^2 \omega\Delta t/2}{(c\Delta t)^2} = \frac{\sin^2 k_x\Delta x/2}{(\Delta x)^2} + \frac{\sin^2 k_y\Delta y/2}{(\Delta y)^2} + \frac{\sin^2 k_z\Delta z/2}{(\Delta z)^2}. \tag{5.32}$$

This is a natural generalization of the result in one dimension (5.3). Taylor expansion of the sine functions shows that $\omega^2 = c^2(k_x^2 + k_y^2 + k_z^2)[1 + O(k^2h^2)]$, so that the deviation from the correct dispersion relation for electromagnetic waves is $O(k^2h^2)$ for a cubic grid with $\Delta x = \Delta y = \Delta z = h$. Note that the dispersion is anisotropic. The wave propagation is the slowest along the coordinate directions, and faster (and closer to the correct result) in oblique directions.

The maximum time-step for stability follows from the requirement

$$\sin^2(\omega\Delta t/2) \leq 1 \text{ for all } \boldsymbol{k},$$

just as in one dimension, and this gives

$$c\Delta t \leq \left[\frac{1}{(\Delta x)^2} + \frac{1}{(\Delta y)^2} + \frac{1}{(\Delta z)^2}\right]^{-1/2}. \tag{5.33}$$

For a cubic grid with $\Delta x = \Delta y = \Delta z = h$, the stability condition simplifies to

$$\Delta t \leq \frac{h}{c\sqrt{3}}. \tag{5.34}$$

In comparison with the 1D case, the maximum time-step has been reduced by a factor $\sqrt{3}$. Because of this stability requirement, the spatial discretization error is generally larger than the temporal discretization error for the FDTD scheme in three dimensions, but they cancel each other to some extent. This means, for example, that there is no magic time-step in this case. (Actually, for fields varying equally fast in all directions, $|k_x h| = |k_y h| = |k_z h|$, the stability limit (5.34) *is* the magic time-step, but this works only for propagation in those particular directions.)

Waves propagating along the coordinate axes suffer most from numerical dispersion. To quantify the effects of the numerical disperions, we consider a wave propagating in the x-direction, i.e., $k_x = k$ and $k_y = k_z = 0$. Further, we assume that $\Delta x = \Delta y = \Delta z = h$ and $c\Delta t/h = 1/\sqrt{3}$. In this case (5.32) simplifies to

$$\sin(\omega \Delta t/2) = \frac{1}{\sqrt{3}} \sin(kh/2). \tag{5.35}$$

An expression for the phase velocity $v_p = \omega/k$ of this wave can be derived from a series expansion of (5.35):

$$v_p = \frac{\omega}{k} = c\left(1 - \frac{k^2h^2}{36} + O(k^4h^4)\right). \tag{5.36}$$

If we demand the relative error in phase velocity to be less than 1%, we require $(kh)^2 < 36/100$, which, since $k \equiv 2\pi/\lambda$, leads to $\lambda/h < \pi/\sqrt{0.09} \approx 10.5$, that is, at least 10.5 cells per wavelength. This takes account of the partial cancellation of the spatial and temporal errors in (5.32).

The same assumptions as in the preceding paragraph yield the following expression for the group velocity:

$$v_g = \frac{\partial \omega}{\partial k} = c\left(1 - \frac{k^2h^2}{12} + O(k^4h^4)\right). \tag{5.37}$$

From this we find that a resolution of about 18 cells per wavelength is required to reduce the relative error of the group velocity to 1%. This is a stricter requirement on the resolution as compared to the result derived from (5.36). Typically, about 18 cells per wavelength is used as a rule of thumb for problems that involve only a few wavelengths and engineering accuracy requirements.

The FDTD often requires even higher resolutions if one asks for a fixed *absolute* phase error across the whole computational domain, in particular for problems that are large in terms of wavelengths, since the phase errors accumulate. The absolute phase error is

$$e_{\text{phase}} = (\tilde{k} - k)L = \left(\frac{\omega}{c(1 - (kh)^2/36 + \cdots)} - \frac{\omega}{c} \right) L \approx \frac{k^3 h^2 L}{36} \tag{5.38}$$

for a system with fixed size L. To keep e_{phase} constant, the cell size must scale with frequency as $\omega^{-3/2}$, and consequently, the computational time is proportional to $1/(h^3 \Delta t) \propto \omega^6$.

The error associated with the numerical dispersion relation provides important understanding for one of the contributions to the total error. It must be emphasized that convergence studies or other means of estimating the actual error are, in general, necessary to achieve reliable results for real-world problems.

Review Questions

5.2-1 Draw the unit cell for the FDTD scheme in three space dimensions and add all the field components for both the electric field and the magnetic field.

5.2-2 Reduce the FDTD scheme for the full Maxwell's equations to one and two dimensions. Derive the corresponding wave equations by eliminating the magnetic (or the electric) field.

5.2-3 How many time-levels of the electric and magnetic fields must be stored in the computer's memory for the FDTD scheme?

5.2-4 Derive the Yee scheme from the integral representation of Maxwell's equations.

5.2-5 Show that (5.27) can be derived from (5.15)–(5.20).

5.2-6 Derive the stability condition given the numerical dispersion relation. Motivate the steps in your derivation.

5.3 Boundary Conditions for Open Regions

The FDTD is often applied to microwave problems such as calculation of:

- Radiation patterns from antennas
- Radar cross sections (RCS) for different targets, e.g., aircraft

These problems involve *open regions*, and in principle, the computational domain extends to infinity. Of course it is not practical to discretize an infinite region, and instead, special boundary conditions can be applied to terminate the computational region. Such boundary conditions serve to absorb outgoing waves, and are called

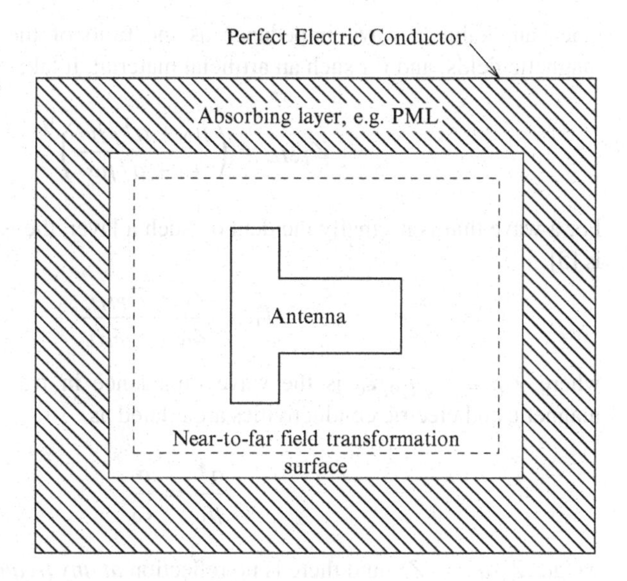

Fig. 5.13 Typical setup for computing the radiation pattern of an antenna with the FDTD

absorbing boundary conditions (ABC). Then, the fields in the *near zone* can be transformed to the *far zone*, several wavelengths or more from the antenna, by means of a so-called near-to-far-field transformation (NTF). Figure 5.13 illustrates its use in an FDTD calculation of the radiation pattern of an antenna.

5.3.1 The Perfectly Matched Layer

A popular set of absorbing boundary conditions is the *perfectly matched layer* (PML) invented by Bérenger [9]. The PML is a layer of artificial material surrounding the computational region and designed to damp waves propagating in the normal direction. The region is then terminated by a PEC. If the waves are sufficiently damped out in the absorbing layer, very little reflection will occur at this PEC surface. The thicker the absorbing layer is, the more efficient is the damping that can be achieved.

Here, we indicate how Bérenger's PML works. The basic idea behind the method is to introduce both an electric conductivity σ and a magnetic conductivity σ^* in the absorbing layer:

$$\epsilon_0 \frac{\partial \boldsymbol{E}}{\partial t} + \sigma \boldsymbol{E} = \nabla \times \boldsymbol{H}, \tag{5.39}$$

$$\mu_0 \frac{\partial \boldsymbol{H}}{\partial t} + \sigma^* \boldsymbol{H} = -\nabla \times \boldsymbol{E}. \tag{5.40}$$

One can define a wave impedance as the ratio of the transversal electric and magnetic fields, and for such an artificial material, it takes the value

$$Z_{PML} = \left(\frac{\mu_0 + \sigma^*/j\omega}{\epsilon_0 + \sigma/j\omega} \right)^{1/2}.$$

For a wave that is normally incident on such a layer, the wave reflection coefficient is [5]

$$\Gamma_0 = \frac{Z_0 - Z_{PML}}{Z_0 + Z_{PML}},$$

where $Z_0 \equiv \sqrt{\mu_0/\epsilon_0}$ is the wave impedance in free space. Evidently, if the magnetic and electric conductivities are related as

$$\frac{\sigma^*}{\mu_0} = \frac{\sigma}{\epsilon_0}, \tag{5.41}$$

we get $Z_{PML} = Z_0$, and there is no reflection *at any frequency*.

For oblique incidence, things become more complicated, and it is harder to avoid reflection. However, Bérenger found a trick that achieves this. It consists in splitting each component of $\boldsymbol{E}$ and $\boldsymbol{H}$ into two parts, for instance, $E_x = E_{xy} + E_{xz}$, according to the direction of the curl operator that contributes to $\partial \boldsymbol{E}/\partial t$. Then, one uses nonzero σ and σ^* only for the derivative in the direction normal to the absorbing layer.

As an example, let us assume that the PML has $\hat{z}$ as the normal direction. The two equations for E_x and E_y are split into four:

$$\epsilon \frac{\partial E_{xy}}{\partial t} = \frac{\partial (H_{zx} + H_{zy})}{\partial y}, \tag{5.42}$$

$$\epsilon \frac{\partial E_{xz}}{\partial t} = -\frac{\partial (H_{yz} + H_{yx})}{\partial z} - \sigma_z E_{xz}, \tag{5.43}$$

$$\epsilon \frac{\partial E_{yz}}{\partial t} = \frac{\partial (H_{xy} + H_{xz})}{\partial z} - \sigma_z E_{yz}, \tag{5.44}$$

$$\epsilon \frac{\partial E_{yx}}{\partial t} = -\frac{\partial (H_{zx} + H_{zy})}{\partial x}. \tag{5.45}$$

The evolution equation for E_z is not modified for a layer with $\hat{z}$ as normal. The magnetic field is treated in a similar way. What is achieved with this trick is that the layer modifies the propagation only in the z-direction, which is the normal direction of the PML, not in the tangential directions x and y. Therefore, no reflection occurs even for waves obliquely incident on the Bérenger PML.

In practice, some reflection occurs if σ varies strongly on the scale of the grid. Therefore, one often chooses profiles for the conductivity, such as parabolic $\sigma(z) = \sigma_0[(z-z_0)/L_z]^2$, for a layer that extends from $z = z_0$ to $z = z_0 + L_z$. Such layers are very good absorbers; 6–8 cells can give a reflection coefficient of –60 to –80 dB.

The PML works well, even when placed very close to the radiating structure or scatterer. This means that it is effective in decreasing the number of cells and consequently reducing the computational cost.

There are alternatives to Bérenger's PML. One that gives the same dispersion properties, without splitting the field components, uses anisotropic, tensorial $\overline{\overline{\epsilon}}_r$ and $\overline{\overline{\mu}}_r$ [55]:

$$\overline{\overline{\epsilon}}_r = \overline{\overline{\mu}}_r = \begin{pmatrix} 1 - j\sigma/\omega\epsilon & 0 & 0 \\ 0 & 1 - j\sigma/\omega\epsilon & 0 \\ 0 & 0 & (1 - j\sigma/\omega\epsilon)^{-1} \end{pmatrix}. \tag{5.46}$$

This involves modifications of the time-stepping.

5.3.2 Near-to-Far-Field Transformation

Figure 5.13 shows a typical setup for computing the radiation pattern of an antenna. The result of main interest is the fields in the *far zone*, several wavelengths from the antenna. This can be computed without extending the computational domain to the far zone by using a *near-to-far-field transformation* (NTF) close to the antenna and adding an ABC just outside the NTF surface. Formulas for the NTF can be found in the book on FDTD by Taflove [80]. Without going through the derivation, we state the formulas for the far field in frequency domain based on the Fourier transform of the near field computed by the FDTD scheme. (The Fourier transform can be computed as part of the time-stepping procedure for selected frequencies.) The field can be expressed in terms of the electric ($\boldsymbol{A}$) and magnetic ($\boldsymbol{F}$) vector potentials as

$$\boldsymbol{E} = -\frac{j\omega}{k^2}\nabla \times \nabla \times \boldsymbol{A} - \frac{1}{\epsilon_0}\nabla \times \boldsymbol{F},$$

$$\boldsymbol{H} = -\frac{j\omega}{k^2}\nabla \times \nabla \times \boldsymbol{F} + \frac{1}{\mu_0}\nabla \times \boldsymbol{A}.$$

The potentials can be calculated from the equivalent electric current $\boldsymbol{J}_s = \hat{\boldsymbol{n}} \times \boldsymbol{H}$ and magnetic current $\boldsymbol{M}_s = -\hat{\boldsymbol{n}} \times \boldsymbol{E}$ on the NTF surface ($\hat{\boldsymbol{n}}$ denotes the outward normal of the NTF surface $\partial\Omega$):

$$\boldsymbol{A} = \frac{\mu_0}{4\pi}\oint_{\partial\Omega} \boldsymbol{J}_s(\boldsymbol{r}')\frac{\exp(-jkR)}{R}dS',$$

$$\boldsymbol{F} = \frac{\epsilon_0}{4\pi}\oint_{\partial\Omega} \boldsymbol{M}_s(\boldsymbol{r}')\frac{\exp(-jkR)}{R}dS'. \tag{5.47}$$

Here R denotes the distance between the source point, $\boldsymbol{r}'$, and the point where we observe the field, $\boldsymbol{r}$. For large distances, one can approximate R in the denominators of (5.47) as a constant, R_0, and in the argument of the exponential as $R \approx R_0 - r' \cos\psi$, where ψ is the angle between $\boldsymbol{r}$ and $\boldsymbol{r}'$. The fields in the radiation zone are

$$\boldsymbol{E} \approx j\omega(\hat{\boldsymbol{r}} \times \hat{\boldsymbol{r}} \times \boldsymbol{A} + Z_0 \hat{\boldsymbol{r}} \times \boldsymbol{F}),$$

$$\boldsymbol{H} \approx j\omega\left(\hat{\boldsymbol{r}} \times \hat{\boldsymbol{r}} \times \boldsymbol{F} - \frac{1}{Z_0}\hat{\boldsymbol{r}} \times \boldsymbol{A}\right).$$

Review Questions

5.3-1 What is meant by an open-region problem and how are these problems handled by FDTD programs?
5.3-2 Use the wave impedance to explain why a normally incident wave is not reflected by the PML at any frequency.
5.3-3 How did Bérenger avoid reflections by the PML for oblique incidence?
5.3-4 How are the electric and magnetic conductivity profiles usually chosen for the PML in an FDTD implementation? What reflection coefficients can be achieved with a PML that is 6–8 cells thick?
5.3-5 Outline a technique for the computation of the fields in the far zone given an FDTD solution in the near zone. Mention some practical situations in which this technique can be used.

Summary

- The FDTD is a standard tool for microwave problems in which the geometrical dimensions are comparable to the wavelength. Its main advantage is that it is both efficient and simple to implement.
- Although the FDTD scheme is very popular, the method suffers from some drawbacks:
  - A main drawback of the FDTD is the way it deals with curved and oblique boundaries, where the standard FDTD solution, known as "staircasing," does not give very accurate results. In this respect, finite elements can do much better.
  - Another disadvantage of the FDTD (in common with finite elements) is that the phase error can become significant when the computational domain is many wavelengths. In this respect, the method of moments is better.
  - Furthermore, the time-step is limited by $\Delta t \leq h/(c\sqrt{3})$, which means that the FDTD cannot be used for eddy current problems.

- The time-dependent system of two first-order equations (Faraday's and Ampère's laws) allows for staggering in both space and time. The discretization of this system exploits centered differences and offers explicit time-stepping. In 1D, we discretize

$$\frac{\partial E_x}{\partial z} = -\mu \frac{\partial H_y}{\partial t}, \qquad -\frac{\partial H_y}{\partial z} = \epsilon \frac{\partial E_x}{\partial t},$$

with $E_x = E_x(r, n)$ and $H_y = H_y(r + \frac{1}{2}, n + \frac{1}{2})$, where r is an integer space index and n is an integer time index. [The corresponding wave equation $\partial^2 E/\partial t^2 = c^2 \partial^2 E/\partial x^2$ can be treated by centered second-order differences and explicit time-stepping.]
- Staggering in three dimensions:

$$E_x|^n_{p+\frac{1}{2},q,r}, \qquad E_y|^n_{p,q+\frac{1}{2},r}, \qquad E_z|^n_{p,q,r+\frac{1}{2}},$$

$$H_x|^{n+\frac{1}{2}}_{p,q+\frac{1}{2},r+\frac{1}{2}}, \qquad H_y|^{n+\frac{1}{2}}_{p+\frac{1}{2},q,r+\frac{1}{2}}, \qquad H_z|^{n+\frac{1}{2}}_{p+\frac{1}{2},q+\frac{1}{2},r}.$$

Electric field components are placed on the midpoint of the edges aligned with the field components. Magnetic field components are centered on the surfaces normal to the field components.
- Numerical dispersion relations (relations between ω and k for $E \propto \exp[j(\omega t - \boldsymbol{k} \cdot \boldsymbol{r})]$) are derived from the finite-difference equations. In three dimensions, we get

$$\frac{\sin^2(\omega \Delta t/2)}{(c\Delta t)^2} = \frac{\sin^2(k_x \Delta x/2)}{(\Delta x)^2} + \frac{\sin^2(k_y \Delta y/2)}{(\Delta y)^2} + \frac{\sin^2(k_z \Delta z/2)}{(\Delta z)^2}.$$

- The stability condition (Courant condition) $c\Delta t/h < 1/\sqrt{n}$ in n dimensions. This can be derived from the numerical dispersion relation.
- Several extensions of the FDTD, such as absorbing boundary conditions, near-to-far-field transformation, and subgrid models for thin wires and slots have been developed, and these allow the FDTD to be applied to a wide range of problems.

Problems

P.5-1 For finite difference computations on unbounded domains, the finite grid must be terminated by boundary conditions that mimic a free-space problem. Use (1.13) to derive boundary conditions for (5.2) when $R = 1$.

P.5-2 Consider a specific point z_0 at a specific time t_0 in Fig. 5.1. A perturbation of the field at this point and time influences the field at later times $t > t_0$ in the region $z_0 - c(t - t_0) < z < z_0 + c(t - t_0)$. Similarly, the field values at earlier times $t < t_0$ within the region $z_0 - c(t_0 - t) < z < z_0 + c(t_0 - t)$ will have an influence on the field at $z = z_0$ and $t = t_0$, and this region is referred to as the light-collecting sector. Relate the stability condition for the 1D FDTD scheme to the light-collecting sector. What happens when the light-collecting sector covers a larger angle than the stencil in (5.2)?

P.5-3 Show that the dispersion relation of the 1D wave equation (5.1) can be expanded as

$$\omega = ck\left[1 - \frac{(k\Delta z)^2}{24}(1 - R^2) + O((k\Delta z)^4)\right]. \tag{5.48}$$

How many points per wavelength are required to get the frequency correct (a) to 1%, (b) to 0.1% if $R = 1/\sqrt{3}$?

P.5-4 Consider the case in which the coupled first-order system shown in (5.12) and (5.13) is applied to solve a problem with continuously varying material parameters. Where should $\epsilon(z)$ and $\mu(z)$ be evaluated on the grid? How would the corresponding problem be treated when the wave equation

$$\frac{\partial}{\partial z}\left(\frac{1}{\mu(z)}\frac{\partial E_x}{\partial z}\right) - \epsilon(z)\frac{\partial^2 E_x}{\partial t^2} = 0$$

is used instead? Where should $\epsilon(z)$ and $\mu(z)$ be evaluated in this case?

P.5-5 Consider the case in which the coupled first-order system shown in (5.12) and (5.13) is applied to a problem with piecewise continuous materials; i.e., there are material discontinuities. Let the grid points associated with an electric field tangential to the material interface be placed on the material interface. How should $\epsilon(z)$ and $\mu(z)$ be evaluated in order to maintain an $O(h^2)$ error? How would the corresponding problem be treated when the wave equation

$$\frac{\partial}{\partial z}\left(\frac{1}{\mu(z)}\frac{\partial E_x}{\partial z}\right) - \epsilon(z)\frac{\partial^2 E_x}{\partial t^2} = 0$$

is used instead? Where are $\epsilon(z)$ and $\mu(z)$ evaluated in this case? Can optimal convergence be maintained?

P.5-6 Suppose that a current-carrying and electrically perfectly conducting wire with radius $r_0 \ll h$, where $h = \Delta x = \Delta y = \Delta z$ denotes the grid spacing, runs along the z-axis. Use the near-field approximations $H_\varphi \propto 1/r$ and $E_r \propto 1/r$ (in cylindrical coordinates) to derive appropriate difference approximations taking into account the wire.

P.5-7 Maxwell's equations can be written in terms of the scalar potential ϕ and the vector potential $\boldsymbol{A}$:

$$\boldsymbol{E} = -\nabla\phi - \frac{\partial \boldsymbol{A}}{\partial t}, \qquad \boldsymbol{B} = \mu\boldsymbol{H} = \nabla\times\boldsymbol{A}.$$

How should the potentials be placed on the grid in order to match Yee's locations for the fields?

P.5-8 In two dimensions (say the solution is independent of z), one can separate electromagnetic fields into TE components, with $E_z = 0$, and TM, with $H_z = 0$. The simplest way to compute these is to use the wave equations for H_z and E_z, respectively, in two dimensions. However, it is also possible to describe TE polarization by a set of first-order equations for E_x, E_y, and H_z, while TM polarization can be described by first-order equations for H_x, H_y, and E_z. Write down the relevant sets of equations and show how suitable staggered finite difference schemes can be found, e.g., as subsets of the 3D Yee scheme.

P.5-9 Derive the finite difference equation for updating $E_x|_{p+\frac{1}{2},q,r}$ starting from the integral form of Ampère's law (5.22).

P.5-10 Show that about 11 points per wavelength gives 1% error in the numerical dispersion relation for a cubic grid by Taylor expanding the dispersion relation (5.32) for $\omega^2(k)$ to order k^4h^4 and using the approximation $\omega^2 = c^2(k_x^2 + k_y^2 + k_z^2)$ in the term $\propto \omega^4$. When the time-step is at the stability limit of (5.34), the result can be written

$$\frac{\omega^2}{c^2} = k_x^2 + k_y^2 + k_z^2 - \frac{h^2}{72}[(k_x^2 - k_y^2)^2 + (k_y^2 - k_z^2)^2 + (k_z^2 - k_x^2)^2].$$

[Thus, when the time-step is at the stability limit, only solutions that propagate maximally obliquely have zero dispersion. In all other directions, the spatial dispersion dominates, and the phase speed is below c. For smaller time steps, the phase speed is less than c in all directions.]

P.5-11 The curl-curl equation (5.28) also has electrostatic solutions that are linear functions of time, i.e. $\boldsymbol{E}(\boldsymbol{r}, t) = t\nabla\phi(\boldsymbol{r})$. Can such a solution appear in an FDTD simulation without sources?

P.5-12 Does the FDTD scheme preserve the electric charge if there are no electric currents?

P.5-13 Carry out the derivation of the numerical dispersion relation for the 3D FDTD scheme by rewriting (5.28) in matrix form and setting the determinant of this matrix to zero.

P.5-14 Derive the impedance Z_{PML} from (5.39)–(5.40) by assuming that the field components vary as $\exp(-j\boldsymbol{k} \cdot \boldsymbol{r})$ and that $\boldsymbol{E}$ and $\boldsymbol{H}$ are perpendicular to $\boldsymbol{k}$.

P.5-15 Consider the computation of an electrical motor at $f = 50$ Hz and a spatial resolution of $h = 5$ mm. How many time-steps are needed if we want to time-step 5 wave periods, or 0.1 s?

P.5-16 We note that 1% relative phase error is obtained with about 10 points per wavelength. How much does the computation time for a 3D problem increase if we want to reduce the relative phase error by a factor 10?

Computer Projects

C.5-1 Propose some different ways of visualizing the numerical dispersion relation (5.32). Write a program that given the different parameters needed implements your ideas for the visualization. Experiment with different resolutions, spatial and temporal. It can be beneficial to use $k_x = k \sin\theta \cos\phi$, $k_y = k \sin\theta \sin\phi$, and $k_z = k \cos\theta$. How do the results depend on the direction of propagation?

C.5-2 Implement the 1D FDTD scheme for $0 \le z \le a$; see Sect. 5.2.1. Extend your program to include the losses shown in (5.39)–(5.40). Let $\epsilon = \epsilon_0$, $\mu = \mu_0$ and introduce a conductive region for $a - w \le z \le a$ where the losses satisfy the condition shown in (5.41) and w is the width of the conductive region. Where should $\sigma(z)$ and $\sigma^*(z)$ be evaluated on the staggered grid? Set up a numerical experiment so that you can study the reflection coefficient for the electric field $E_x(z,t)$, which satisfies the boundary conditions $E_x(0,t) = g(t)$ and $E_x(a,t) = 0$. Let $g(t) = \exp[-(t - t_0)^2/d_0^2] \sin[2\pi f_0(t - t_0)]$ and choose appropriate values for t_0, d_0, and f_0. Experiment with different conductivity profiles $\sigma(z)$ and $\sigma^*(z)$ given by (5.41). Try a constant conductivity profile and optimize the value σ_{const} for the conductivity. A very common choice is the quadratic profile $\sigma(z) = \sigma_{\max}[(z - (a - w))/w]^2$, where $\sigma_{\max}$ is a constant to be optimized. Plot the reflection coefficient as a function of σ_{const} and $\sigma_{\max}$. Explain your findings. What happens if the condition (5.41) is violated? How does the reflection coefficient depend on frequency?

C.5-3 Write a program that implements the 2D FDTD scheme. Use it to compute the resonant frequencies of a circular cavity with metal boundaries. How do you represent the circular boundary on the Cartesian grid? How do you excite the problem? Suggest and implement some different excitations and compare the approaches.

C.5-4 Modify the program in Sect. 5.2.3 so that inhomogeneous materials $\epsilon(\boldsymbol{r})$ and $\mu(\boldsymbol{r})$ can be considered. Extend the implementation so that also a source current $\boldsymbol{J}(\boldsymbol{r})$ can be included. Let the electric and magnetic field be identically zero as an initial condition. Is the condition of solenoidal magnetic flux density (1.4) preserved numerically at all times? Does the solution computed by your program satisfy the equation of continuity for electric charge?

Chapter 6
The Finite Element Method

The finite element method (FEM) is a standard tool for solving differential equations in many disciplines, e.g., electromagnetics, solid and structural mechanics, fluid dynamics, acoustics, and thermal conduction. Jin [40, 41] and Peterson [54] give good accounts of the FEM for electromagnetics. More mathematical treatments of the same topic are given in [12, 48]. This chapter gives an introduction to FEM in general and FEM for Maxwell's equations in particular. Practical issues, such as how to handle unstructured grids and how to write FEM programs, will be discussed in some detail.

In the FEM, the domain with the sought electromagnetic field is subdivided into small subdomains of simple shape and, as an initial example, we may consider a circular domain in two dimensions that is subdivided into triangular subdomains. The collection of triangular subdomains cover the original circular domain and the triangular subdomains do not overlap each other. (Thus, the circular boundary is approximated by a polygon that consists of the outermost edges of the triangular subdomains.) The field solution is expressed in terms of a low-order polynomial (for example a linear polynomial) on each of these subdomains and, consequently, we have a *piecewise* low-order polynomial representation of the field on the circular domain. In general, such a representation of the field is not flexible enough to be able to exactly fulfill the differential equation and its boundary conditions in a pointwise manner. The FEM relaxes this requirement slightly and, instead, it attempts to find a field solution that fulfills the differential equation and its boundary conditions in an averaged sense. There are different approaches to how to construct this relaxed fulfillment of the differential equation and its boundary condition and two important methods are (i) to set the weighted average of the residual to zero and (ii) to exploit a variational method to find a stationary point of a quadratic form. For a FEM that works correctly, the approximate solution tends to the exact solution as the size of the subdomains tends to zero and, consequently, the number of subdomains tend to infinity. Clearly, the smaller subdomains allows for a better approximation of both

T. Rylander et al., *Computational Electromagnetics*, Texts in Applied Mathematics 51, DOI 10.1007/978-1-4614-5351-2_6,

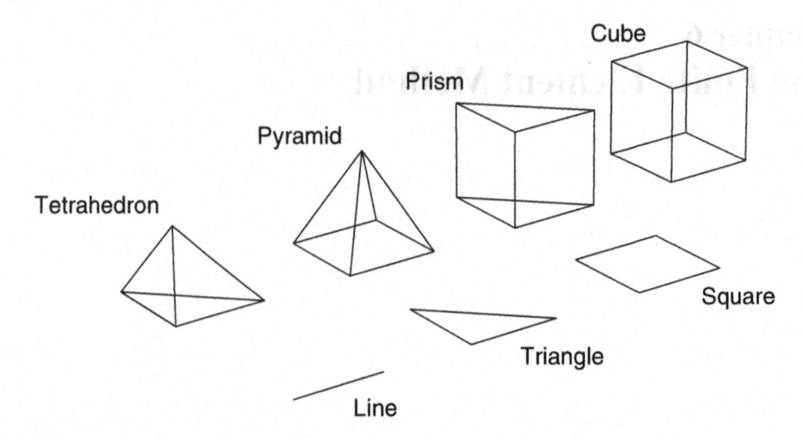

Fig. 6.1 Different element shapes: a line in one dimension, a triangle and square in two dimensions, and a tetrahedron, prism, pyramid, and cube in three dimensions

the circular domain and the exact field solution. This chapter describes the FEM in terms of the weighted residual method for a range of situations and, towards the end of the chapter, also the variational methods are introduced and contrasted to the weighted residual method.

A very strong point of the FEM, and the main reason why it is a favorite method in many branches of engineering, is its ability to deal with complex geometries. Typically, this is done using unstructured grids, which are commonly referred to as (unstructured) meshes. These meshes may consist of triangles in two dimensions and tetrahedra in three dimensions. However, there are several types of element shapes, as shown in Fig. 6.1: triangles and quadrilaterals in two dimensions, tetrahedra, prisms, pyramids, and hexahedra in three dimensions.

Unstructured meshes with, for instance, tetrahedra allow good representations of curved objects, which are hard to represent on the Cartesian grids used by finite difference methods. Moreover, unstructured meshes allow for higher resolution locally in order to resolve fine structures of the geometry and rapid variations of the solution. Another nice property of the FEM is that the method provides a well-defined representation of the sought function everywhere in the solution domain. This makes it possible to apply many mathematical tools and prove important properties concerning stability and convergence.

A disadvantage of the FEM, compared to the FDTD, is that explicit formulas for updating the fields in time-domain simulations cannot be derived in the general case. Instead, a linear system of equations has to be solved in order to update the fields. Consequently, provided that the same number of cells are used for the two methods, the FEM requires more computer resources, both in terms of CPU time and memory.

Normally, the FEM is used to solve differential equations. However, it is also possible to apply the FEM to integral equations, where the unknown field is part of the integrand. In CEM, the FEM applied to integral equations is referred to as the Method of Moments and this technique is discussed in Chapter 7.

6.1 General Recipe

We start by giving the general recipe for how to solve a differential equation by the FEM. The equation is written as $L[f] = s$, where L is an operator, s the source, and f the unknown function to be computed in the region Ω.

- Subdivide the solution domain Ω into cells, or *elements*. For example, a 2D domain can be subdivided into triangles or quadrilaterals.
- Approximate the solution by an expansion in a finite number of *basis functions*, i.e., $f(\mathbf{r}) \approx \sum_{i=1}^{n} f_i \varphi_i(\mathbf{r})$, where f_i are (unknown) coefficients multiplying the basis functions $\varphi_i(\mathbf{r})$. The basis functions are generally low-order polynomials that are nonzero only in a few adjacent elements.
- Form the residual $r = L[f] - s$, which we want to make as small as possible. In general, it will not be zero pointwise, but we require it to be zero in the so-called weak sense by setting a weighted average of it to zero.
- Choose *test*, or *weighting*, functions w_i, $i = 1, 2, \ldots, n$ (as many as there are unknown coefficients) for weighting the residual r. Often, the weighting functions are the same as the basis functions, $w_i = \varphi_i$, and this method is then called Galerkin's method.
- Set the weighted residuals to zero and solve for the unknowns f_i; i.e., solve the set of equations $\langle w_i, r \rangle = \int_\Omega w_i \, r \, d\Omega = 0$, $i = 1, 2, \ldots, n$.

In mathematical definitions, the term *finite element* usually refers to an element (e.g., a triangle) together with a polynomial space defined in this element (e.g., the space of linear functions) and a set of degrees of freedom defined on this space (e.g., the values of the linear functions in the corners (nodes) of the triangle). This definition is seldom used in electrical engineering, where one tends to focus on the basis functions used to expand the solution instead.

Review Questions

6.1-1 List some pros and cons of the finite element method.

6.1-2 Compare the steps of the general recipe for the FEM to the typical discretization procedure employed for finite difference methods. Identify similarities and differences.

6.1-3 What is a finite element?

6.2 1D Finite Element Analysis

As the first model problem we choose a second-order ordinary differential equation, namely the 1D Helmholtz equation:

$$-\frac{d}{dx}\left(\alpha\frac{df}{dx}\right)+\beta f = s, \quad a < x < b, \tag{6.1}$$

$$f(a) = f_a, \tag{6.2}$$

$$f(b) = f_b. \tag{6.3}$$

Here $f = f(x)$ is the sought solution, and the material properties $\alpha = \alpha(x)$ and $\beta = \beta(x)$ and the source $s = s(x)$ are prescribed functions of x.

There are many physical systems that are modeled by (6.1), for example, a transversal wave in a 1D medium, such as a light wave propagating and being reflected in dielectric layers. In this case we have $f(x) = E_z(x)$, and the coefficients are $\alpha(x) = 1/\mu(x), \beta(x) = j\omega\sigma(x) - \omega^2\epsilon(x)$, where ω is the angular frequency, and $s(x) = -j\omega J_z(x)$ (which vanishes, unless there are current-carrying conductors).

We seek the function $f(x)$ on the interval $a < x < b$. According to the general recipe for the FEM, we first divide this interval into subintervals (elements). Let us assume, for example, $a = -2$ and $b = 5$ and divide the x-axis into 7 equally large elements. We call the endpoints of each element *nodes*, and they have the coordinates $x_i = i - 3$ where $i = 1, 2, \ldots, 8$. We introduce the *nodal basis functions* $\varphi_i(x)$, which are linear on each interval, one at node i and zero at all other nodes, as shown in Fig. 6.2. These basis functions are often called "tent functions."

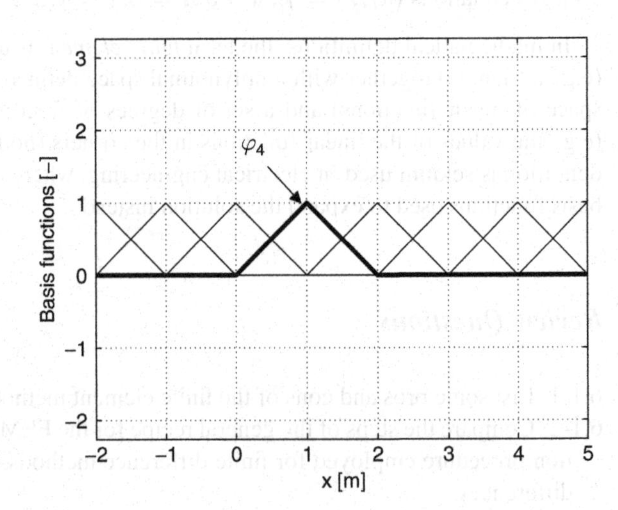

Fig. 6.2 1D linear elements. In particular, the basis function $\varphi_4(x)$ is emphasized by a thick line

We seek approximate solutions that are expanded in the basis functions (in the following, f will denote this approximate solution):

$$f(x) = \sum_{j=1}^{8} f_j \varphi_j(x). \tag{6.4}$$

Note that $f(x_i) = f_i$, so that the expansion coefficients are the values of f at the nodes. Since $f(a) = f_a$ and $f(b) = f_b$ are known, we set $f_1 = f_a$ and $f_8 = f_b$.

In the next step, we follow Galerkin's method and choose the test functions $w_i(x) = \varphi_i(x)$, where $i = 2, 3, \ldots, 7$ (the endpoints are excluded because the corresponding function values are known). We multiply the residual of (6.1) by the test function $w_i(x)$ and integrate from $x = a$ to $x = b$. To move one of the derivatives from f to the test function w_i, we use integration by parts. This gives the *weak form* of the original problem, which is the weighted average of the residual:

$$\int_a^b \left(\alpha w_i' f' + \beta w_i f - w_i s\right) dx = 0. \tag{6.5}$$

In this case, the boundary term $[w_i \alpha f']_a^b$ vanishes, since $w_i(a) = w_i(b) = 0$.

By substituting (6.4) into the weak form (6.5) and choosing $w_2(x) = \varphi_2(x)$, we generate an equation involving six unknowns: the coefficients f_j for the interior nodes x_j, where $j = 2, 3, \ldots, 7$. Next, we pick $w_3(x) = \varphi_3(x)$ to generate a second equation, and so on. In the end, we have six equations and six unknowns, and this is formulated as a system of linear equations $\mathbf{Az} = \mathbf{b}$ with

$$A_{ij} = \int_a^b \left(\alpha \varphi_i' \varphi_j' + \beta \varphi_i \varphi_j\right) dx, \tag{6.6}$$

$$z_j = f_j, \tag{6.7}$$

$$b_i = \int_a^b \varphi_i s \, dx. \tag{6.8}$$

Here, $i = 2, 3, \ldots, 7$ (for the equations) and $j = 1, 2, \ldots, 8$ (for the coefficients), so $\mathbf{A}$ has 8 columns and 6 rows, $\mathbf{z}$ has 8 rows, and $\mathbf{b}$ has 6 rows. The coefficients f_1 and f_8 are known from the boundary conditions and can be moved to the right-hand side:

$$\begin{pmatrix} A_{22} & A_{23} & \ldots & A_{27} \\ A_{32} & A_{33} & \ldots & A_{37} \\ \vdots & \vdots & \ddots & \vdots \\ A_{72} & A_{73} & \ldots & A_{77} \end{pmatrix} \begin{bmatrix} f_2 \\ f_3 \\ \vdots \\ f_7 \end{bmatrix} = \begin{bmatrix} b_2 \\ b_3 \\ \vdots \\ b_7 \end{bmatrix} - \begin{bmatrix} A_{21} f_1 + A_{28} f_8 \\ A_{31} f_1 + A_{38} f_8 \\ \vdots \\ A_{71} f_1 + A_{78} f_8 \end{bmatrix}.$$

The part of the system matrix $\mathbf{A}$ that remains on the left-hand side is square; that is, we have as many unknowns as equations. In the present case, the function values at the endpoints are known, and we do not use the corresponding weighting functions. The matrix **A** is sparse because the basis functions give only nearest-neighbor coupling of the unknowns. Also note that **A** is symmetric, $A_{ij} = A_{ji}$. This is related to the fact that the Helmholtz operator is self-adjoint and we used Galerkin's method.

The boundary conditions (6.2) and (6.3) specify the value of the function $f(x)$ at the boundary. Other types of boundary conditions can specify the derivative of $f(x)$ or a linear combination of $f(x)$ and its derivative. At either boundary, for instance the left one $x = a$, we can apply conditions of the following standard types:

$$f(a) = p \tag{6.9}$$

or

$$f'(a) + \gamma f(a) = q. \tag{6.10}$$

Equation (6.9) is called a *Dirichlet* boundary condition, and it eliminates an unknown. Equation (6.10) is called a *Neumann* boundary condition when $\gamma = 0$ and a *Robin* boundary condition when $\gamma \neq 0$. For the Neumann and Robin boundary conditions, $f(a)$ must be introduced as an extra unknown. We generate the extra equation by testing with $w_1(x) = \varphi_1(x)$. Dirichlet boundary conditions are referred to as essential, whereas Neumann and Robin boundary conditions are called natural. Further, if q or p is zero, the boundary conditions are called homogeneous.

Review Questions

6.2-1 Write down an explicit expression for the nodal basis function $\varphi_i(x)$ and its derivative for a nonuniform discretization in one space dimension.
6.2-2 Explain the terms *Galerkin's method* and *weak form*.
6.2-3 How many test functions are needed for a 1D finite element problem?
6.2-4 Explain the difference between Dirichlet, Neumann, and Robin boundary conditions.
6.2-5 Are the numbers of basis functions and test functions always the same?

6.3 2D Finite Element Analysis

We extend the model problem (6.1) to two dimensions, but still f is a scalar-valued function:

$$-\nabla \cdot (\alpha \nabla f) + \beta f = s \text{ in } S, \tag{6.11}$$

$$f = p \text{ on } L_1, \tag{6.12}$$

$$\hat{\boldsymbol{n}} \cdot (\alpha \nabla f) + \gamma f = q \text{ on } L_2. \tag{6.13}$$

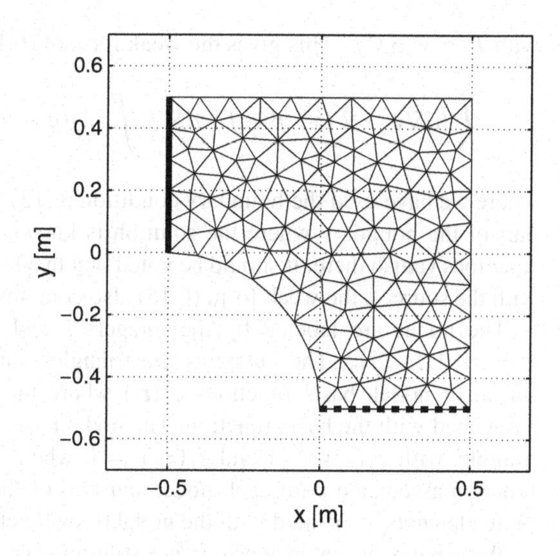

Fig. 6.3 A 2D conducting plate. The computational domain S, i.e., the plate, is divided into triangular elements. Automatically, the boundary of S is discretized into line segments. This boundary divides into the two parts, denoted by L_1 and L_2, with different types of boundary condition according to (6.12) and (6.13)

The boundary of the solution domain S has two parts, L_1 and L_2, with different types of boundary conditions.

Analogously to the 1D model problem, there are many physical situations that can be modeled by (6.11). Let us consider a specific example where we wish to compute the resistance between the left and bottom edges of the conducting plate shown in Fig. 6.3. In this case, f is the electrostatic potential, α the conductivity, $\beta = 0$, and $s = 0$. The electric potential along the thick solid line on the boundary is set to 10 V, i.e., a Dirichlet boundary condition $f = 10$. Along the thick dashed line the potential is set to 0 V. The remaining part of the boundary is an insulating material. On this part of the boundary, we use a Neumann boundary condition, $\hat{\boldsymbol{n}} \cdot \nabla f = 0$, which means no flux of charge across the boundary. We now continue with the derivation based on the general model problem, and at the end of this section, we will show the solution for the specific example concerning the resistance computation.

We multiply (6.11) by a test function w_i and integrate over S:

$$\int_S w_i \left[-\nabla \cdot (\alpha \nabla f) + \beta f\right] dS = \int_S w_i s \, dS.$$

Next, integrate by parts using the identity

$$\nabla \cdot [w_i (\alpha \nabla f)] = \alpha \nabla w_i \cdot \nabla f + w_i \nabla \cdot (\alpha \nabla f) \tag{6.14}$$

and Gauss's theorem in 2D:

$$\int_S \nabla \cdot \boldsymbol{F} \, dS = \int_{L_1 + L_2} \hat{\boldsymbol{n}} \cdot \boldsymbol{F} \, dl,$$

with $\boldsymbol{F} = w_i \alpha \nabla f$. This gives the weak form of (6.11)–(6.13):

$$\int_S (\alpha \nabla w_i \cdot \nabla f + \beta w_i f)\, dS - \int_{L_2} w_i (q - \gamma f)\, dl = \int_S w_i s\, dS, \tag{6.15}$$

where we have used the boundary condition (6.13). The boundary integral over the part of the boundary where the solution is known (L_1) vanishes because the test functions vanish there. It should be noted that in addition to the differential equation with the sources, the weak form (6.15) also contains the boundary conditions.

The nodes are labeled by the integers i and they are located at $\boldsymbol{r}_i$, where $i = 1, 2, \ldots, N_n$. The elements are triangles, and again, we choose piecewise linear, or nodal, basis functions $\varphi_i(\boldsymbol{r})$ where the subindex i refers to the node associated with the basis function. The nodal basis functions are linear inside each triangle, with $\varphi_i(\boldsymbol{r}_i) = 1$ and $\varphi_i(\boldsymbol{r}_j) = 0$ when $i \neq j$. There is one such basis function associated with each node, and two of them are shown in Fig. 6.4. The finite elements associated with the nodal basis functions are called nodal elements.

We expand the, again approximate, solution $f(\boldsymbol{r})$ in terms of the basis functions:

$$f(\boldsymbol{r}) = \sum_{j=1}^{N_n} f_j \varphi_j(\boldsymbol{r}). \tag{6.16}$$

Next, we substitute (6.16) into the weak form (6.15) and use Galerkin's method, i.e., choose $w_i(\boldsymbol{r}) = \varphi_i(\boldsymbol{r})$ for all nodes where f is unknown. This gives a linear system of equations $\mathbf{Az} = \mathbf{b}$, where the elements are given by

$$A_{ij} = \int_S (\alpha \nabla \varphi_i \cdot \nabla \varphi_j + \beta \varphi_i \varphi_j)\, dS + \int_{L_2} \gamma \varphi_i \varphi_j\, dl, \tag{6.17}$$

$$z_j = f_j, \tag{6.18}$$

$$b_i = \int_S \varphi_i s\, dS + \int_{L_2} \varphi_i q\, dl. \tag{6.19}$$

Here, the index j runs over all nodes, and i only over those nodes where f is unknown (not those on the boundary L_1 with the Dirichlet condition). The variables are reordered to collect those where f is known in the vector $\mathbf{z}_e$, while $\mathbf{z}_n$ denotes the remaining unknowns,

$$\left(\mathbf{A}_e \;\middle|\; \mathbf{A}_n \right) \begin{bmatrix} \mathbf{z}_e \\ \mathbf{z}_n \end{bmatrix} = \mathbf{A}_e \mathbf{z}_e + \mathbf{A}_n \mathbf{z}_n = \mathbf{b}.$$

The matrix $\mathbf{A}$ is partitioned in the same way. This results in a square matrix $\mathbf{A}_n$ and a rectangular part $\mathbf{A}_e$ accounting for the Dirichlet boundary condition. The final

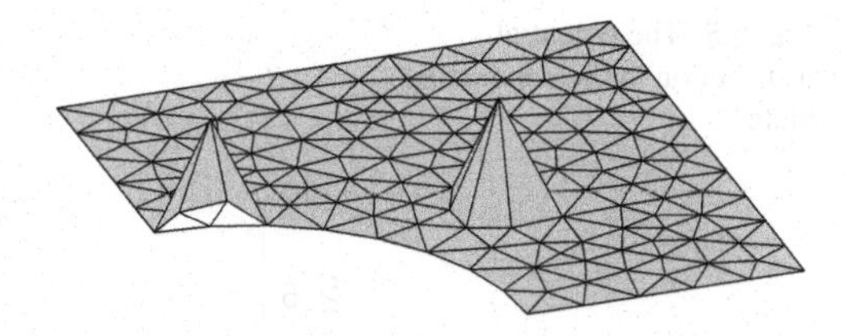

Fig. 6.4 Illustration of two nodal basis functions, one on the boundary and one in the interior of the solution domain

system of equations to be solved for $\mathbf{z}_n$ is $\mathbf{A}_n\mathbf{z}_n = \mathbf{b} - \mathbf{A}_e\mathbf{z}_e$, where $\mathbf{A}_n$ and $\mathbf{b} - \mathbf{A}_e\mathbf{z}_e$ contain only known numbers. In Sect. 6.2, this procedure is shown at a very detailed level. Here, it is expressed in terms of matrices and vectors, which is more convenient for 2D and 3D problems.

Finally, we return to the specific example in which we wanted to compute the resistance of the metal plate, where the thickness of the plate is denoted by h. The numerical solution, i.e., the approximate electrostatic potential, is shown in Fig. 6.5. Based on the potential, the resistance can be computed in two ways:

- Integrate the normal component of the current density $\boldsymbol{J} = -\sigma\nabla\phi$ over a cross-section of the plate to obtain the total current that flows through the plate. For example,

$$I = \int_{z=0}^{h}\int_{x=0}^{0.5} \sigma \frac{\partial\phi}{\partial y}\bigg|_{y=-0.5} dx. \tag{6.20}$$

  The resistance is then obtained from $R = U/I$, where $U = \Delta\phi = 10$ V.
- Compute the total power dissipation in the plate (see Sect. 6.3.3 for a similar approach used for a capacitance computation);

$$P = \int_V \boldsymbol{J}\cdot\boldsymbol{E}\,dV = \int_V \sigma|\nabla\phi|^2\,dV = h\mathbf{z}^{\mathrm{T}}\mathbf{A}\mathbf{z} = h\mathbf{z}^{\mathrm{T}}\mathbf{b}, \tag{6.21}$$

  and then calculate the resistance from $P = U^2/R$, which gives $R = U^2/P$.

The latter approach is generally preferred, since it is trivial to compute and often leads to better accuracy.

6.3.1 The Assembling Procedure

In practice, the matrix and vector components in (6.17)–(6.19) are computed by assembling contributions from all elements. To illustrate the assembling procedure, we consider the capacitance calculation in Sect. 3.1. The differential equation is again $\nabla^2\phi = 0$, and only the boundary conditions differ from the previous example.

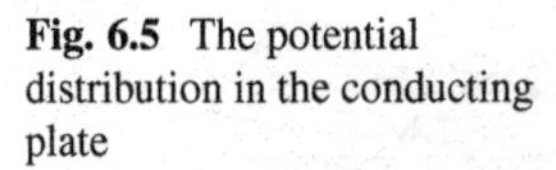
Fig. 6.5 The potential distribution in the conducting plate

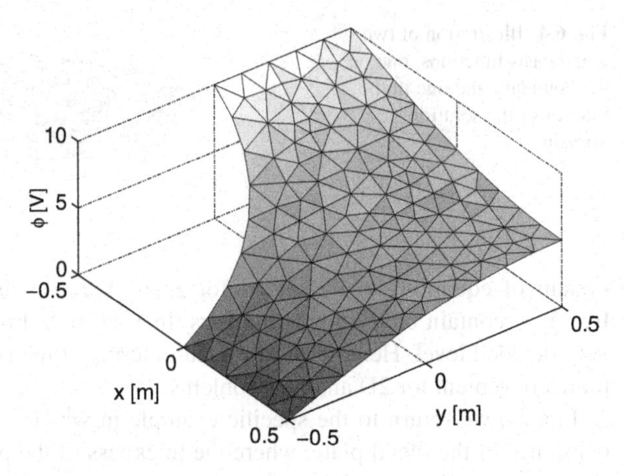

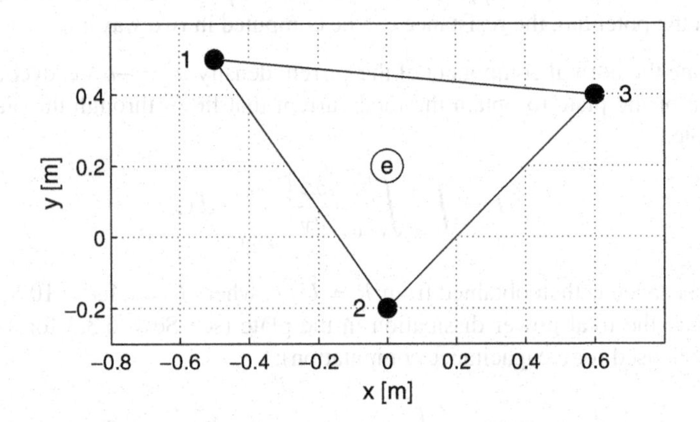

Fig. 6.6 The numbering of local nodes for the element e

The elements A_{ij} of the system matrix are computed by evaluating the integral $\int_S \nabla\varphi_i \cdot \nabla\varphi_j \, dS$ over the domain S between the inner and outer conductors. In the assembling procedure, we break up this integral into integrals over each element S^e, and sum the contributions from all the elements, i.e.,

$$A_{ij} = \int_S \nabla\varphi_i \cdot \nabla\varphi_j \, dS = \sum_{e=1}^{N_e} \int_{S^e} \nabla\varphi_i \cdot \nabla\varphi_j \, dS, \tag{6.22}$$

where N_e is the total number of elements.

Now we will concentrate on evaluating the integrals restricted to a single element. We use a local numbering of the nodes for the element e, as shown in Fig. 6.6, and denote the coordinates of the nodes by $\boldsymbol{r}_1^e$, $\boldsymbol{r}_2^e$, and $\boldsymbol{r}_3^e$, respectively.

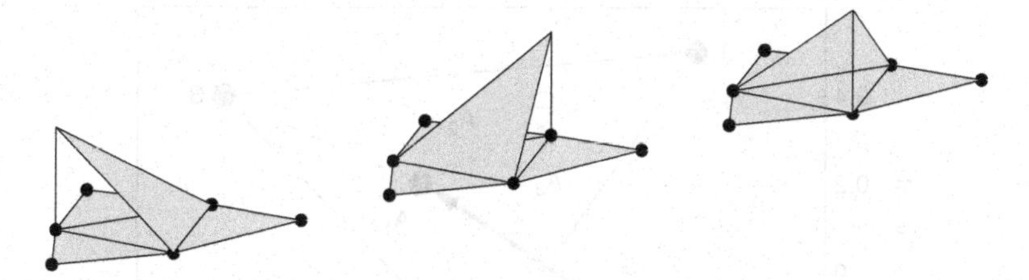

Fig. 6.7 The three basis functions for element e. The adjacent elements sharing an edge with element e are also shown

6.3.1.1 The Nodal Basis Functions

The local basis functions (i.e., the basis functions restricted to one element) are denoted by $\varphi_i^e(x, y)$, where the superindex labels the element ($e = 1, \ldots, N_e$) and the subindex the local node number ($i = 1, 2, 3$). There is one local basis function associated with each node of the element, and these are shown in Fig. 6.7. The global basis function associated with node i is built up by the local basis functions associated with that particular node in the surrounding elements.

The basis functions have the following properties:

- Inside each element, they are linear in x and y, i.e.,

$$\varphi_i^e(x, y) = a_i^e + b_i^e x + c_i^e y. \tag{6.23}$$

- They equal unity on one node and vanish on the others:

$$\varphi_i^e(x_i^e, y_i^e) = 1, \quad \varphi_i^e(x_j^e, y_j^e) = 0, \forall i \neq j. \tag{6.24}$$

We will now construct explicit expressions for $\varphi_i^e(x, y)$ with these properties. To do this, we divide the element e into three triangles as shown in Fig. 6.8. Here, A_i^e is the area of subtriangle i, opposing vertex i of the element, and $A_{\text{tot}}^e = A_1^e + A_2^e + A_3^e$. The point inside the element, where we evaluate $\varphi_i^e(x, y)$, has the position $\boldsymbol{r} = x\hat{\boldsymbol{x}} + y\hat{\boldsymbol{y}}$.

The basis functions $\varphi_i^e(x, y)$ can be constructed by means of the area coordinates A_i^e as

$$\varphi_i^e(x, y) = \frac{A_i^e}{A_{\text{tot}}^e}. \tag{6.25}$$

[We note that the functions φ_i^e also are called simplex coordinates and barycentric coordinates.] It is easy to verify that these elements satisfy the requirements (6.23)–(6.24). A_i^e can be written as

$$A_1^e = \frac{1}{2}\hat{\boldsymbol{z}} \cdot \left(\boldsymbol{r}_3^e - \boldsymbol{r}_2^e\right) \times \left(\boldsymbol{r} - \boldsymbol{r}_2^e\right),$$

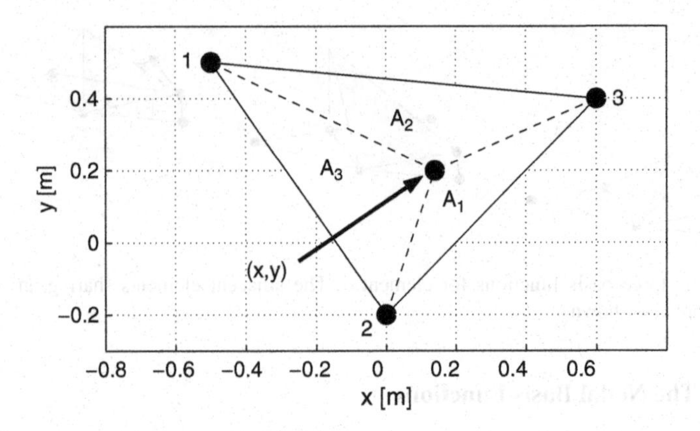

Fig. 6.8 Partition used to construct $\varphi_i^e(x, y)$

$$A_2^e = \frac{1}{2}\hat{z} \cdot \left(\boldsymbol{r}_1^e - \boldsymbol{r}_3^e\right) \times \left(\boldsymbol{r} - \boldsymbol{r}_3^e\right),$$

$$A_3^e = \frac{1}{2}\hat{z} \cdot \left(\boldsymbol{r}_2^e - \boldsymbol{r}_1^e\right) \times \left(\boldsymbol{r} - \boldsymbol{r}_1^e\right),$$

or more compactly

$$A_i^e = \frac{1}{2}(\boldsymbol{r} - \boldsymbol{r}_{i+1}^e) \cdot \hat{z} \times \boldsymbol{s}_i, \tag{6.26}$$

where

$$\boldsymbol{s}_i = \boldsymbol{r}_{i-1}^e - \boldsymbol{r}_{i+1}^e \tag{6.27}$$

is the edge in the counterclockwise direction opposing node i. The total area of the element is

$$A_{\text{tot}}^e = \frac{1}{2}\hat{z} \cdot \boldsymbol{s}_2 \times \boldsymbol{s}_3. \tag{6.28}$$

Now it is simple to find the gradients of the local basis functions,

$$\nabla\varphi_i^e = \frac{\hat{z} \times \boldsymbol{s}_i}{2A_{\text{tot}}^e}, \tag{6.29}$$

and these are, of course, constant inside each element. Therefore, the integral over one element e, contributing to the system matrix in (6.22), can be evaluated by multiplying the scalar product of the local basis functions by the area of the element:

$$A_{ij}^e = \int_{S^e} \nabla\varphi_i^e \cdot \nabla\varphi_j^e \, dS = \frac{\boldsymbol{s}_i \cdot \boldsymbol{s}_j}{4A_{\text{tot}}^e}. \tag{6.30}$$

Notice that we need to relate the three *local* node numbers of element e to their corresponding *global* node numbers before we add the element contributions A^e_{ij} to the global system matrix **A**.

6.3.1.2 The Element Matrix

Here we give a MATLAB function that computes all the contributions to **A** from a single finite element described by its coordinates given in the argument `xy`. Since there are three basis functions in each element, we can store all its contributions in a 3×3 matrix, which we will refer to as the *element matrix*. We name the MATLAB function `CmpElMtx`, and for the element shown in Fig. 6.6, this should be called with the argument `xy = [-0.5 0.0 0.6; 0.5 -0.2 0.4]`.

```
% ---------------------------------------------------------------
% Compute element matrix for a triangle and its node basis
% ---------------------------------------------------------------
function Ae = CmpElMtx(xy)

% Arguments:
% xy = the coordinates of the nodes of the triangle
% Returns:
%    Ae = element matrix corresponding to the Laplace operator

% Edges
s1 = xy(:,3)-xy(:,2);
s2 = xy(:,1)-xy(:,3);
s3 = xy(:,2)-xy(:,1);

% Area of the triangle
Atot = 0.5*(s2(1)*s3(2)-s2(2)*s3(1));

% Check whether area is negative (nodes given counterclockwise)

if (Atot < 0)
  error('The nodes of the element given in wrong order')
end

% Compute the gradient of the vectors.
grad_phi1e = [-s1(2);s1(1)]/(2*Atot);
grad_phi2e = [-s2(2);s2(1)]/(2*Atot);
grad_phi3e = [-s3(2);s3(1)]/(2*Atot);

grad_phi = [grad_phi1e grad_phi2e grad_phi3e];

% Compute all the integrals for this particular element.
for iIdx = 1:3
  for jIdx = 1:3
    Ae(iIdx,jIdx) = grad_phi(:,iIdx)' * grad_phi(:,jIdx) * Atot;
  end
end
```

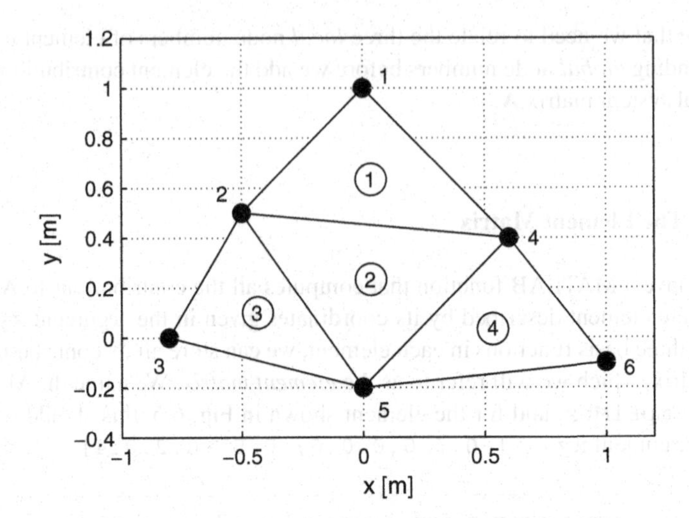

Fig. 6.9 A 2D mesh. The node numbers are shown next to the corresponding nodes and the element numbers in the center of the corresponding triangles

The right-hand side **b** is constructed following the same assembling procedure, i.e., by summing the contributions b_i^e from each element. Often, A_{ij}^e and b_i^e are evaluated by numerical rather than analytical integration.

Now, we have one row in **A** and **b** for every node in the mesh, since we have tested the differential equation at all the nodes, including those where the solution is known from the Dirichlet boundary condition. This is not exactly what we want, since the test function must be zero along the Dirichlet boundary. We correct this by removing the rows in **A** and **b** corresponding to nodes where the solution is known. A more efficient approach, in particular for large problems, is to compute the local contribution for each element but assemble only the rows that are not associated with a Dirichlet boundary.

6.3.2 *Unstructured Meshes in Practice*

When writing FEM programs it is important to treat unstructured meshes in an efficient and well-organized way. The most common way is explained here for the small mesh shown in Fig. 6.9. The mesh consists of 6 nodes and 4 triangular elements.

We will use the fact that a triangle is built up by three nodes. Therefore, we store the coordinates of the nodes in a table `no2xy`; i.e., given a global node number the table `no2xy` provides its coordinates. Next, we construct the triangles by listing the nodes that are the vertices of each triangle in another table `el2no`; i.e., given an

Table 6.1 Given a global node number, this table (no2xy) provides its coordinates

Node	1	2	3	4	5	6
x	0.0	−0.5	−0.8	0.6	0.0	1.0
y	1.0	0.5	0.0	0.4	−0.2	−0.1

Table 6.2 Given an element number, this table (el2no) provides its global node numbers

Element	1	2	3	4
Node 1	1	4	3	5
Node 2	2	2	5	6
Node 3	4	5	2	4

element number the table el2no provides its global node numbers. For the mesh shown in Fig. 6.9, the information in no2xy is given in Table 6.1, and el2no in Table 6.2.

This is how it looks in MATLAB:

```
>> no2xy

no2xy =

         0   -0.5000   -0.8000    0.6000         0    1.0000
    1.0000    0.5000         0    0.4000   -0.2000   -0.1000

>> el2no

el2no =

     1     4     3     5
     2     2     5     6
     4     5     2     4
```

The same idea can be used to store other types of elements such as lines and quadrilaterals.

6.3.3 MATLAB: 2D FEM Using Nodal Basis Functions

We will present a program showing the assembling procedure for the capacitance calculation in Sect. 3.1. However, first it is useful to show how the mesh can be generated and used for computation in MATLAB.

6.3.3.1 Generate a Mesh of Triangles

Mesh generation is a discipline in itself, and it is an active field of research. An overview of both commercial and free mesh generators (a program that creates a FEM mesh) is available at the Meshing Research Corner [53]. Many of these programs use their own input and output format. However, most likely the output is based on the ideas presented in Sect. 6.3.2. Thus, if we understand the basic principles of how an unstructured mesh is organized, we can extract the necessary information from most mesh generators. Of course, we will need the documentation of the mesh generator and to work with simple examples in the beginning.

6.3.3.2 Solving the Laplace Equation

Now we are ready to write the program solving for the potential $\phi(\boldsymbol{r}) = \sum_{j=1}^{N_n} \phi_j \varphi_j(\boldsymbol{r})$ at the nodes (vector z). Once the potential is known, the capacitance per unit length C can be computed from the energy relation $C = 2W/U^2$, where W is the electrostatic energy per unit length and U is the potential difference between the inner and outer conductors. The electrostatic energy per unit length can be computed using the following quadratic form (see Sect. 6.9):

$$\begin{aligned} W[\phi] &= \frac{1}{2}\int_S \boldsymbol{E}\cdot\boldsymbol{D}\,dS = \frac{1}{2}\int_S \epsilon_0|\nabla\phi|^2 dS \\ &= \frac{\epsilon_0}{2}\sum_{i=1}^{N_n}\sum_{j=1}^{N_n}\phi_i\left[\int_S \nabla\varphi_i\cdot\nabla\varphi_j\,dS\right]\phi_j \\ &= \frac{\epsilon_0}{2}\mathbf{z}^T\mathbf{A}\mathbf{z}. \end{aligned}$$

The MATLAB calculation can be done as follows:

```
% Physical constants
mu0 = 4*pi*1e-7;             % Permeability in vacuum
c0 = 299792456;              % Speed of light in vacuum
eps0 = 1/(mu0*c0*c0);        % Permittivity in vacuum

% Voltage between inner and outer conductor.
U = 1;

% Read the grid from the file 'unimesh0.mat'.
% This file contains the variables no2xy, el2no, noInt, noExt
load unimesh0
noNum = size(no2xy,2);
elNum = size(el2no,2);

% Scale the domain to measure 2cm x 2cm.
% The initial mesh fitted the unit square:
% -1 < x < 1 and -1 < y < 1.
no2xy = 1e-2*no2xy;

% Assemble the matrix A and vector b.
A = zeros(noNum);
b = zeros(noNum,1);
```

```
for elIdx = 1:elNum
  % Get the nodes and their coordinates
  % for the element 'elIdx'.
  no = el2no(:,elIdx);
  xy = no2xy(:,no);

  % Compute the element matrix and add
  % the contribution to the global matrix.
  A_el = CmpElMtx(xy);
  A(no,no) = A(no,no) + A_el;
end

% Get the indices of the nodes.
no_ess = union(noInt, noExt);
no_all = 1:noNum;
no_nat = setdiff(no_all, no_ess);

% Pick out the parts of the matrix and the vectors
% needed to solve the problem.
A_ess     = A(no_nat,no_ess);
A_nat     = A(no_nat,no_nat);
b         = b(no_nat);

z         = zeros(length(no_all),1);
z(noInt)  = U*ones(length(noInt),1);
z_ess     = z(no_ess);

% Solve the system of linear equations.
z_nat = A_nat\(b - A_ess*z_ess);

% Build up the total solution.
z = zeros(length(no_all),1);
z(no_ess) = z_ess;
z(no_nat) = z_nat;

% Compute the capacitance.
W = 0.5*eps0*(z'*A*z);
C = 2*W/U^2;

disp(['C per unit length [pF/m] = ' num2str(C/1e-12)])
```

The potential distribution computed by the MATLAB program is shown in Fig. 6.10, and the calculated value of the capacitance is 91.47360 pF/m. Not all these digits are correct, and we will discuss how to improve the accuracy in the next section. Note that there are large gradients near the reentrant corners of the inner conductor where the electric field is singular, but these gradients are not well resolved on the rather coarse mesh in Fig. 6.10.

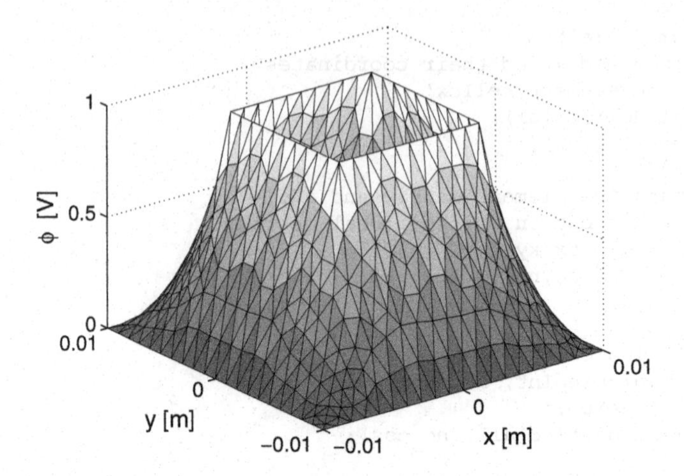

Fig. 6.10 The potential distribution between the inner conductor ($\phi = 1$ V) and the outer conductor ($\phi = 0$ V)

Review Questions

6.3-1 Explain how Dirichlet, Neumann, and Robin boundary conditions are incorporated into the system of linear equations for a FEM.

6.3-2 Derive the weak forms of the 2D Helmholtz equation with homogeneous Dirichlet and homogeneous Neumann boundary conditions. What are the differences between the two weak forms?

6.3-3 What is done in the assembly procedure?

6.3-4 Explain the difference between local and global node numbers.

6.3-5 Is a solution expanded in nodal basis functions φ_i guaranteed to be continuous?

6.3-6 How are unstructured finite element meshes constructed, represented, and stored by computers?

6.3-7 List the steps involved in computing the capacitance for a coaxial cable by the FEM.

6.4 Adaptivity

Triangular elements allow for local refinement of the mesh. Hence high resolution can be used where it is required, for example close to singularities and fine geometrical features, whereas lower resolution can be used where that is sufficient. This allows us to use the computational power where it contributes the most to the overall accuracy.

In general, one does not know a priori how to refine the mesh in order to get optimal efficiency. Therefore, adaptive schemes are usually based on a posteriori error estimates or error indicators (see, e.g., [48, 69]). A typical adaptive algorithm repeats the following steps until a satisfactory solution is obtained:

1. Compute the numerical solution on the current mesh.
2. Compute a posteriori error indicators for all individual elements.
3. Refine the mesh by splitting the elements with largest errors into smaller elements.

Algorithms for splitting selected elements into smaller elements may be quite complicated (see, e.g., [10]). However, software for mesh generation often includes this functionality.

To illustrate the advantages of adaptivity, we return to the capacitance calculation that we have already used as an illustration in Sects. 3.1 and 6.3 (see also Appendix C.3). For uniform meshes, this singularity reduces the convergence from $O(h^2) = O(1/N_n)$ to $O(h^{4/3}) = O(N_n^{-2/3})$, where N_n is the number of nodes in the mesh. By using FEM with *adaptively* generated meshes it is possible to restore quadratic convergence so that the error scales as $O(1/N_n)$ despite the singularity. We use the code shown in Sect. 6.3.3 and two sets of meshes. The first set of meshes is generated by uniform refinement (all elements are split into smaller elements), and the second set is generated by adaptive refinement (only selected elements are refined). A close-up of one of the adaptively refined meshes is shown in Fig. 6.11.

The relative error of the computed capacitance is shown in Fig. 6.12 for both uniformly and adaptively refined meshes. The horizontal axis shows the total number of nodes N_n in the mesh. The circles show the relative error $|C(N_n) - C_0|/C_0$ of the computed capacitances for different N_n. The exact value of C_0 is unknown in this case, but a sufficiently accurate reference solution ($C_0 = 90.6145$ pF with 6 correct digits) is obtained by careful extrapolation of the computed values.

The solid curves in Fig. 6.12 fit the error model $e(N_n) = \alpha/N_n^p$ to computed values of the capacitance $C(N_n)$. With uniform mesh refinement we find that the capacitance converges as $N_n^{-0.7} \propto h^{1.4}$. This is quite close to the theoretical asymptotic convergence rate $h^{4/3}$. With adaptive mesh refinement we find that the convergence rate is restored to $N^{-1} \propto h^2$, which is the rate we get for uniformly refined meshes when the solution is smooth (sufficiently regular).

Review Questions

6.4-1 Why and when is adaptivity useful? List advantages and disadvantages of adaptivity. Write down a general formulation, in words, for the objective of an adaptive computation. How could you achieve this objective?

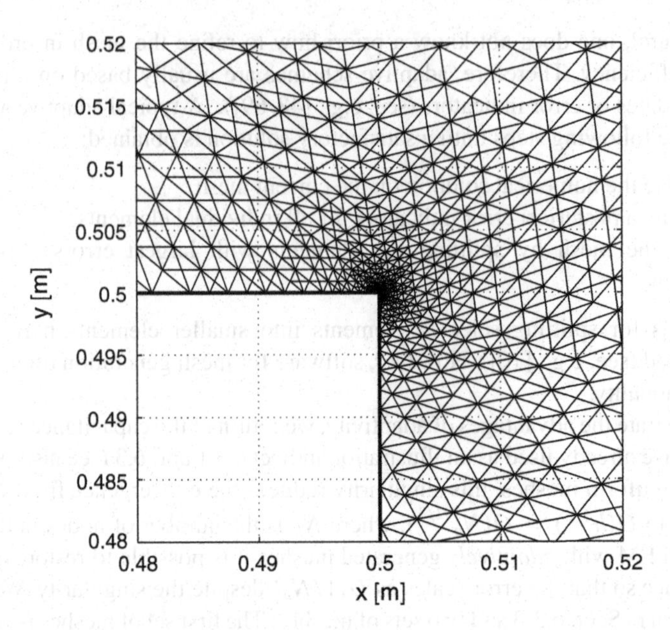

Fig. 6.11 The mesh after adaptive mesh refinement at one of the corners of the inner conductor where the potential changes rapidly. The smallest triangles at the corner measure approximately 40 μm

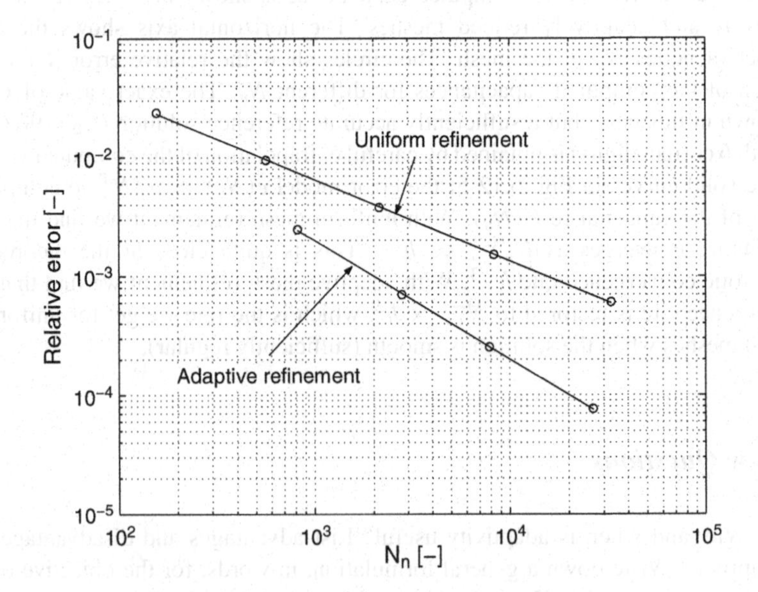

Fig. 6.12 The relative error in the capacitance as a function of the number of nodes in the mesh

6.4-2 Can adaptivity restore the nominal order of convergence even if the solution is singular? What implications does this have for the error as a function of the number of degrees of freedom?

6.5 Vector Equations

In this section, we will discuss a vector equation: the curl-curl equation of electromagnetics. However, as an intermediate step, we will first see how to choose elements for the 1D Maxwell equations written in terms of two variables, one component each of $\boldsymbol{E}$ and $\boldsymbol{H}$.

6.5.1 Mixed-Order FEM for Systems of First-Order Equations

In Sect. 6.2 we studied the model problem (6.1), i.e., the second-order equation for the electric field E in one dimension:

$$\frac{d}{dx}\left(\frac{1}{\mu}\frac{dE}{dx}\right) + \omega^2 \epsilon E = 0. \tag{6.31}$$

The second-order equation can be split into two first-order equations involving also the magnetic field H (a factor of j is removed in order to avoid complex variables):

$$\frac{dE}{dx} - \omega\mu H = 0, \tag{6.32}$$

$$\frac{dH}{dx} + \omega\epsilon E = 0. \tag{6.33}$$

To solve this pair of first-order equations, we first seek finite element representations for E and H that are suited for this. Somewhat arbitrarily, we choose to expand E, as before, in piecewise linear functions $l_i(x)$ (often referred to as "tent functions"). This gives

$$E(x) = \sum_{i=0}^{N} E_i l_i(x). \tag{6.34}$$

Equation (6.32) then leads us to expand H in the same class of functions as dE/dx, that is, in piecewise constants $c_i(x)$ ("top-hat functions"). This gives

$$H(x) = \sum_{i=0}^{N-1} H_{i+\frac{1}{2}} c_{i+\frac{1}{2}}(x), \tag{6.35}$$

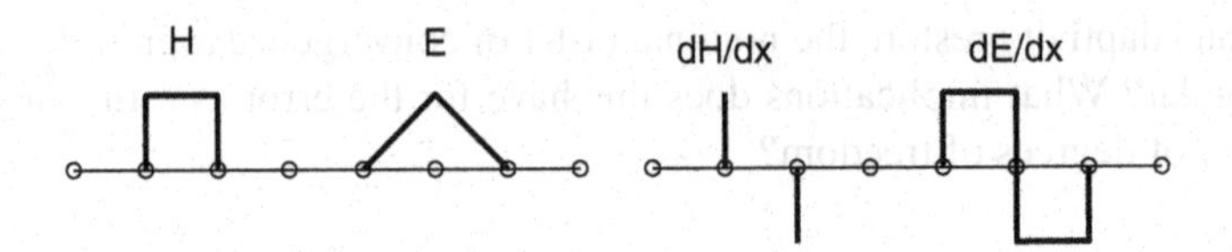

Fig. 6.13 Basis functions for the electric and magnetic fields together with their derivatives

where $c_{i+\frac{1}{2}}(x) = 1$ if $x_i < x < x_{i+1}$, and otherwise, $c_{i+\frac{1}{2}}(x) = 0$. Fig. 6.13 shows the tent and top-hat functions together with their derivatives.

To solve the set of first-order equations (6.32)–(6.33), we try a form of Galerkin's method. Since (6.32) contains H and dE/dx, which are both piecewise constant, we multiply (6.32) by piecewise constant weighting functions $c_{i+\frac{1}{2}}(x)$ and integrate over x. After division by the step length h this gives

$$\frac{E_{i+1} - E_i}{h} - \omega\mu H_{i+\frac{1}{2}} = 0, \tag{6.36}$$

which is exactly the simplest finite difference approximation for (6.32) on a staggered grid.

Equation (6.33), on the other hand, contains E. Therefore, we multiply it by a piecewise linear weighting function and integrate over x:

$$\int_{x_{i-1}}^{x_{i+1}} \left(\frac{dH}{dx} + \omega\epsilon E\right) l_i(x)dx = 0. \tag{6.37}$$

We substitute the representations (6.34) and (6.35) into (6.37) and obtain

$$\int_{x_{i-1}}^{x_{i+1}} (H_{i+\frac{1}{2}} - H_{i-\frac{1}{2}})\delta(x - x_i)l_i(x)dx + \omega\epsilon\left[\int_{x_{i-1}}^{x_i} E_{i-1}l_{i-1}(x)l_i(x)dx\right.$$
$$\left. + \int_{x_{i-1}}^{x_{i+1}} E_i l_i^2(x)dx + \int_{x_i}^{x_{i+1}} E_{i+1}l_i(x)l_{i+1}(x)dx\right] = 0. \tag{6.38}$$

Evaluation of the integrals and division by h gives

$$\frac{H_{i+\frac{1}{2}} - H_{i-\frac{1}{2}}}{h} + \omega\epsilon\left[\frac{2}{3}E_i + \frac{1}{6}(E_{i-1} + E_{i+1})\right] = 0. \tag{6.39}$$

Thus, the FEM equations corresponding to the coupled system (6.32)–(6.33) of first-order equations are

$$\frac{E_{i+1} - E_i}{h} = \omega\mu H_{i+\frac{1}{2}}, \tag{6.40}$$

$$\frac{H_{i+\frac{1}{2}} - H_{i-\frac{1}{2}}}{h} = -\omega\epsilon\left[\frac{2}{3}E_i + \frac{1}{6}(E_{i-1} + E_{i+1})\right]. \tag{6.41}$$

This FEM-discretized system looks almost the same as the finite difference approximation of (6.32)–(6.33) with staggered meshes. It differs only in the form for E on the right-hand side of (6.41). The similarity comes from the choice of basis and test functions. E was expanded in piecewise linear functions that are centered on the nodes: the integer mesh. H was expanded in piecewise constant functions that are centered on the midpoints or the half-mesh. Furthermore, (6.40) is centered on the half-grid. We constructed it this way by multiplying (6.32) by the piecewise constants $c_{i+\frac{1}{2}}(x)$ before integration. Similarly, (6.41) is centered on the integer grid, because we multiplied (6.33) by the piecewise linear functions $l_i(x)$.

This is a simple example of *mixed elements*. We can make the following analogy with staggered meshes for finite differences:

- A variable expanded in piecewise linear functions (FEM) is placed on the integer mesh (FD).
- A variable expanded in piecewise constant functions (FEM) is placed on the half mesh (FD).
- An equation multiplied by piecewise linear functions (FEM) is evaluated on the integer mesh (FD).
- An equation multiplied by piecewise constant functions (FEM) is evaluated on the half mesh (FD).

To emphasize the similarity between finite element and finite difference methods, we mention that if the integration in (6.38) is made by the trapezoidal rule, $\int_{x_i}^{x_{i+1}} f(x)dx \approx (h/2)[f(x_i) + f(x_{i+1})]$, the $\omega\epsilon E$ term becomes "lumped", $(4E_i + E_{i-1} + E_{i+1})/6 \to E_i$, and the FEM scheme becomes identical to the finite difference scheme.

One can see that the discretization (6.40)–(6.41) is in fact a Galerkin method, because the equation for ωE has been tested with the basis functions for E and the equation for ωH has been tested with the basis functions for H. It may also be noted that Faraday's law is identically satisfied by the FEM representation for E and H, while Ampère's law (6.33) is satisfied only in the weak sense, that is, as a weighted average.

6.5.2 The Curl-Curl Equation and Edge Elements

So far, we have discussed basis functions only for *scalar* equations, and used piecewise linear (nodal) and piecewise constant basis functions. To deal with *vector* quantities, such as the electric field, a first attempt might be to expand each vector component separately in nodal basis functions. It turns out that such an approach leads to nonphysical solutions, referred to as spurious modes.

This can be avoided by using *edge elements* [51], which are very well suited for approximating electromagnetic fields. The (basis functions for) edge elements are constructed such that their tangential components are continuous across element borders, whereas their normal components are allowed to be discontinuous. Edge elements are also called curl-conforming because the continuous tangential components imply that the curl of an edge element does not contain delta functions at the element boundaries. Thus, an electric field that is expanded in terms of edge elements has a curl that is square integrable.

In this section, we will show how edge elements can be applied to solve the curl-curl equation for $\boldsymbol{E}$:

$$\nabla \times \left(\mu^{-1} \nabla \times \boldsymbol{E}\right) - \left(\omega^2 \epsilon - j\omega\sigma\right) \boldsymbol{E} = -j\omega \boldsymbol{J}^s \text{ in } S, \tag{6.42}$$

$$\hat{\boldsymbol{n}} \times \boldsymbol{E} = \boldsymbol{P} \text{ on } L_1, \tag{6.43}$$

$$\hat{\boldsymbol{n}} \times \left(\mu^{-1} \nabla \times \boldsymbol{E}\right) + \gamma \hat{\boldsymbol{n}} \times \hat{\boldsymbol{n}} \times \boldsymbol{E} = \boldsymbol{Q} \text{ on } L_2. \tag{6.44}$$

Again, we have both Dirichlet and Robin boundary conditions, and $\boldsymbol{J}^s$ is an imposed source current.

We proceed along similar lines as in the scalar problem (6.11)–(6.13). Thus, we take the scalar product of (6.42) and the test function $\boldsymbol{W}_i$ and integrate over the computational domain S using the vector identity (4.4):

$$\begin{aligned} \nabla \cdot \left[\boldsymbol{W}_i \times \left(\mu^{-1} \nabla \times \boldsymbol{E}\right)\right] = \mu^{-1} \left(\nabla \times \boldsymbol{W}_i\right) \cdot \left(\nabla \times \boldsymbol{E}\right) \\ - \boldsymbol{W}_i \cdot \nabla \times \left(\mu^{-1} \nabla \times \boldsymbol{E}\right). \end{aligned} \tag{6.45}$$

The divergence term in (6.45) is integrated using Gauss's law in two dimensions $\int_S \nabla \cdot \boldsymbol{F} \, dS = \oint_{L_1 + L_2} \boldsymbol{F} \cdot \hat{\boldsymbol{n}} \, dl$, which gives the weak form of the vector Helmholtz equation

$$\begin{aligned} \int_S \left[\mu^{-1} \left(\nabla \times \boldsymbol{W}_i\right) \cdot \left(\nabla \times \boldsymbol{E}\right) - \left(\omega^2 \epsilon - j\omega\sigma\right) \boldsymbol{W}_i \cdot \boldsymbol{E}\right] dS \\ + \int_{L_2} \boldsymbol{W}_i \cdot \left(\boldsymbol{Q} - \gamma \hat{\boldsymbol{n}} \times \hat{\boldsymbol{n}} \times \boldsymbol{E}\right) dl = -j\omega \int_S \boldsymbol{W}_i \cdot \boldsymbol{J}^s dS. \end{aligned} \tag{6.46}$$

The major difference from the scalar problem lies in the choice of basis functions, where we use the edge element basis functions $\boldsymbol{N}_i(\boldsymbol{r})$ instead of nodal basis functions φ_i in this case.

Edge elements associate the degrees of freedom to the edges of the mesh rather than the nodes (this is why they are usually referred to as edge elements in the first place). Therefore, we have to number all the edges in the mesh and also give them reference directions. We will discuss the basis functions in more detail later. The edges are labeled by integers $1, 2, \ldots, N_e$. We expand the solution $\boldsymbol{E}(\boldsymbol{r})$ in terms of the basis functions:

$$\boldsymbol{E}(\boldsymbol{r}) = \sum_{j=1}^{N_e} E_j \boldsymbol{N}_j(\boldsymbol{r}), \tag{6.47}$$

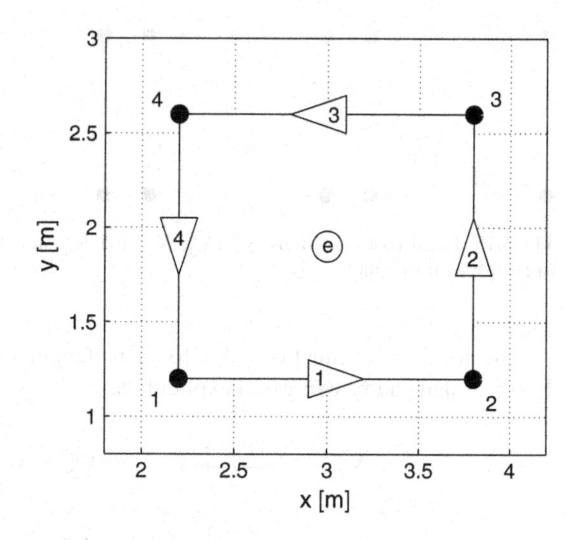

Fig. 6.14 Local numbering for the element e. The local reference directions for the edges are indicated by arrows, and the corresponding local edge numbers are shown inside the arrows

where E_j is the tangential electric field along the jth edge, in the direction of its reference direction.

We follow Galerkin's method, choose the test functions $\boldsymbol{W}_i(\boldsymbol{r}) = \boldsymbol{N}_i(\boldsymbol{r})$, and substitute (6.47) and the test functions into the weak form (6.46). This gives a linear system of equations $\mathbf{Az} = \mathbf{b}$ with

$$A_{ij} = \int_S \left[\mu^{-1} (\nabla \times \boldsymbol{N}_i) \cdot (\nabla \times \boldsymbol{N}_j) - (\omega^2 \epsilon - j\omega\sigma)\, \boldsymbol{N}_i \cdot \boldsymbol{N}_j\right] dS + \int_{L_2} \gamma\, (\hat{\boldsymbol{n}} \times \boldsymbol{N}_i) \cdot (\hat{\boldsymbol{n}} \times \boldsymbol{N}_j)\, dl, \tag{6.48}$$

$$z_j = E_j, \tag{6.49}$$

$$b_i = -j\omega \int_S \boldsymbol{N}_i \cdot \boldsymbol{J}^s\, dS - \int_{L_2} \boldsymbol{N}_i \cdot \boldsymbol{Q}\, dl. \tag{6.50}$$

The index j labels all edges and i all edges where $\boldsymbol{E}$ is unknown, i.e., all edges excluding those on the boundary L_1.

6.5.3 Edge Elements on Cartesian Grids

Here, we give explicit expressions for the edge basis functions $\boldsymbol{N}_i$. For simplicity, we first study those on a rectangular element that occupies the region defined by $x_a^e \leq x \leq x_b^e$ and $y_a^e \leq y \leq y_b^e$. The local numbering of the nodes and the edges is shown in Fig. 6.14 together with the local reference directions of the edges.

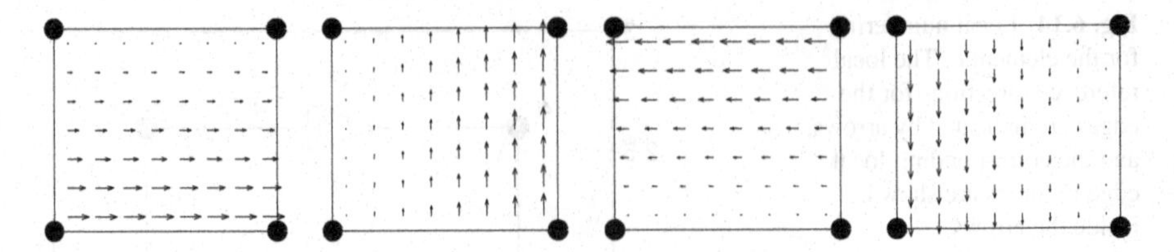

Fig. 6.15 Local basis functions N_1^e, N_2^e, N_3^e, and N_4^e on a rectangular element, shown in this order from left to right

The local basis functions N_i^e for a rectangular finite element are shown in Fig. 6.15 and can be expressed explicitly as

$$N_1^e = +\frac{y_b^e - y}{y_b^e - y_a^e}\hat{\boldsymbol{x}}, \qquad N_2^e = +\frac{x - x_a^e}{x_b^e - x_a^e}\hat{\boldsymbol{y}},$$

$$N_3^e = -\frac{y - y_a^e}{y_b^e - y_a^e}\hat{\boldsymbol{x}}, \qquad N_4^e = -\frac{x_b^e - x}{x_b^e - x_a^e}\hat{\boldsymbol{y}}. \tag{6.51}$$

The global basis functions must be chosen such that the tangential components of $\boldsymbol{E}$ are continuous across element boundaries. However, the normal component is allowed to be discontinuous [since $\nabla \cdot \boldsymbol{E}$ does not appear in the FEM matrix (6.48)]. Therefore, it is natural to associate the basis functions with the value of the electric field along the edges. The required representation is simply

$$E_x(x, y) = \sum_{ij} E_x|_{i,j}\; c_{i+\frac{1}{2}}(x) l_j(y),$$

$$E_y(x, y) = \sum_{ij} E_y|_{i,j}\; l_i(x) c_{j+\frac{1}{2}}(y). \tag{6.52}$$

Two such global basis functions are shown in Fig. 6.16.

Note that the edge elements have a mixed order of representation. Within each cell, E_x is constant in x and linear in y, and vice versa for E_y. The edge elements are not complete to first order, but represent a subset that is suitable for the curl-curl equation.

6.5.3.1 Edge Elements on Bricks and Hexahedra

We extend edge elements on rectangles to brick elements (hexahedra) in three dimensions. The electric field is represented as

$$E_x(x, y, z) = \sum_{ijk} E_x|_{i,j,k}\; c_{i+\frac{1}{2}}(x) l_j(y) l_k(z),$$

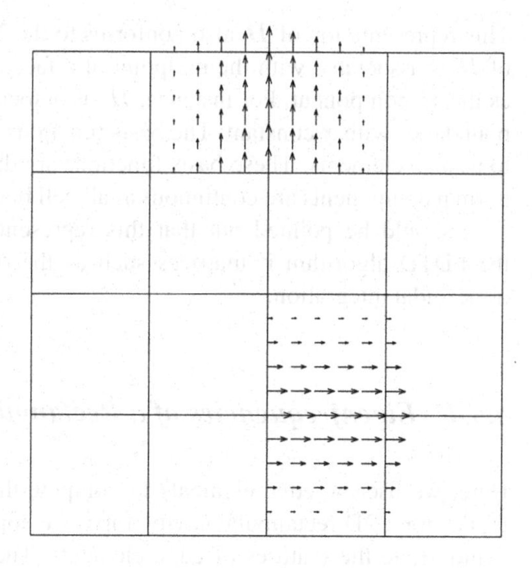

Fig. 6.16 Two global basis functions for rectangular edge elements on a grid

$$E_y(x,y,z) = \sum_{ijk} E_y|_{i,j,k}\, l_i(x) c_{j+\frac{1}{2}}(y) l_k(z),$$

$$E_z(x,y,z) = \sum_{ijk} E_z|_{i,j,k}\, l_i(x) l_j(y) c_{k+\frac{1}{2}}(z). \quad (6.53)$$

These edge elements are the FEM equivalent of the Yee cell. For instance, E_x in the Yee cell is located at the midpoint of the element in the x-direction, and the FEM basis function is the piecewise constant $c_{i+\frac{1}{2}}(x)$, also associated with the midpoint in x. In the y and z directions, the Yee cell puts E_x on the integer grid, and the FEM representation is in terms of piecewise linears, which are also associated with the integer grid.

For the magnetic field, we choose a representation that corresponds to the curl of the electric field. For instance, from the x-component of Faraday's law, $j\omega\mu H_x = \partial E_y/\partial z - \partial E_z/\partial y$, and the edge element representation (6.53) for $\boldsymbol{E}$, we see that the equation can be satisfied exactly if H_x is expanded with piecewise linears in x, and piecewise constants in y and z. Thus, for $\boldsymbol{H}$, we choose the representation

$$H_x(x,y,z) = \sum_{ijk} H_x|_{i,j,k}\, l_i(x) c_{j+\frac{1}{2}}(y) c_{k+\frac{1}{2}}(z),$$

$$H_y(x,y,z) = \sum_{ijk} H_y|_{i,j,k}\, c_{i+\frac{1}{2}}(x) l_j(y) c_{k+\frac{1}{2}}(z),$$

$$H_z(x,y,z) = \sum_{ijk} H_z|_{i,j,k}\, c_{i+\frac{1}{2}}(x) c_{j+\frac{1}{2}}(y) l_k(z). \quad (6.54)$$

This representation of $\boldsymbol{H}$ also conforms to the Yee arrangement. Each component of $\boldsymbol{H}$ is associated with the midpoint of a face that has the same normal direction as the $\boldsymbol{H}$ component. For instance, H_x is associated with the midpoints of the cell boundaries with x constant. The basis functions we have chosen for $\boldsymbol{H}$ are referred to as *face* elements. These basis functions are divergence-conforming, because the normal components are continuous at all cell boundaries.

It should be pointed out that this representation of $\boldsymbol{E}$ and $\boldsymbol{H}$ gives exactly the FDTD algorithm if matrices such as the one in (6.48) are assembled using trapezoidal integration.

6.5.4 Eigenfrequencies of a Rectangular Cavity

Here, we use the edge elements to compute the eigenfrequencies and the eigenmodes for a 2D rectangular cavity. First, we consider a 2×2-element resonator to demonstrate the features of edge elements. Then, we increase the resolution and study a more realistic case.

6.5.4.1 2 x 2-Element Resonator

We choose a square domain with width $a_x = 2$ m and height $a_y = 2$ m. The cavity resonator is discretized by 2×2 square elements, which is the smallest possible system that gives meaningful results. The mesh with numbering of nodes, edges, and elements is shown in Fig. 6.17. The positive reference directions (in this case chosen arbitrarily) for the edges are indicated by the arrows.

The numbering is systematically organized in Table 6.3 for the nodes, Table 6.4 for the edges, and Table 6.5 for the elements.

The boundary of the computational domain is metal and the interior S is air, i.e., $\sigma = 0$, $\mu = \mu_0$ and $\epsilon = \epsilon_0$. Thus, the eigenvalue problem is stated as

$$\nabla \times \nabla \times \boldsymbol{E} = \omega^2 \epsilon_0 \mu_0 \boldsymbol{E} \text{ in } S, \tag{6.55}$$

$$\hat{\boldsymbol{n}} \times \boldsymbol{E} = \boldsymbol{0} \text{ on } L_1. \tag{6.56}$$

We use (6.45) to arrive at the weak form

$$\int_S (\nabla \times \boldsymbol{W}_i) \cdot (\nabla \times \boldsymbol{E})\, dS = k^2 \int_S \boldsymbol{W}_i \cdot \boldsymbol{E}\, dS, \tag{6.57}$$

where $k^2 = \omega^2 \epsilon_0 \mu_0$. We expand the electric field in terms of the basis functions, i.e., approximate the electric field by (6.47), and test with $\boldsymbol{W}_i = \boldsymbol{N}_i$. Then, we get a generalized eigenvalue problem $\mathbf{Sz} = k^2 \mathbf{Mz}$, from which we solve for the eigenvalues k^2 and the eigenvectors $\mathbf{z} = [z_2, z_4, z_{11}, z_{12}]$, where z_2, z_4, z_{11}, and z_{12}

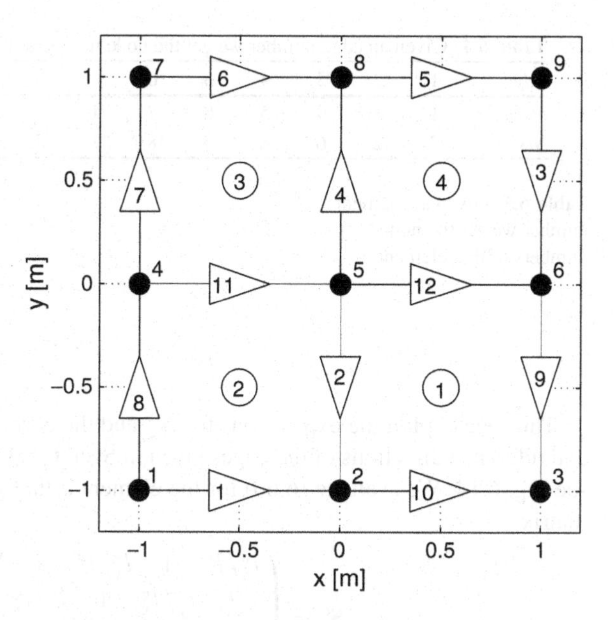

Fig. 6.17 Grid for 2×2-element resonator. The nodes (with numbers) are shown by black dots, and the edges (with numbers and positive directions) are indicated by the arrows centered on the edges of the grid. The element numbers are shown in the circles, centered in the corresponding elements

Table 6.3 Given a node number we get the coordinates of that node

Node	1	2	3	4	5	6	7	8	9
x	−1.0	0.0	1.0	−1.0	0.0	1.0	−1.0	0.0	1.0
y	−1.0	−1.0	−1.0	0.0	0.0	0.0	1.0	1.0	1.0

correspond to edges in the interior of the cavity. The remaining coefficients in (6.47) are zero because of the PEC boundary. The elements in **S** and **M** are given by

$$S_{ij} = \int_S (\nabla \times \boldsymbol{N}_i) \cdot (\nabla \times \boldsymbol{N}_j)\, dS, \tag{6.58}$$

$$M_{ij} = \int_S \boldsymbol{N}_i \cdot \boldsymbol{N}_j\, dS, \tag{6.59}$$

where the indices i and j run over all edges except those on the metal boundary, i.e., $i = 2, 4, 11, 12$ and $j = 2, 4, 11, 12$. By terminology borrowed from mechanical engineering, **S** is called the stiffness matrix and **M** is called the mass matrix.

For realistic cases, however, we do not evaluate S_{ij} and M_{ij} by (6.58) and (6.59). It is more convenient to use the assembling procedure described in Sect. 6.3.1. Consequently, we evaluate the element matrices S^e_{ij} and M^e_{ij} by

$$S^e_{ij} = \int_{y^e_a}^{y^e_b} \int_{x^e_a}^{x^e_b} (\nabla \times \boldsymbol{N}^e_i) \cdot \left(\nabla \times \boldsymbol{N}^e_j\right) dx\, dy, \tag{6.60}$$

$$M^e_{ij} = \int_{y^e_a}^{y^e_b} \int_{x^e_a}^{x^e_b} \boldsymbol{N}^e_i \cdot \boldsymbol{N}^e_j\, dx\, dy. \tag{6.61}$$

Table 6.4 Given an edge number we get the node numbers of that edge

Edge	1	2	3	4	5	6	7	8	9	10	11	12
Node 1	1	5	9	5	8	7	4	1	6	2	4	5
Node 2	2	2	6	8	9	8	7	4	3	3	5	6

Table 6.5 Given an element number we get the node numbers of that element.

Element	1	2	3	4
Node 1	6	1	7	6
Node 2	5	2	4	9
Node 3	2	5	5	8
Node 4	3	4	8	5

Thus, we exploit the expressions for $\mathbf{N}_i^e$ and the corresponding *local* numbering and reference directions of the edges given in Sect. 6.5.3 for an arbitrary element e; see Fig. 6.14. We evaluate (6.60) for the element e that gives the element stiffness matrix

$$\mathbf{S}^e = \begin{pmatrix} l_x^e/l_y^e & 1 & l_x^e/l_y^e & 1 \\ 1 & l_y^e/l_x^e & 1 & l_y^e/l_x^e \\ l_x^e/l_y^e & 1 & l_x^e/l_y^e & 1 \\ 1 & l_y^e/l_x^e & 1 & l_y^e/l_x^e \end{pmatrix}, \tag{6.62}$$

where the edges of the rectangle have lengths $l_x^e = x_b^e - x_a^e$ and $l_y^e = y_b^e - y_a^e$ along the x- and y-axes, respectively. Evaluation of (6.61) for the element e gives the corresponding element mass matrix

$$\mathbf{M}^e = \frac{l_x^e l_y^e}{6} \begin{pmatrix} 2 & 0 & -1 & 0 \\ 0 & 2 & 0 & -1 \\ -1 & 0 & 2 & 0 \\ 0 & -1 & 0 & 2 \end{pmatrix}, \tag{6.63}$$

The assembling procedure gives the global matrices $\mathbf{S}$ and $\mathbf{M}$ shown below, where the subindices in brackets show the index of the element that contributed to the matrix element. For edge elements, the reference direction of the edges must be compared for the local and global elements. If one of the two edges is reversed between the local and global ordering, the sign of the corresponding row and column in the element matrix must be changed before it is added to the global matrix:

$$\mathbf{S} = \underbrace{\begin{pmatrix} +1 & 0 & 0 & -1 \\ 0 & 0 & 0 & 0 \\ 0 & 0 & 0 & 0 \\ -1 & 0 & 0 & +1 \end{pmatrix}}_{\texttt{Element 1}} + \underbrace{\begin{pmatrix} +1 & 0 & +1 & 0 \\ 0 & 0 & 0 & 0 \\ +1 & 0 & +1 & 0 \\ 0 & 0 & 0 & 0 \end{pmatrix}}_{\texttt{Element 2}}$$

$$
+\underbrace{\begin{pmatrix} 0 & 0 & 0 & 0 \\ 0 & +1 & +1 & 0 \\ 0 & +1 & +1 & 0 \\ 0 & 0 & 0 & 0 \end{pmatrix}}_{\texttt{Element 3}} + \underbrace{\begin{pmatrix} 0 & 0 & 0 & 0 \\ 0 & +1 & 0 & -1 \\ 0 & 0 & 0 & 0 \\ 0 & -1 & 0 & +1 \end{pmatrix}}_{\texttt{Element 4}}
$$

$$
= \underbrace{\begin{pmatrix} +1_{[1]} + 1_{[2]} & 0 & +1_{[2]} & -1_{[1]} \\ 0 & +1_{[3]} + 1_{[4]} & +1_{[3]} & -1_{[4]} \\ +1_{[2]} & +1_{[3]} & +1_{[2]} + 1_{[3]} & 0 \\ -1_{[1]} & -1_{[4]} & 0 & +1_{[1]} + 1_{[4]} \end{pmatrix}}_{\texttt{Global matrix}}
$$

$$
\mathbf{M} = \underbrace{\frac{1}{6}\begin{pmatrix} +2 & 0 & 0 & 0 \\ 0 & 0 & 0 & 0 \\ 0 & 0 & 0 & 0 \\ 0 & 0 & 0 & +2 \end{pmatrix}}_{\texttt{Element 1}} + \underbrace{\frac{1}{6}\begin{pmatrix} +2 & 0 & 0 & 0 \\ 0 & 0 & 0 & 0 \\ 0 & 0 & +2 & 0 \\ 0 & 0 & 0 & 0 \end{pmatrix}}_{\texttt{Element 2}}
$$

$$
+\underbrace{\frac{1}{6}\begin{pmatrix} 0 & 0 & 0 & 0 \\ 0 & +2 & 0 & 0 \\ 0 & 0 & +2 & 0 \\ 0 & 0 & 0 & 0 \end{pmatrix}}_{\texttt{Element 3}} + \underbrace{\frac{1}{6}\begin{pmatrix} 0 & 0 & 0 & 0 \\ 0 & +2 & 0 & 0 \\ 0 & 0 & 0 & 0 \\ 0 & 0 & 0 & +2 \end{pmatrix}}_{\texttt{Element 4}}
$$

$$
= \underbrace{\frac{1}{6}\begin{pmatrix} +2_{[1]} + 2_{[2]} & 0 & 0 & 0 \\ 0 & +2_{[3]} + 2_{[4]} & 0 & 0 \\ 0 & 0 & +2_{[2]} + 2_{[3]} & 0 \\ 0 & 0 & 0 & +2_{[1]} + 2_{[4]} \end{pmatrix}}_{\texttt{Global matrix}}
$$

To summarize, we solve the eigenvalue problem

$$
\underbrace{\begin{pmatrix} 2 & 0 & 1 & -1 \\ 0 & 2 & 1 & -1 \\ 1 & 1 & 2 & 0 \\ -1 & -1 & 0 & 2 \end{pmatrix}}_{=\mathbf{S}} \underbrace{\begin{bmatrix} z_2 \\ z_4 \\ z_{11} \\ z_{12} \end{bmatrix}}_{=\mathbf{z}} = k^2 \underbrace{\begin{pmatrix} 2/3 & 0 & 0 & 0 \\ 0 & 2/3 & 0 & 0 \\ 0 & 0 & 2/3 & 0 \\ 0 & 0 & 0 & 2/3 \end{pmatrix}}_{=\mathbf{M}} \underbrace{\begin{bmatrix} z_2 \\ z_4 \\ z_{11} \\ z_{12} \end{bmatrix}}_{=\mathbf{z}}.
$$

Table 6.6 shows the eigenvalues and eigenvectors for this particular setting. *Analytical* treatment of this particular problem shows that there is an infinitely degenerate eigenvalue $k^2 = 0$ that corresponds to electrostatic modes $\boldsymbol{E} = -\nabla\phi$.

Table 6.6 Numerical eigenvalues and eigenvectors for the four-element cavity

Mode	k^2	z_2	z_4	z_{11}	z_{12}
1	0	$+1/2$	$+1/2$	$-1/2$	$+1/2$
2	3	$-1/\sqrt{2}$	$+1/\sqrt{2}$	0	0
3	3	0	0	$+1/\sqrt{2}$	$+1/\sqrt{2}$
4	6	$+1/2$	$+1/2$	$+1/2$	$-1/2$

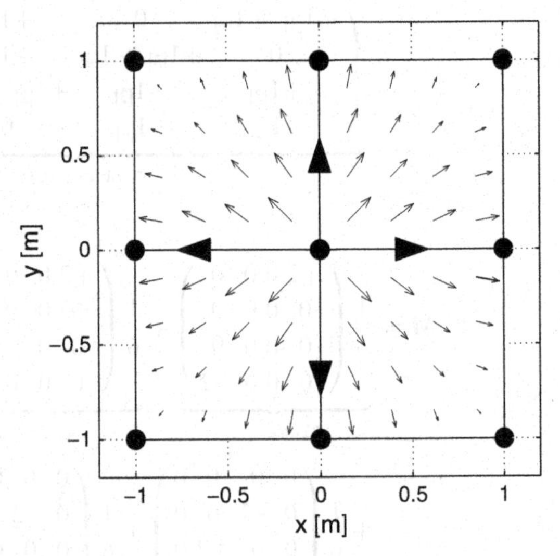

Fig. 6.18 Electric field for mode 1 with $k^2 = 0$. This is a static field that can be expressed in terms of a scalar potential, i.e., $\boldsymbol{E} = -\nabla\phi$

The electromagnetic modes have $k^2 = (\pi/2)^2(n_x^2 + n_y^2)$ for $n_x = 0, 1, \ldots$ and $n_y = 0, 1, \ldots$, where the combination $n_x = n_y = 0$ is excluded. The lowest nonzero eigenvalues are associated with the two (degenerate) modes with $k^2 = (\pi/2)^2 \approx 2.5$ and one mode with $k^2 = 2(\pi/2)^2 \approx 5.0$.

The figures below show the four numerical eigenmodes computed on the 2×2-element discretization. Fig. 6.18 shows the electrostatic mode on this mesh. It can be expressed in terms of a scalar potential, i.e., $\boldsymbol{E} = -\nabla\phi$, where the electric potential ϕ is expanded in piecewise bilinear nodal based finite elements, with $\phi = 0$ on the metal boundary and $\phi \neq 0$ on the central node. This static mode has the eigenvalue $k^2 = 0$.

The next two modes are shown in Fig. 6.19, and they correspond to the physical modes with the lowest resonance frequency. The two modes of the discretized system have the same eigenvalue $k^2 = 3$ and are therefore said to be degenerate. The corresponding analytical eigenvalue is $k^2 = (\pi/2)^2 \approx 2.5$.

Fig. 6.20 shows the third resonance of the cavity. It has the eigenvalue $k^2 = 6$, and the corresponding analytical eigenvalue is $k^2 = 2(\pi/2)^2 \approx 5.0$.

Observe that a linear combination of the four numerical eigenmodes can represent *any* solution on the 2×2-element discretization that satisfies the boundary condition.

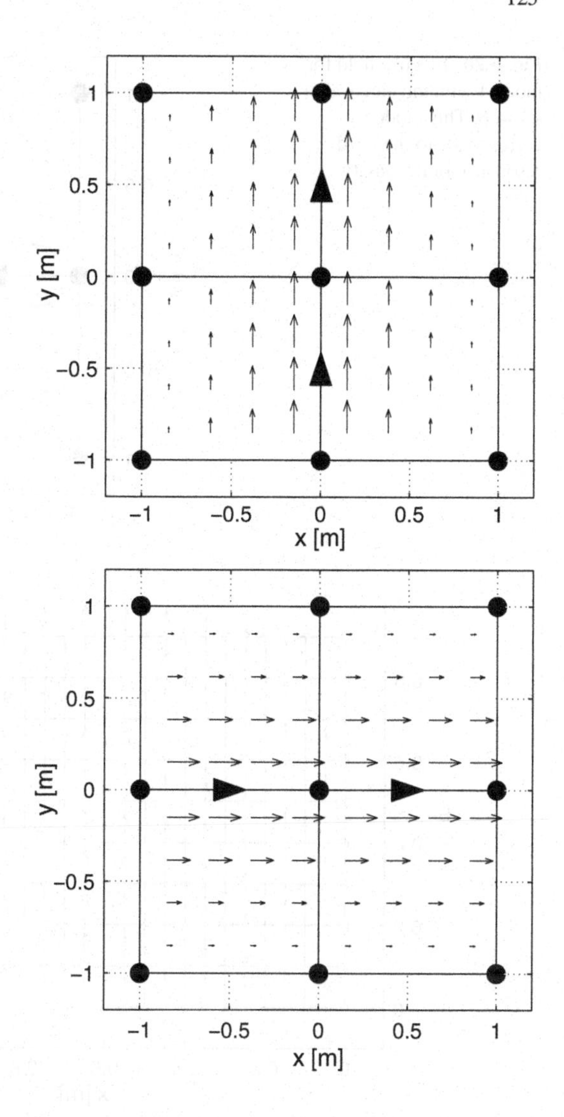

Fig. 6.19 Electric field for mode 2 and 3. The two have the eigenvalue $k^2 = 3$, and thus, they are degenerate. They correspond to the two degenerate fundamental resonances of the cavity

6.5.4.2 Better-Resolved Resonator

Next, we study a rectangular domain with width $a_x = 1.3$ m and height $a_y = 0.9$ m. We choose square cells of side 0.1 m, which gives a grid with 13×9 elements. We follow the approach outlined above, and the fundamental eigenmode, which corresponds to the lowest resonance frequency, is shown in Fig. 6.21. The corresponding analytic eigenmode is $\boldsymbol{E} = E_0 \sin(\pi x/a_x)\hat{\boldsymbol{y}}$.

The numerical eigenvalues k^2 are shown in Fig. 6.22 by circles and the analytical eigenvalues $k^2 = (\pi n_x/a_x)^2 + (\pi n_y/a_y)^2$ by crosses. Again, we have $n_x = 0, 1, 2, \ldots$ and $n_y = 0, 1, 2, \ldots$, where the combination $n_x = n_y = 0$ is excluded.

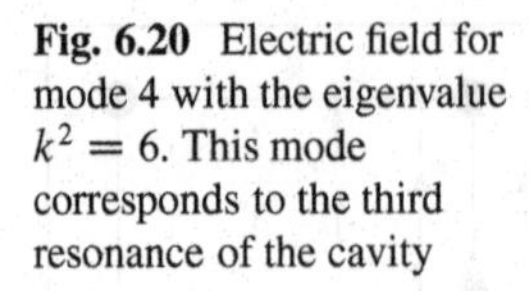

Fig. 6.20 Electric field for mode 4 with the eigenvalue $k^2 = 6$. This mode corresponds to the third resonance of the cavity

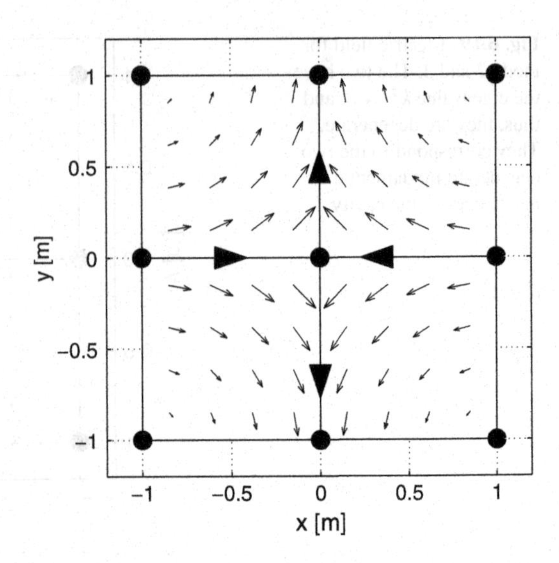

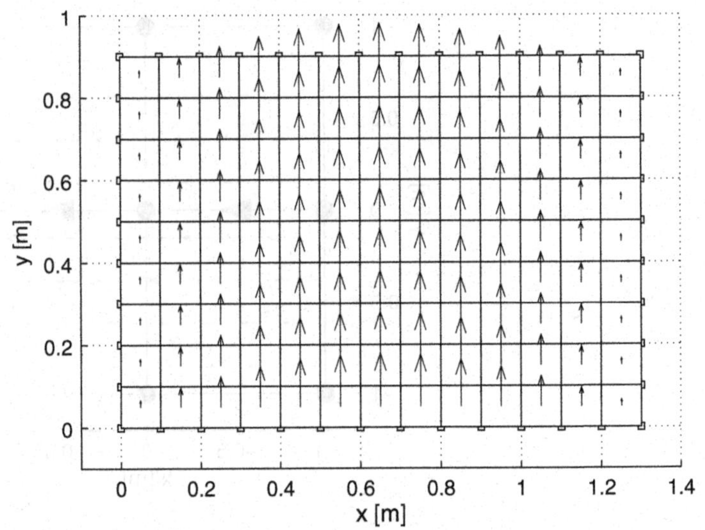

Fig. 6.21 The fundamental eigenmode on a rectangle with width $a_x = 1.3$ m and height $a_y = 0.9$ m

An important and very good property of the edge elements is that there is a one-to-one correspondence between the lowest nonzero numerical eigenmodes and the lowest nonzero analytical eigenmodes. This can be seen in Fig. 6.22 for our particular problem. The nodal elements, which we do not use for vector-valued electromagnetic fields, do not share this property, and the drawbacks of nodal elements can be clearly seen by examining the spectrum of the curl-curl operator. Instead of exact zero eigenvalues for the $\nabla \times \nabla \times$-operator corresponding

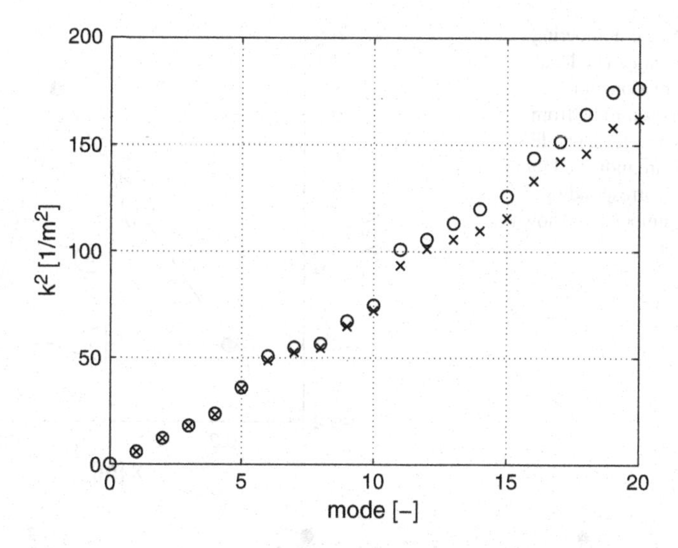

Fig. 6.22 Spectrum of eigenvalues for a rectangle with width $a_x = 1.3$ m and height $a_y = 0.9$ m. The numerically computed eigenvalues are shown by circles and their analytical counterparts by crosses

to electrostatic modes $\boldsymbol{E} = -\nabla\phi$, the nodal elements produce many eigenvalues between 0 and the smallest physical one. This is called spectral pollution, because it adds nonphysical eigenvalues in between the correct eigenvalues shown in Fig. 6.22. The eigenfunctions of the spurious solutions have rapid space variation associated with nonzero divergence. The nodal elements also cause much dispersion at short wavelengths (similar to the 1D result for first-order derivatives on nonstaggered meshes, discussed in Sect. 3.2), and this phenomenon also contributes to the spectral pollution.

By contrast, the edge elements produce exactly one zero eigenvalue for each interior node. Each such eigenvalue corresponds to a mode $\boldsymbol{E} = -\nabla\phi$, which has a zero eigenvalue, since $\nabla \times \nabla \times (-\nabla\phi) = \boldsymbol{0} = k^2(-\nabla\phi)$ gives $k^2 = 0$. With edge elements, this important property is preserved by the discrete representation, because the modes $\boldsymbol{E} = -\nabla\phi$, where ϕ is piecewise bilinear, belong to the set of edge elements. In our problem with the rectangular cavity, there are $12 \times 8 = 96$ interior nodes and therefore 96 zero eigenvalues, and these are given the mode number zero in Fig. 6.22.

It is in particular with respect to the electrostatic modes that the node-based elements fail for electromagnetic problems. Node-based elements do not contain the proper null-space for the curl-operator. The reason for this is that the potential modes $\boldsymbol{E} = -\nabla\phi$ for continuous, piecewise linear ϕ do not have continuous normal components and therefore do not belong to the node-based elements for $\boldsymbol{E}$, which are divergence conforming. The edge elements are not divergence conforming but allow jumps in the normal component at cell boundaries.

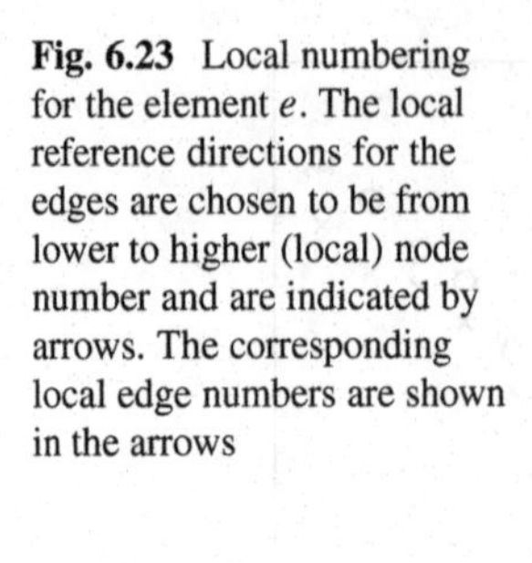

Fig. 6.23 Local numbering for the element e. The local reference directions for the edges are chosen to be from lower to higher (local) node number and are indicated by arrows. The corresponding local edge numbers are shown in the arrows

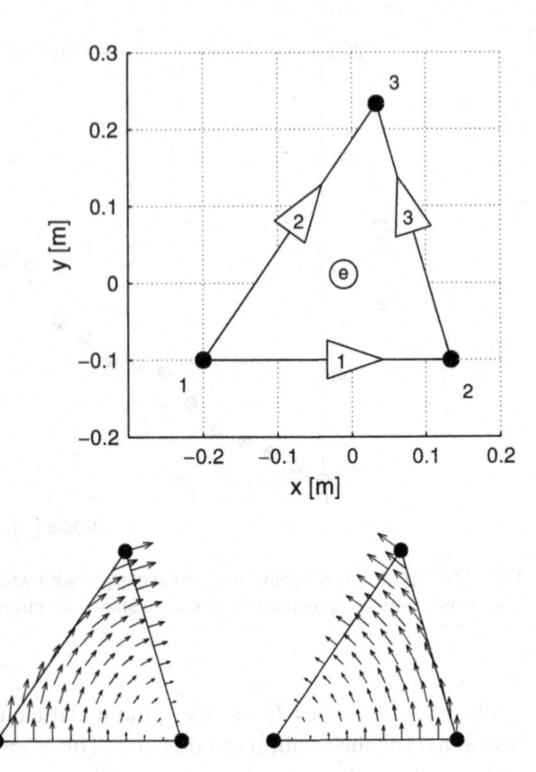

Fig. 6.24 $N_1^e(\boldsymbol{r})$ is shown to the left, $N_2^e(\boldsymbol{r})$ in the middle, and $N_3^e(\boldsymbol{r})$ to the right

6.5.5 Edge Elements on Triangles

Edge elements can also be formulated on triangles, tetrahedra, pyramids, and prisms. Fig. 6.23 shows the local numbering of the nodes and the edges of a triangle.

The edge element basis functions on a triangle can be expressed in the nodal basis functions φ_i^e:

$$
\begin{aligned}
\boldsymbol{N}_1^e &= \varphi_1^e \nabla \varphi_2^e - \varphi_2^e \nabla \varphi_1^e, \\
\boldsymbol{N}_2^e &= \varphi_1^e \nabla \varphi_3^e - \varphi_3^e \nabla \varphi_1^e, \\
\boldsymbol{N}_3^e &= \varphi_2^e \nabla \varphi_3^e - \varphi_3^e \nabla \varphi_2^e.
\end{aligned}
\tag{6.64}
$$

Fig. 6.24 shows the local basis functions. These basis functions are proportional to the vector field $r\hat{\boldsymbol{\phi}}$, where r and ϕ are local polar coordinates around the node opposite to the edge on which the basis function has a nonzero tangential component. The magnitudes of the basis functions are made such that the tangent line integral of the basis function along the edge it is associated with is 1.

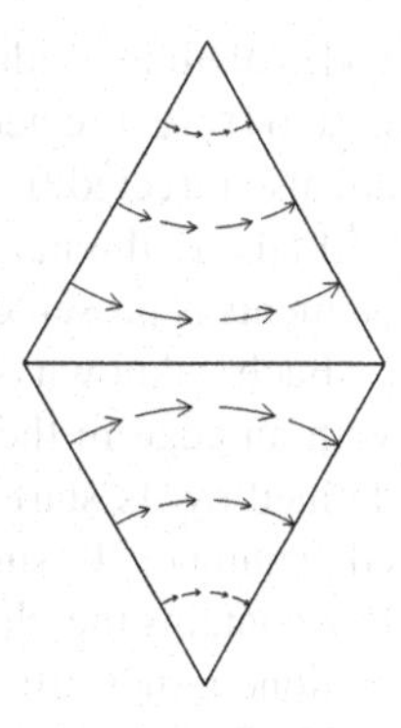

Fig. 6.25 Global edge basis function in 2D, spanning two triangles

Some important properties of the edge elements on triangles are worth pointing out. Just as for the rectangular edge elements, one constructs global basis functions such that the tangential component of $\boldsymbol{E}$ is continuous over element interfaces.

A global edge basis function is shown in Fig.6.25. Note that the normal component is discontinuous at the edges. Similar to the edge elements on rectangles, the tangential component is constant along one edge and zero along all the other edges of the rectangle.

Also similar to their rectangular counterparts, the edge elements on triangles have mixed order. One can add three more functions, constructed in a similar way as those in (6.64), but with the minus signs replaced by plus, to make the basis complete to first order. The "missing" first-order edge elements are gradients of scalar functions. Whether or not it is useful to include these gradients depends on the problem. Since the gradients do not contribute to $\nabla \times \boldsymbol{E}$, it is often more efficient not to leave them out. The edge elements we have discussed here are often referred to as order (0, 1), where 1 refers to those components that contribute to the curl, and 0 to the gradient part. There are also higher-order edge elements available [39, 89], which often can be more economical to use. However, these are not considered in this book.

In practice, the administration of edge elements requires certain special techniques, which are nonstandard in the context of the conventional FEM with node-based elements. These issues can to some extent be avoided on structured meshes of squares or cubes. For unstructured meshes, however, it is necessary to have efficient and reliable techniques, to for example, number the edges in the mesh and associate a reference direction with each edge. It is useful to remember the field representation $\boldsymbol{E}(\boldsymbol{r}) = \sum_{j=1}^{N_e} E_j \boldsymbol{N}_j(\boldsymbol{r})$ when such techniques are designed.

The reference direction is usually based on the global node numbers at the endpoints of the edge under consideration; for example, the vector field of an edge element basis function $\boldsymbol{N}_i$ is directed from the lower to the higher global node number when the coefficient for the basis function is positive. One or several of the basis functions on the local elements that share an edge may be defined in the reverse direction. One way to deal with this problem is to multiply all local basis functions with reverse direction by -1; i.e., the local basis function $\boldsymbol{N}_i^e$ relates to the global basis function as $\boldsymbol{N}_i = -\boldsymbol{N}_i^e$. Another way to deal with this problem is to sort the nodes of all individual element in ascending order. Since the basis functions defined in (6.64) are directed from lower to higher *local* node number; this implies that they are also directed from lower to higher *global* node number. This is the approach we will take in the next section, where a MATLAB program based on triangular edge elements is presented.

Each unknown (or coefficient E_j and its basis function) must also be associated with an edge in the unstructured mesh. We assume that all edges in the mesh are defined by its start and end nodes and that they also have been assigned a global edge number. To simplify the assembly procedure, we want to create a table `el2ed` that contains the global edge numbers for the three edges of each element. This can be done rather efficiently based on sorting techniques; see [41] for a more details. In MATLAB, this can be done by the function `unique`.

6.5.6 MATLAB: FEM with Triangular Edge Elements

We will here present a MATLAB function that given a triangular mesh on the form presented in Sect. 6.3.2, computes the mass and stiffness matrices **M** and **S**. A routine for plotting a field, given the vector with coefficients E_j that corresponds to the field, is also provided.

We begin by sorting the nodes of the individual elements in ascending order. Together with the definition of the basis functions in (6.64), this ensures that the edges–and therefore the tangential components of the basis functions–always are directed from lower to higher global *and local* node numbers.

Next we rewrite the basis functions in (6.64) using $\varphi_1 = 1 - \varphi_2 - \varphi_3$. The basis function can then be expressed as $\boldsymbol{N}_i^e = N_{i2}(\varphi_2, \varphi_3)\nabla\varphi_2 + N_{i3}(\varphi_2, \varphi_3)\nabla\varphi_3$. Noting that $\nabla\varphi_i$, $i = 1, 2, 3$, are constant within each element, we can write the local mass matrix of element e as

$$M_{ij}^e = \iint_e \boldsymbol{N}_i^e \cdot \boldsymbol{N}_j^e \, dx \, dy = \sum_{k=2}^{3} \sum_{l=2}^{3} \nabla\varphi_k \cdot \nabla\varphi_l \iint_e N_{ik} N_{jl} \, dx \, dy. \quad (6.65)$$

The integrals $\iint_e N_{ik}^e N_{jl}^e \, dx \, dy$ are scalar and can be computed through a mapping to a reference element with nodes $(0, 0)$, $(1, 0)$, and $(0, 1)$. The determinant of this mapping is

$$\det(\mathbf{J}^e) = (\boldsymbol{l}_1^e \times \boldsymbol{l}_2^e) \cdot \hat{\mathbf{z}}, \quad (6.66)$$

where $\boldsymbol{l}_i^e$ refers to edge i of element e. Depending on the order of the nodes, $\det(\mathbf{J}^e)$ is equal to plus or minus $2A^e$, where A^e is the area of element e. We then get the following expression for M_{ij}^e:

$$M_{ij}^e = |\det(\mathbf{J}^e)| \left(\nabla\varphi_2 \cdot \nabla\varphi_2 M_{ij}^{22} + \nabla\varphi_2 \cdot \nabla\varphi_3 M_{ij}^{23} + \nabla\varphi_3 \cdot \nabla\varphi_3 M_{ij}^{33} \right), \quad (6.67)$$

where M_{ij}^{kl} are independent of the shape of the triangles and therefore can be precomputed:

$$M_{ij}^{kk} = \int_{\varphi_2=0}^{1} \int_{\varphi_3=0}^{1-\varphi_2} N_{ik} N_{jk} \, d\varphi_2 \, d\varphi_3, \tag{6.68}$$

$$M_{ij}^{kl} = \int_{\varphi_2=0}^{1} \int_{\varphi_3=0}^{1-\varphi_2} [N_{ik} N_{jl} + N_{il} N_{jk}] \, d\varphi_2 \, d\varphi_3, \quad k \neq l. \tag{6.69}$$

Here δ_{kl} denotes the Kronecker delta. With the basis functions in (6.64) we get the following matrices $\mathbf{M}^{kl}$:

$$\mathbf{M}^{22} = \frac{1}{12} \begin{bmatrix} +3 & +1 & -1 \\ +1 & +1 & -1 \\ -1 & -1 & +1 \end{bmatrix}, \mathbf{M}^{23} = \frac{1}{12} \begin{bmatrix} +3 & +3 & +1 \\ +3 & +3 & -1 \\ +1 & -1 & -1 \end{bmatrix}, \mathbf{M}^{33} = \frac{1}{12} \begin{bmatrix} +1 & +1 & +1 \\ +1 & +3 & +1 \\ +1 & +1 & +1 \end{bmatrix}. \tag{6.70}$$

The stiffness matrix is also computed using a mapping to the same reference element. First we use the chain rule:

$$\begin{aligned} \nabla \times \boldsymbol{N}_i^e &= \nabla \times (N_{i2} \nabla \varphi_2 + N_{i3} \nabla \varphi_3) = \nabla N_{i2} \times \varphi_2 + \nabla N_{i3} \times \varphi_3 \\ &= \frac{\partial N_{i2}}{\partial \varphi_3} \nabla \varphi_3 \times \nabla \varphi_2 + \frac{\partial N_{i3}}{\partial \varphi_2} \nabla \varphi_2 \times \nabla \varphi_3 = \frac{\hat{z}}{\det(\mathbf{J}^e)} \left(\frac{\partial N_{i2}}{\partial \varphi_3} - \frac{\partial N_{i3}}{\partial \varphi_2} \right). \end{aligned}$$

Then we obtain

$$\begin{aligned} S_{ij}^e &= \iint_e (\nabla \times \boldsymbol{N}_i^e) \cdot (\nabla \times \boldsymbol{N}_j^e) \, dx \, dy \\ &= \frac{1}{|\det(\mathbf{J}^e)|^2} \iint_e \left(\frac{\partial N_{i2}}{\partial \varphi_3} - \frac{\partial N_{i3}}{\partial \varphi_2} \right) \left(\frac{\partial N_{j2}}{\partial \varphi_3} - \frac{\partial N_{j3}}{\partial \varphi_2} \right) dx \, dy \\ &= \frac{1}{|\det(\mathbf{J}^e)|} \int_{\varphi_2=0}^{1} \int_{\varphi_3=0}^{1-\varphi_2} \left(\frac{\partial N_{i2}}{\partial \varphi_3} - \frac{\partial N_{i3}}{\partial \varphi_2} \right) \left(\frac{\partial N_{j2}}{\partial \varphi_3} - \frac{\partial N_{j3}}{\partial \varphi_2} \right) d\varphi_2 \, d\varphi_3 \\ &= \frac{S_{ij}^{00}}{|\det(\mathbf{J}^e)|}, \end{aligned}$$

where $\mathbf{S}^{00}$ is independent of the shape of the element and can be precomputed:

$$\mathbf{S}^{00} = \begin{bmatrix} +2 & -2 & +2 \\ -2 & +2 & -2 \\ +2 & -2 & +2 \end{bmatrix}. \tag{6.71}$$

```
% -------------------------------------------------------------
% Compute the stiffness and mass matrix for edge elements on
% a triangular grid
% -------------------------------------------------------------
function [M, S, el2ed] = edgeFEM2D(no2xy, el2no)

% Arguments:
%    no2xy = x- and y-coordinates of the nodes
%    el2no = node indices of the triangles
% Returns:
%    M     = Mass matrix
%    S     = Stiffness matrix
%    el2ed = a table that contain the three edge numbers related
%            to each element

% Sort the nodes of each element
el2no = sort(el2no);

% Assign a number to each edge in the grid and create el2ed
n1 = el2no([1 1 2],:);
n2 = el2no([2 3 3],:);
[ed2no,trash,el2ed] = unique([n1(:) n2(:)],'rows');
el2ed = reshape(el2ed,3,size(el2no,2));

% Compute det(J^e), grad phi_2 and grad phi_3
e1 = no2xy(:,el2no(2,:)) - no2xy(:,el2no(1,:)); % 1st edge in
                                                % all elements
e2 = no2xy(:,el2no(3,:)) - no2xy(:,el2no(1,:)); % 2nd edge in
                                                % all elements
detJ = e1(1,:).*e2(2,:) - e1(2,:).*e2(1,:);     % det(J^e) for
                                                % all elements
g2 = [+e2(2,:)./detJ; -e2(1,:)./detJ];          % grad phi_2
g3 = [-e1(2,:)./detJ; +e1(1,:)./detJ];          % grad phi_3

% Define element shape independent matrices
m22 = [+3 +1 -1; +1 +1 -1; -1 -1 +1] / 12;
m23 = [+3 +3 +1; +3 +3 -1; +1 -1 -1] / 12;
m33 = [+1 +1 +1; +1 +3 +1; +1 +1 +1] / 12;
s00 = [+2 -2 +2; -2 +2 -2; +2 -2 +2];

% Compute local matrices and indices for all elements
mloc = m22(:) * (abs(detJ).*sum(g2.*g2)) + ...
       m23(:) * (abs(detJ).*sum(g2.*g3)) + ...
       m33(:) * (abs(detJ).*sum(g3.*g3));
sloc = s00(:) * abs(1./detJ);
rows = el2ed([1 2 3 1 2 3 1 2 3],:);
cols = el2ed([1 1 1 2 2 2 3 3 3],:);

% Assemble.
S = sparse(rows,cols,sloc);
M = sparse(rows,cols,mloc);
```

The presented MATLAB function assumes that the material parameters are constant in the entire mesh. It also assumes homogeneous Neumann boundary conditions, i.e., $\hat{\boldsymbol{n}} \times \nabla \times \boldsymbol{E} = \boldsymbol{0}$, which corresponds to a perfectly magnetic conducting (PMC) boundary. If we instead solved for the magnetic field $\boldsymbol{H}$, we would have $\hat{\boldsymbol{n}} \times \nabla \times \boldsymbol{H} = \boldsymbol{0}$, which corresponds to a PEC boundary. The function `edgeFEM2D` can easily be extended to treat problems where the material parameters vary between elements, but are constant within each element, and problems with homogeneous Dirichlet boundary conditions. However, this is left as a computer exercise.

A function for plotting a solution, expressed as a (real) vector with coefficients, is given below. The field is plotted on a finer mesh than the mesh that was used to compute the solution. The reason for this is to see how the field varies within, and on the interface between, elements. Arrows and color are used to visualize the field itself and its curl respectively.

```
% -----------------------------------------------------------------
% Plot a 2D vector field described by edge elements
% -----------------------------------------------------------------
function plotfield(no2xy, el2no, el2ed, sol)

% Arguments:
%     no2xy = x- and y-coordinates of the nodes
%     el2no = node indices for all triangles
%     el2ed = edge indices for all elements
%     sol   = Coefficient vector (each entry in the vector
%             corresponds to one edge in the mesh)
% Returns:
%     -

% Sort the nodes of each element
el2no = sort(el2no);

% Local coordinates for subgrid plotting
phi_1 = [4 3 2 1 0 3 2 1 0 2 1 0 1 0 0]' / 4;
phi_2 = [0 1 2 3 4 0 1 2 3 0 1 2 0 1 0]' / 4;
phi_3 = [0 0 0 0 0 1 1 1 1 2 2 2 3 3 4]' / 4;

% Gradients of the simplex functions
% (constant within each element)
edge1 = no2xy(:,el2no(2,:)) - no2xy(:,el2no(1,:));
edge2 = no2xy(:,el2no(3,:)) - no2xy(:,el2no(1,:));
detJ = edge1(1,:).*edge2(2,:) - edge1(2,:).*edge2(1,:);
grad_phi_2x =  edge2(2,:)./ detJ;
grad_phi_2y = -edge2(1,:)./ detJ;
grad_phi_3x = -edge1(2,:)./ detJ;
grad_phi_3y =  edge1(1,:)./ detJ;
grad_phi_1x = 0 - grad_phi_2x - grad_phi_3x;
grad_phi_1y = 0 - grad_phi_2y - grad_phi_3y;

% Solution values associated to the 1st, 2nd, and
% 3rd edges in each element
sol1 = sol(el2ed(1,:)).';
sol2 = sol(el2ed(2,:)).';
sol3 = sol(el2ed(3,:)).';
```

```
% Field values
Ex = phi_1 * ( grad_phi_2x.*sol1 + grad_phi_3x.*sol2) + ...
     phi_2 * (-grad_phi_1x.*sol1 + grad_phi_3x.*sol3) + ...
     phi_3 * (-grad_phi_1x.*sol2 - grad_phi_2x.*sol3);

Ey = phi_1 * ( grad_phi_2y.*sol1 + grad_phi_3y.*sol2) + ...
     phi_2 * (-grad_phi_1y.*sol1 + grad_phi_3y.*sol3) + ...
     phi_3 * (-grad_phi_1y.*sol2 - grad_phi_2y.*sol3);

Hz = (sol1 - sol2 + sol3)./detJ;

% Create subgrid
p1 = no2xy(:,el2no(1,:));
p2 = no2xy(:,el2no(2,:));
p3 = no2xy(:,el2no(3,:));
psub = kron(p1,phi_1') + kron(p2,phi_2') + kron(p3,phi_3');

% Initiate plotting
ih = ishold;
ax = newplot;

% Plot the curl of the field (constant within each element)
patch('faces',el2no','vertices',no2xy','facevertexcdata',Hz(:), ...
      'facecolor',get(ax,'defaultsurfacefacecolor'), ...
      'edgecolor',get(ax,'defaultsurfaceedgecolor'));
axis equal, hold on

% Plot the field itself as arrows
quiver(psub(1,:),psub(2,:),Ex(:)',Ey(:)','k');

% Plot the mesh
xy1 = no2xy(:,el2no(1,:));
xy2 = no2xy(:,el2no(2,:));
xy3 = no2xy(:,el2no(3,:));
xy = [xy1; xy2; xy3; xy1; NaN*xy1];
plot(xy(1:2:end),xy(2:2:end),'k')

% Create a new colormap
mrz = max(abs(Hz(:)));
caxis([-mrz, mrz]);
c = (0:64)'/64; d = [c c ones(size(c))];
colormap([d ;1 1 1; d(end:-1:1,end:-1:1)]);

if ~ih, hold off, end
```

We exploit this implementation to compute the eigenmodes $\boldsymbol{H}$ and eigenvalues k^2 for a cavity resonator with a circular metal boundary of radius $a = 1$ m. The solution satisfies the eigenvalue problem $\nabla \times \nabla \times \boldsymbol{H} = k^2 \boldsymbol{H}$ with boundary condition $\hat{\boldsymbol{n}} \times \nabla \times \boldsymbol{H} = \boldsymbol{0}$, where $\boldsymbol{H} = \hat{\boldsymbol{x}} H_x(x, y) + \hat{\boldsymbol{y}} H_y(x, y)$. A relatively coarse grid is used to compute the fundamental mode shown in Fig. 6.26. The numerical mode has $ka = 2.4412$, and this computed value compares well with the analytical

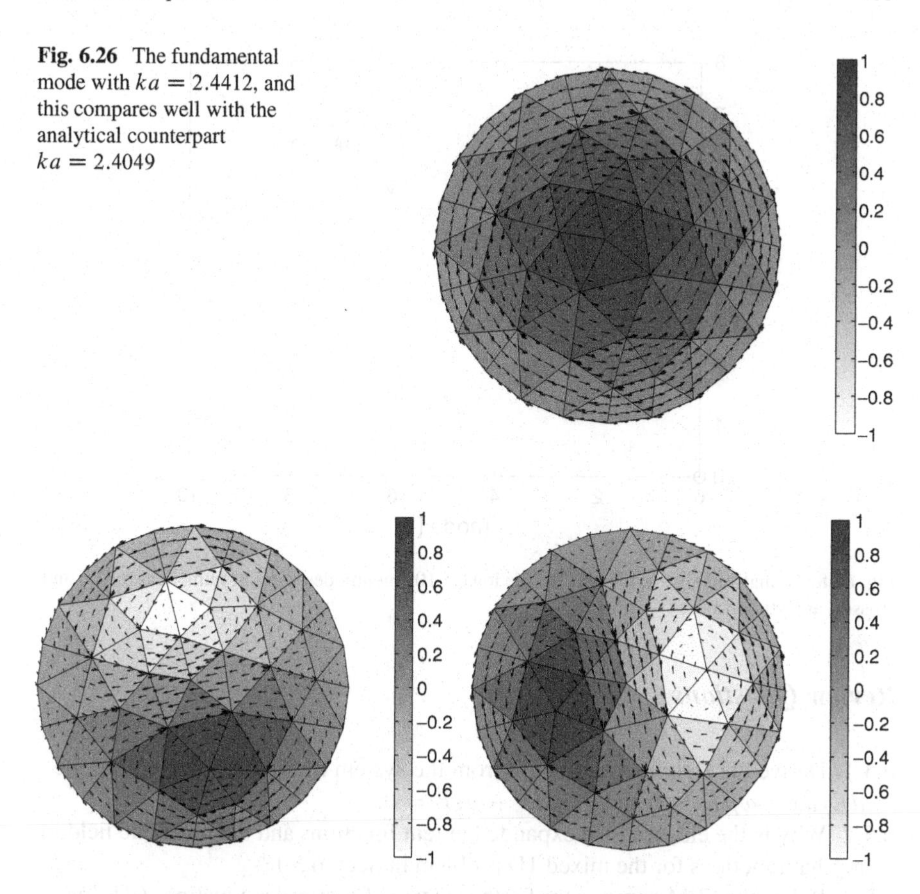

Fig. 6.26 The fundamental mode with $ka = 2.4412$, and this compares well with the analytical counterpart $ka = 2.4049$

Fig. 6.27 Two degenerate modes associated with the second-smallest $ka = 3.8831$ and $ka = 3.8846$, which compares well with the analytical counterpart $ka = 3.8318$

counterpart, i.e., the first zero $ka = 2.4049$ of the Bessel function $J_0(ka)$. The next mode is degenerated, and analytically it has $ka = 3.8318$, which corresponds to the first zero of $J_1(ka)$. The two numerically computed eigenmodes are shown in Fig. 6.27, and they have $ka = 3.8831$ and $ka = 3.8846$. The ten lowest eigenvalues are shown in Fig. 6.28, where the crosses indicate the analytical solution and the circles the numerical result. We note that there are no spurious modes, the multiplicity of the lowest modes is correct, and the error for the higher-order modes is surprisingly small. There are 48 zero eigenvalues and $N_n = 49$ nodes in the mesh, which includes all the nodes on the boundary. The zero eigenvalues correspond to modes $\boldsymbol{H} = \nabla\psi$, where the potentials ψ are different linear combinations of nodal basis functions φ_i. However, while there are 49 linearly independent potentials ψ, there are only 48 linearly independent modes $\boldsymbol{H} = \nabla\psi$, since a constant (but nonzero) ψ corresponds to zero magnetic field.

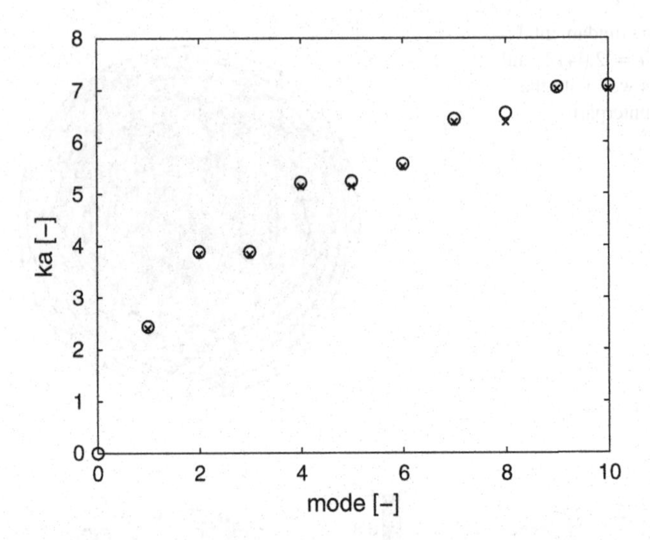

Fig. 6.28 Nomalized eigenvalues ka for the lowest 10 eigenmodes: circles, numerical result; and crosses, analytical values

Review Questions

6.5-1 Derive the Helmholtz equation from the system of first-order equations, i.e., $dE/dx - \omega\mu H = 0$ and $dH/dx + \omega\epsilon E = 0$.

6.5-2 Why is the electric field expanded in tent functions and the magnetic field in top-hat functions for the mixed 1D problem in Sect. 6.5.1?

6.5-3 Relate the FEM expressions for the system of first-order equations ($dE/dx - \omega\mu H = 0$ and $dH/dx + \omega\epsilon E = 0$) to the corresponding finite-difference approximations. Do you need to apply special techniques for a one-to-one correspondence?

6.5-4 How do tent and top-hat functions relate to the integer and half-mesh used for finite difference approximations?

6.5-5 Describe the differences and similarities between the FEM for scalar and vector equations.

6.5-6 Why are edge elements needed? Why are they called edge elements? Why are they referred to as curl-conforming elements? List some of the characteristic properties of edge elements.

6.5-7 What is the physical meaning of the degrees of freedom for a vector field expanded in terms of edge elements? How does this translate to an electric field that can be represented as the gradient of a scalar potential?

6.5-8 Derive the weak form of the vector Helmholtz equation, $\nabla \times (\mu^{-1} \nabla \times \boldsymbol{E}) - (\omega^2\epsilon - j\omega\sigma)\boldsymbol{E} = -j\omega \boldsymbol{J}^s$, with some suitable boundary conditions.

6.5-9 Write down the explicit expressions for the edge elements on a rectangle.

6.5-10 Describe the functions (with respect to x, y, and z) that are used for the x-components of the electric and magnetic fields, respectively, on a grid of brick elements.

6.5-11 Derive explicit expressions for the matrix elements in (6.60) and (6.61) by evaluating the integrals by hand. Use the expressions in (6.51) for the basis and test functions.

6.5-12 How many static modes are supported by the mesh in Fig. 6.21 and why? How many static modes are supported by the mesh in Fig. 6.26?

6.5-13 Write down explicit expressions for edge elements on triangles in terms of (a) polar coordinates and (b) nodal basis functions.

6.5-14 Show that for triangles, the tangential component of a given basis function is constant along one edge and zero along the other edges of the element. Does this also hold for rectangular edge elements?

6.6 Practical Implementations of the FEM

Up to this point, we have exploited analytical evaluation of the integrals in the weak formulations. For cases with inhomogeneous material parameters, analytical evaluation is typically not feasible and, then, numerical integration is exploited instead of analytical integration. There are also other situations where it is complicated or undesirable to use analytical evaluation of the integrals in the weak formulation. Here, we introduce numerical integration in combination with a technique that involves a so-called reference element, which is used in most FEM codes.

Consider a finite element in physical space that is described in terms of its Cartesian coordinates (x, y, z). Rather than integrating directly with respect to the physical coordinates (x, y, z), it is common to perform the numerical integration on a so-called reference element, which is described in terms of another coordinate system with the Cartesian coordinates (u, v, w). The physical element in (x, y, z)-space is related to the reference element in (u, v, w)-space by a transformation, which is referred to as the mapping from the reference element to the physical element. The basis functions are defined on the reference element and, then, transformed to the physical element by means of the mapping. Thus, an integral that is originally formulated in (x, y, z)-space can be reformulated and evaluated in (u, v, w)-space, i.e. on the reference element, regardless of the shape of the physical element as long as the physical element is not degenerated or incorrectly shaped in some other way.

For parts of the discussion that follows, we use triangular elements in order to provide an explicit demonstration of the techniques presented and, hopefully, this makes it simpler to understand the concept of the reference element. However, this recipe is general and can be applied to other types of finite elements used for both two- and three-dimensional problems. Appendix B contains information about the reference elements and their basis functions for the triangle and other typical finite element shapes: quadrilaterals; tetrahedrons; prisms and hexahedrons. It is

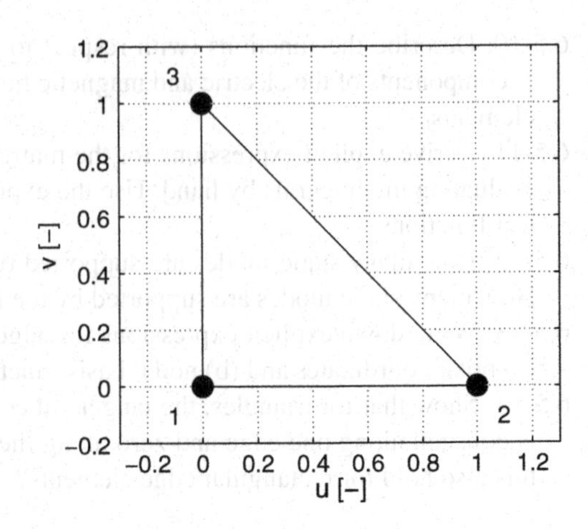

Fig. 6.29 Reference triangle defined in the (u, v)-space

possible to implement all the examples demonstrated in this chapter by means of numerical integration on the reference element. (The analytical treatment presented in the previous sections in this chapter is mainly useful for special cases that feature piecewise constant material parameters. Also, it can be easier to get a basic understanding of the FEM based on such formulations.)

6.6.1 *The Reference Element*

The reference element for the triangle is shown in Fig. 6.29. It is defined in the (u, v)-space on the domain $0 \leq u \leq 1 - v$ and $0 \leq v \leq 1$. It is useful to notice that the reference element does not change its shape or size as the elements in the physical finite element mesh. We provide a relation between the reference element and the physical elements in Sect. 6.6.2.

On this reference triangle, we have the nodal basis functions

$$\varphi_1(u, v) = 1 - u - v$$
$$\varphi_2(u, v) = u$$
$$\varphi_3(u, v) = v$$

and the linear edge element basis functions

$$\mathbf{N}_1(u, v) = \varphi_1 \tilde{\nabla} \varphi_2 - \varphi_2 \tilde{\nabla} \varphi_1 = \hat{\mathbf{u}}(1 - v) + \hat{\mathbf{v}} u$$
$$\mathbf{N}_2(u, v) = \varphi_2 \tilde{\nabla} \varphi_3 - \varphi_3 \tilde{\nabla} \varphi_2 = -\hat{\mathbf{u}} v + \hat{\mathbf{v}} u$$
$$\mathbf{N}_3(u, v) = \varphi_3 \tilde{\nabla} \varphi_1 - \varphi_1 \tilde{\nabla} \varphi_3 = -\hat{\mathbf{u}} v + \hat{\mathbf{v}}(u - 1),$$

where $\tilde{\nabla} = \hat{\mathbf{u}} \partial/\partial u + \hat{\mathbf{v}} \partial/\partial v + \hat{\mathbf{w}} \partial/\partial w$ differentiates with respect to the reference element coordinate system.

6.6.2 Mapping From the Reference Element to the Physical Element

Consider a triangle with straight edges in the physical two-dimensional space (x, y). An example is given in Fig.6.6 that shows a physical triangle with index e, where the nodes are located at $\boldsymbol{r}_1^e = -0.5\hat{\boldsymbol{x}} + 0.5\hat{\boldsymbol{y}}$, $\boldsymbol{r}_2^e = -0.2\hat{\boldsymbol{y}}$ and $\boldsymbol{r}_3^e = 0.6\hat{\boldsymbol{x}} + 0.4\hat{\boldsymbol{y}}$.

In order to have a relation between the reference element and the physical element, we define the vector $\boldsymbol{r}^e = \boldsymbol{r}^e(u, v)$ to be given by

$$\boldsymbol{r}^e(u, v) = \hat{\boldsymbol{x}}\, x^e(u, v) + \hat{\boldsymbol{y}}\, y^e(u, v)$$
$$= \sum_{i=1}^{N} \boldsymbol{r}_i^e \varphi_i(u, v),$$

where $N = 3$ for the triangle. (We use the same type of expression for quadrilaterals, where $N = 4$ since the quadrilateral has four nodes and four nodal basis functions.)

Thus, we have

$$\boldsymbol{r}^e(u, v) = (1 - u - v)\boldsymbol{r}_1^e + u\boldsymbol{r}_2^e + v\boldsymbol{r}_3^e$$

for an arbitrary point (u, v) in the domain $0 \le u \le 1 - v$ and $0 \le v \le 1$.

In order to investigate the mapping, we introduce the parameter ξ that is zero at the start point of an edge of the triangle and one at the end point of that same edge. As we move along the edge, the parameter ξ increases from zero to one and the entire length of the edge is spanned by $0 \le \xi \le 1$. Now, we consider the first edge of the triangle with $u = \xi$ and $v = 0$, where we have

$$\boldsymbol{r}^e(\xi, 0) = (1 - \xi)\boldsymbol{r}_1^e + \xi\boldsymbol{r}_2^e.$$

The second edge with $u = 1 - \xi$ and $v = \xi$ yields

$$\boldsymbol{r}^e(1 - \xi, \xi) = (1 - \xi)\boldsymbol{r}_2^e + \xi\boldsymbol{r}_3^e$$

and the third edge with $u = 0$ and $v = 1 - \xi$ gives

$$\boldsymbol{r}^e(0, 1 - \xi) = (1 - \xi)\boldsymbol{r}_3^e + \xi\boldsymbol{r}_1^e.$$

Consequently, the edges of the reference element are mapped to the edges of the physical element. Similarly, the vertices of the reference element are mapped to the vertices of the physical element. The same procedure can be applied to all points interior to the reference element, which are mapped to the corresponding points interior to the physical element.

For three dimensional finite elements, we have in an analogous manner

$$\begin{aligned} \boldsymbol{r}^e(u,v,w) &= \hat{\boldsymbol{x}}\, x^e(u,v,w) + \hat{\boldsymbol{y}}\, y^e(u,v,w) + \hat{\boldsymbol{z}}\, z^e(u,v,w) \\ &= \sum_{i=1}^{N} \boldsymbol{r}_i^e \varphi_i(u,v,w), \end{aligned}$$

where N is the number of nodes of the element.

These mappings are referred to as linear mappings since they exploit the linear nodal basis functions φ_i. As a consequence, the edges of the elements are straight lines. It is also feasible to use higher-order basis functions such as quadratic basis functions in the mapping, which gives curved edges and faces in the general case. A higher-order mapping improves the geometric approximation of smoothly curved boundaries.

6.6.2.1 Integration on the Reference Element

Next, we introduce the Jacobian

$$\mathbf{J}^e = \begin{pmatrix} \partial x^e/\partial u & \partial y^e/\partial u & \partial z^e/\partial u \\ \partial x^e/\partial v & \partial y^e/\partial v & \partial z^e/\partial v \\ \partial x^e/\partial w & \partial y^e/\partial w & \partial z^e/\partial w \end{pmatrix}, \qquad (6.72)$$

which is evaluated given the mapping $\boldsymbol{r}^e = \boldsymbol{r}^e(u,v,w)$ for the physical element with index e. We notice that the mapping and its Jacobian depend not only on the type of element but also on the shape of this element.

In the 2D case, we have that $x^e = x^e(u,v)$ and $y^e = y^e(u,v)$ as described above. In addition, we remove the influence of the third space coordinate on the mapping by setting $z^e = w$, such that a 2D reference element in the plane $w = 0$ is located in the plane $z^e = 0$. This construction yields the Jacobian

$$\mathbf{J}^e = \begin{pmatrix} \partial x^e/\partial u & \partial y^e/\partial u & 0 \\ \partial x^e/\partial v & \partial y^e/\partial v & 0 \\ 0 & 0 & 1 \end{pmatrix}, \qquad (6.73)$$

which can be used in a convenient manner in the following.

For an infinitesimal volume element, we have the relation $dx\,dy\,dz = \det(\mathbf{J}^e)\,du\,dv\,dw$. Similarly, we have the relation $dx\,dy = \det(\mathbf{J}^e)\,du\,dv$ for the corresponding 2D situation.

Thus, we have the following relation for integration in 3D

$$\int_{V^e} f(x, y, z) dx\, dy\, dz = \int_{\tilde{V}} f(u, v, w) \det(\mathbf{J}^e)\, du\, dv\, dw,$$

where V^e is volume occupied by the physical element e and $\tilde{V}$ is the corresponding domain for the reference element. Similarly, integration in 2D yields the relation

$$\int_{S^e} f(x, y) dx\, dy = \int_{\tilde{S}} f(u, v) \det(\mathbf{J}^e)\, du\, dv$$

where S^e is surface occupied by the physical element e and $\tilde{S}$ is the corresponding domain for the reference element.

The integrand $f(x, y, z)$ in 3D and $f(x, y)$ in 2D must also be related to the reference element and expressed in terms of its coordinate system. For example, the integrand would involve expressions of the type $f(x, y) = \nabla\varphi_i^e(x, y) \cdot \nabla\varphi_j^e(x, y)$ for an electrostatic problem in 2D, where $\varphi_i^e(x, y)$ is the linear nodal basis functions on element e expressed in terms of the physical coordinates (x, y).

6.6.2.2 Nodal Elements

First, we consider the nodal elements $\varphi_i^e(x, y)$ that can be expressed as

$$\varphi_i^e(x, y) = \varphi_i^e(x(u, v), y(u, v)) = \varphi_i(u, v).$$

This relation is supposed to be interpreted and used in the following manner, where we for simplicity consider a triangular element as an example. Given a point (u, v) in the reference triangle, we can evaluate the nodal basis function $\varphi_i(u, v)$, which is one at node i and zero at the two other nodes with a linear variation inside the reference element. In physical space, $\varphi_i^e(x, y)$ is also a linear function with the value one at node i and zero at the two other nodes. Given the linear mapping $\boldsymbol{r}^e = \boldsymbol{r}^e(u, v)$, we get a point $(x^e(u, v), y^e(u, v))$ in physical space that corresponds to the point (u, v) on the reference element. Thus, the nodal basis function $\varphi_i^e(x, y)$ evaluated at the point $(x, y) = (x^e(u, v), y^e(u, v))$ in physical space yields the same value as the nodal basis function $\varphi_i(u, v)$ evaluated at the point (u, v) on the reference element. The same type of result holds for all nodal basis functions listed in Appendix B.

Further, we have the gradient $\nabla\varphi_i^e(x, y)$ of the nodal elements that can be expressed as

$$\nabla\varphi_i^e(x, y) = \nabla\varphi_i^e(x(u, v), y(u, v)) = [\mathbf{J}^e]^{-1}\, \tilde{\nabla}\varphi_i(u, v). \tag{6.74}$$

where we distinguish the operator ∇ that differentiates with respect to the physical space coordinates from the corresponding operator $\tilde{\nabla}$ that differentiates with respect to the coordinates used for the reference element. The relation (6.74) is a result of the chain rule $\tilde{\nabla}\varphi_i(u,v,w) = \mathbf{J}^e\,\nabla\varphi_i^e(x,y,z)$ that is expressed in detail as

$$\begin{bmatrix} \partial\varphi_i/\partial u \\ \partial\varphi_i/\partial v \\ \partial\varphi_i/\partial w \end{bmatrix} = \begin{pmatrix} \partial x^e/\partial u & \partial y^e/\partial u & \partial z^e/\partial u \\ \partial x^e/\partial v & \partial y^e/\partial v & \partial z^e/\partial v \\ \partial x^e/\partial w & \partial y^e/\partial w & \partial z^e/\partial w \end{pmatrix} \begin{bmatrix} \partial\varphi_i^e/\partial x \\ \partial\varphi_i^e/\partial y \\ \partial\varphi_i^e/\partial z \end{bmatrix}, \tag{6.75}$$

where this result holds for all nodal basis functions listed in Appendix B.

Thus, we can now evaluate integrals of the following type

$$\int_{S^e} \alpha(x,y)\nabla\phi_i^e\cdot\nabla\phi_j^e\,dx\,dy =$$

$$= \int_{\tilde{S}} \alpha(x^e(u,v),y^e(u,v))\left([\mathbf{J}^e]^{-1}\,\tilde{\nabla}\phi_i\right)\cdot\left([\mathbf{J}^e]^{-1}\,\tilde{\nabla}\phi_j\right)\det(\mathbf{J}^e)\,du\,dv$$

$$\int_{S^e} \beta(x,y)\phi_i^e\phi_j^e\,dx\,dy =$$

$$= \int_{\tilde{S}} \beta(x^e(u,v),y^e(u,v))\phi_i\phi_j\,\det(\mathbf{J}^e)\,du\,dv$$

This type of treatment is applicable for all the nodal elements listed in Appendix B, which also includes elements for three-dimensional problems.

6.6.2.3 Edge Elements

The edge element basis functions on the triangle are given in (6.64) and they can be expressed as

$$\boldsymbol{N}_i^e = \varphi_{i_1}^e\nabla\varphi_{i_2}^e - \varphi_{i_2}^e\nabla\varphi_{i_1}^e$$

for the edge i that starts at node i_1 and ends at node i_2. Consequently, we can express these basis functions as

$$\begin{aligned} \boldsymbol{N}_i^e &= \varphi_{i_1}^e\nabla\varphi_{i_2}^e - \varphi_{i_2}^e\nabla\varphi_{i_1}^e \\ &= [\mathbf{J}^e]^{-1}\left(\varphi_{i_1}\tilde{\nabla}\varphi_{i_2} - \varphi_{i_2}\tilde{\nabla}\varphi_{i_1}\right) \\ &= [\mathbf{J}^e]^{-1}\,\boldsymbol{N}_i \end{aligned}$$

where $\boldsymbol{N}_i^e = \boldsymbol{N}_i^e(x,y)$ is the basis function on the physical triangle and $\boldsymbol{N}_i = \boldsymbol{N}_i(u,v)$ is the basis function on the reference triangle.

Similarly, we would like to relate the curl of the edge element basis functions on the physical element to the corresponding quantity on the reference element. Thus, we consider

$$\begin{aligned}\nabla \times \boldsymbol{N}_i^e &= \nabla \times \left(\varphi_{i_1}^e \nabla \varphi_{i_2}^e - \varphi_{i_2}^e \nabla \varphi_{i_1}^e\right) = 2\nabla \varphi_{i_1}^e \times \nabla \varphi_{i_2}^e \\ &= 2\left([\mathbf{J}^e]^{-1}\,\tilde{\nabla}\varphi_{i_1}\right) \times \left([\mathbf{J}^e]^{-1}\,\tilde{\nabla}\varphi_{i_1}\right) \\ &= \frac{[\mathbf{J}^e]^T}{\det(\mathbf{J}^e)}\left(2\tilde{\nabla}\varphi_{i_1} \times \tilde{\nabla}\varphi_{i_1}\right) = \frac{[\mathbf{J}^e]^T}{\det(\mathbf{J}^e)}\tilde{\nabla} \times \boldsymbol{N}_i\end{aligned}$$

where we have exploited the identity

$$(\mathbf{A}^{-1}\mathbf{x}) \times (\mathbf{A}^{-1}\mathbf{y}) = \frac{\mathbf{A}^T}{\det(\mathbf{A})}\mathbf{x} \times \mathbf{y}$$

that holds for a general (and invertible) 3×3-matrix $\mathbf{A}$ and arbitrary vectors $\mathbf{x}$ and $\mathbf{y}$ of dimension 3. In this context, we have used $\mathbf{A} = \mathbf{J}^e$ together with $\mathbf{x} = \tilde{\nabla}\varphi_{i_1}$ and $\mathbf{y} = \tilde{\nabla}\varphi_{i_2}$.

Thus, we can now evaluate integrals of the following type

$$\begin{aligned}&\int_{S^e} \frac{1}{\mu(x,y)} \nabla \times \boldsymbol{N}_i^e \cdot \nabla \times \boldsymbol{N}_j^e \, dx\, dy = \\ &= \int_{\tilde{S}} \frac{1}{\mu(x^e(u,v), y^e(u,v))} \left(\frac{[\mathbf{J}^e]^T\,\tilde{\nabla} \times \boldsymbol{N}_i}{\det(\mathbf{J}^e)}\right) \cdot \left(\frac{[\mathbf{J}^e]^T\,\tilde{\nabla} \times \boldsymbol{N}_j}{\det(\mathbf{J}^e)}\right) \det(\mathbf{J}^e)\, du\, dv \\ &= \int_{\tilde{S}} \frac{1}{\mu(x^e(u,v), y^e(u,v))} \frac{1}{\det(\mathbf{J}^e)} \left([\mathbf{J}^e]^T\,\tilde{\nabla} \times \boldsymbol{N}_i\right) \cdot \left([\mathbf{J}^e]^T\,\tilde{\nabla} \times \boldsymbol{N}_j\right) du\, dv\end{aligned}$$

$$\begin{aligned}&\int_{S^e} \epsilon(x,y) \boldsymbol{N}_i^e \cdot \boldsymbol{N}_j^e \, dx\, dy = \\ &= \int_{\tilde{S}} \epsilon(x^e(u,v), y^e(u,v)) \left([\mathbf{J}^e]^{-1} \boldsymbol{N}_i\right) \cdot \left([\mathbf{J}^e]^{-1} \boldsymbol{N}_j\right) \det(\mathbf{J}^e)\, du\, dv\end{aligned}$$

This type of treatment is applicable for all the edge elements listed in Appendix B, which also includes elements for three-dimensional problems.

6.6.3 Numerical Integration

Finally, we wish to evaluate the integrals formulated on the reference element. This is accomplished by numerical integration according to a so-called quadrature rule on the form

$$\int_{\tilde{S}} f(u,v)\, du\, dv = \sum_{q=1}^{N} \xi_q f(u_q, v_q)$$

where (u_q, v_q) is the q-th quadrature point with the weight ξ_q and N is the number of quadrature points. The corresponding relation for 3D integrals is

$$\int_{\tilde{V}} f(u, v, w)\, du\, dv\, dw = \sum_{q=1}^{N} \xi_q f(u_q, v_q, w_q)$$

Appendix B contains quadrature rules, i.e. quadrature points and weights, for the reference triangle, quadrilateral, tetrahedron, prism and hexahedron.

6.6.4 *MATLAB: FEM for the First-Order System in 3D*

We consider a metal cavity with the boundary S that encloses the volume V. Inside the cavity, we have the permittivity $\epsilon = \epsilon(\boldsymbol{r})$ and the conductivity $\sigma = \sigma(\boldsymbol{r})$, whereas the permeability is constant according to $\mu = \mu_0$. Furthermore, we approximate the metal walls by a PEC, which yields

$$\nabla \times \boldsymbol{E} = -j\omega \boldsymbol{B} \quad \text{in } V, \tag{6.76}$$

$$\nabla \times \frac{\boldsymbol{B}}{\mu_0} = (\sigma + j\omega\epsilon)\boldsymbol{E} \quad \text{in } V, \tag{6.77}$$

$$\hat{\boldsymbol{n}} \times \boldsymbol{E} = 0 \quad \text{on } S. \tag{6.78}$$

We are interested in computing the eigenfrequencies ω in combination with the eigenmodes represented by the electric field $\boldsymbol{E}$ and magnetic flux density $\boldsymbol{B}$. For cavities with losses represented by σ, the eigenfrequencies are complex, where the imaginary part corresponds to the damping of the eigenmode. If (6.76)-(6.78) are formulated in terms of only the electric field, we get a non-linear eigenvalue-problem in terms of the eigenfrequency ω. However, the first-order system (6.76)-(6.78) with both the electric field and the magnetic flux density avoids this complication and, thus, we get a linear eigenvalue-problem.

6.6.4.1 Weak Form

Here, we use the dot product to weigh the residual of Faraday's law (6.76) with the weighting function $\boldsymbol{W}_i^{\mathrm{FL}}$ and Ampère's law (6.77) by the weighting function $\boldsymbol{W}_i^{\mathrm{AL}}$, where integration over the entire computational domain yields

$$\int_V \boldsymbol{W}_i^{\mathrm{FL}} \cdot (\nabla \times \boldsymbol{E})\, dV = -j\omega \int_V \boldsymbol{W}_i^{\mathrm{FL}} \cdot \boldsymbol{B}\, dV, \tag{6.79}$$

$$\int_V \boldsymbol{W}_i^{\mathrm{AL}} \cdot (\nabla \times \boldsymbol{B})\, dV = \mu_0 \int_V \sigma \boldsymbol{W}_i^{\mathrm{AL}} \cdot \boldsymbol{E}\, dV + j\omega\mu_0 \int_V \epsilon \boldsymbol{W}_i^{\mathrm{AL}} \cdot \boldsymbol{E}\, dV. \tag{6.80}$$

Next, we proceed in a similar fashion as described in Sect. 6.5.1, where we exploited piecewise linear basis functions for the electric field and piecewise constant basis functions for the magnetic field to construct a mixed-order FEM for the system of first-order equations. Here, we expand the electric field in edge elements $\boldsymbol{N}_j$ as demonstrated previously, which yields $\boldsymbol{E} = \sum_j E_j \boldsymbol{N}_j$. Similarly, the magnetic flux density is expanded in so-called face elements $\boldsymbol{M}_j$ and, thus, we have $\boldsymbol{B} = \sum_j B_j \boldsymbol{M}_j$. (The unknowns B_j are associated with the faces in the mesh, which is analogous with the location of the magnetic field in the Yee cell as described in Chapter 5.) As mentioned before, the edge elements are curl-conforming and they feature a continuous tangential component on the interfaces between elements. Similarly, the face elements are divergence-conforming and they feature a continuous normal component on the interfaces between elements, which is appropriate for the approximation of the magnetic flux density and its boundary condition (1.2.1). Detailed expressions for the divergence-conforming elements can be found in Appendix B. It can be useful to compare this approach to Sect. 6.5.3.1, where the magnetic field is chosen to be linear along the field component and constant in the directions perpendicular to the field component. In fact, the divergence-conforming basis functions for brick-shaped elements yield exactly the approximation shown in Sect. 6.5.3.1. Here, we will use tetrahedrons instead of brick shaped elements but, although the detailed expressions for the basis functions change, the features of the approximation are identical.

As shown in Sect. 6.5.1, we choose the weighting functions for Ampère's law from the set of basis functions that are used to expand the electric field, i.e. $\boldsymbol{W}_i^{\mathrm{AL}} = \boldsymbol{N}_i$. Similarly, the weighting functions for Faraday's law are chosen from the set of basis functions that are used to expand the magnetic flux density, i.e. $\boldsymbol{W}_i^{\mathrm{FL}} = \boldsymbol{M}_i$. Given the boundary condition (6.78), the tangential electric field is known on the external boundary of the computational domain and, therefore, the tangential component of the weighting function $\boldsymbol{W}_i^{\mathrm{AL}}$ is identically zero on S, which should be compared with the corresponding situation described in Sect. 6.5.4.

We apply integration by parts to the left-hand side of (6.80) by means of the vector identity $\nabla \cdot (\boldsymbol{W}_i^{\mathrm{AL}} \times \boldsymbol{B}) = (\nabla \times \boldsymbol{W}_i^{\mathrm{AL}}) \cdot \boldsymbol{B} - \boldsymbol{W}_i^{\mathrm{AL}} \cdot (\nabla \times \boldsymbol{B})$ and Gauss' law, i.e.

$$\begin{aligned}
\int_V \boldsymbol{W}_i^{\mathrm{AL}} \cdot (\nabla \times \boldsymbol{B})\, dV &= \int_V \left[(\nabla \times \boldsymbol{W}_i^{\mathrm{AL}}) \cdot \boldsymbol{B} - \nabla \cdot (\boldsymbol{W}_i^{\mathrm{AL}} \times \boldsymbol{B}) \right] dV \\
&= \int_V (\nabla \times \boldsymbol{W}_i^{\mathrm{AL}}) \cdot \boldsymbol{B}\, dV - \int_S (\boldsymbol{W}_i^{\mathrm{AL}} \times \boldsymbol{B}) \cdot \hat{\boldsymbol{n}}\, dS \\
&= \int_V (\nabla \times \boldsymbol{W}_i^{\mathrm{AL}}) \cdot \boldsymbol{B}\, dV - \int_S (\hat{\boldsymbol{n}} \times \boldsymbol{W}_i^{\mathrm{AL}}) \cdot \boldsymbol{B}\, dS \\
&= \int_V (\nabla \times \boldsymbol{W}_i^{\mathrm{AL}}) \cdot \boldsymbol{B}\, dV,
\end{aligned}$$

where we have exploited the boundary condition (6.78) that implies that $\hat{\boldsymbol{n}} \times \boldsymbol{W}_i^{\mathrm{AL}} = \mathbf{0}$ on the boundary S.

Thus, we have the system

$$\int_V \boldsymbol{W}_i^{\mathrm{FL}} \cdot (\nabla \times \boldsymbol{E})\, dV = -j\omega \int_V \boldsymbol{W}_i^{\mathrm{FL}} \cdot \boldsymbol{B}\, dV, \tag{6.81}$$

$$\int_V (\nabla \times \boldsymbol{W}_i^{\mathrm{AL}}) \cdot \boldsymbol{B}\, dV = \mu_0 \int_V \sigma \boldsymbol{W}_i^{\mathrm{AL}} \cdot \boldsymbol{E}\, dV + j\omega\mu_0 \int_V \epsilon \boldsymbol{W}_i^{\mathrm{AL}} \cdot \boldsymbol{E}\, dV. \tag{6.82}$$

which can be expressed as an eigenvalue problem on the form $\mathbf{Az} = \lambda \mathbf{Bz}$ according to

$$\begin{pmatrix} \mathbf{0} & -\mathbf{C} \\ \mathbf{C}^T & -Z_0 \mathbf{M}^{(\sigma)} \end{pmatrix} \begin{bmatrix} c_0 \mathbf{b} \\ \mathbf{e} \end{bmatrix} = \frac{j\omega}{c_0} \begin{pmatrix} \mathbf{M}^{(1)} & \mathbf{0} \\ \mathbf{0} & \mathbf{M}^{(\epsilon_r)} \end{pmatrix} \begin{bmatrix} c_0 \mathbf{b} \\ \mathbf{e} \end{bmatrix}. \tag{6.83}$$

Here, the unknown coefficients E_j for the electric field are collected in the vector $\mathbf{e}$ and, similarly, the coefficients B_j for the magnetic flux density are collected in the vector $\mathbf{b}$. We have scaled the magnetic flux density and both Faraday's law and Ampère's law in order to achieve a more well-balanced system of equations. Thus, the eigenvalue is $j\omega/c_0$ and the eigenvector $\mathbf{z}$ is the combination of the column vector $c_0\mathbf{b}$ and the electric field coefficients $\mathbf{e}$.

6.6.4.2 Evaluation of Integrals on the Reference Element

The assembling procedure yields the global matrices $\mathbf{C}$, $\mathbf{M}^{(1)}$, $\mathbf{M}^{(\epsilon_r)}$ and $\mathbf{M}^{(\sigma)}$ and, here, the corresponding *element* matrices for element e are given by

$$C_{ij} = \int_{V^e} \boldsymbol{M}_i^e \cdot (\nabla \times \boldsymbol{N}_j^e)\, dx\, dy\, dz$$

$$M_{ij}^{(1)} = \int_{V^e} \boldsymbol{M}_i^e \cdot \boldsymbol{M}_j^e\, dx\, dy\, dz$$

$$M_{ij}^{(\epsilon_r)} = \int_{V^e} \epsilon_r \boldsymbol{N}_i^e \cdot \boldsymbol{N}_j^e\, dx\, dy\, dz$$

$$M_{ij}^{(\sigma)} = \int_{V^e} \sigma \boldsymbol{N}_i^e \cdot \boldsymbol{N}_j^e\, dx\, dy\, dz.$$

Given the derivations in Sect. 6.6.2, the relation between the curl-conforming basis function $\boldsymbol{N}_i^e$ on the physical tetrahedron and the curl-conforming basis function $\boldsymbol{N}_i$ on the reference tetrahedron is given by

$$\boldsymbol{N}_i^e = [\mathbf{J}^e]^{-1} \boldsymbol{N}_i$$

and, similarly, the corresponding relation for its curl is given by

$$\nabla \times \boldsymbol{N}_i^e = \frac{[\mathbf{J}^e]^T}{\det(\mathbf{J}^e)} \tilde{\nabla} \times \boldsymbol{N}_i.$$

The divergence-conforming elements on the tetrahedron are given by

$$\boldsymbol{M}_i^e = 2(\varphi_{i_3}^e \nabla \varphi_{i_2}^e \times \nabla \varphi_{i_1}^e + \varphi_{i_2}^e \nabla \varphi_{i_1}^e \times \nabla \varphi_{i_3}^e + \varphi_{i_1}^e \nabla \varphi_{i_3}^e \times \nabla \varphi_{i_2}^e)$$

where i_1, i_2 and i_3 are the node indices of the three nodes on face i of the tetrahedron. (Further details on the divergence-conforming basis functions for the tetrahedron can be found in Sect. B.2.1.) In Sect. 6.6.2, we exploited the relation $\nabla \varphi_i^e(x, y, z) = [\mathbf{J}^e]^{-1} \tilde{\nabla} \varphi_i(u, v, w)$ and that

$$(\mathbf{A}^{-1}\mathbf{x}) \times (\mathbf{A}^{-1}\mathbf{y}) = \frac{\mathbf{A}^T}{\det(\mathbf{A})} \mathbf{x} \times \mathbf{y},$$

for a general (and invertible) 3×3-matrix $\mathbf{A}$ and arbitrary vectors $\mathbf{x}$ and $\mathbf{y}$ of dimension 3. Thus, we have the following relation between the divergence-conforming basis function $\boldsymbol{M}_i^e$ on the physical element and the divergence-conforming basis function $\boldsymbol{M}_i$ on the reference element

$$\begin{aligned}
\boldsymbol{M}_i^e &= 2(\varphi_{i_3}^e \nabla \varphi_{i_2}^e \times \nabla \varphi_{i_1}^e + \varphi_{i_2}^e \nabla \varphi_{i_1}^e \times \nabla \varphi_{i_3}^e + \varphi_{i_1}^e \nabla \varphi_{i_3}^e \times \nabla \varphi_{i_2}^e) \\
&= \frac{[\mathbf{J}^e]^T}{\det(\mathbf{J}^e)} 2(\varphi_{i_3} \tilde{\nabla} \varphi_{i_2} \times \tilde{\nabla} \varphi_{i_1} + \varphi_{i_2} \tilde{\nabla} \varphi_{i_1} \times \tilde{\nabla} \varphi_{i_3} + \varphi_{i_1} \tilde{\nabla} \varphi_{i_3} \times \tilde{\nabla} \varphi_{i_2}) \\
&= \frac{[\mathbf{J}^e]^T}{\det(\mathbf{J}^e)} \boldsymbol{M}_i.
\end{aligned}$$

This gives the following element matrices evaluated on the reference element

$$\begin{aligned}
C_{ij} &= \int_{\tilde{V}} \left(\frac{[\mathbf{J}^e]^T}{\det(\mathbf{J}^e)} \boldsymbol{M}_i \right) \cdot \left(\frac{[\mathbf{J}^e]^T}{\det(\mathbf{J}^e)} (\nabla \times \boldsymbol{N}_j) \right) \det(\mathbf{J}^e)\, du\, dv\, dw, \\
M_{ij}^{(1)} &= \int_{\tilde{V}} \left(\frac{[\mathbf{J}^e]^T}{\det(\mathbf{J}^e)} \boldsymbol{M}_i \right) \cdot \left(\frac{[\mathbf{J}^e]^T}{\det(\mathbf{J}^e)} \boldsymbol{M}_j \right) \det(\mathbf{J}^e)\, du\, dv\, dw, \\
M_{ij}^{(\epsilon_r)} &= \int_{\tilde{V}} \epsilon_r \left([\mathbf{J}^e]^{-1} \boldsymbol{N}_i \right) \cdot \left([\mathbf{J}^e]^{-1} \boldsymbol{N}_j \right) \det(\mathbf{J}^e)\, du\, dv\, dw, \\
M_{ij}^{(\sigma)} &= \int_{\tilde{V}} \sigma \left([\mathbf{J}^e]^{-1} \boldsymbol{N}_i \right) \cdot \left([\mathbf{J}^e]^{-1} \boldsymbol{N}_j \right) \det(\mathbf{J}^e)\, du\, dv\, dw.
\end{aligned}$$

Here, the permittivity ϵ_r and the conductivity σ are functions of the coordinate $\boldsymbol{r}^e$ that spans the physical element, where $\boldsymbol{r}^e$ is a function of the reference element coordinates u, v and w according to the mapping $\boldsymbol{r}^e = \boldsymbol{r}^e(u, v, w)$ as described in Sect. 6.6.2. Given the quadrature points on the reference element, the mapping yields the corresponding points in the physical element. The points in the physical element can then be used directly to evaluate the permittivity $\epsilon_r = \epsilon_r(x, y, z)$ and the conductivity $\sigma = \sigma(x, y, z)$.

6.6.4.3 Overview of the MATLAB Implementation

The MATLAB implementation uses the main script `mixedFEM3D.m` for the computation of the eigenvalues and eigenvectors. Also, this script visualizes the results of the computation. Before details of the script `mixedFEM3D.m` and other parts of the implementation are given, we describe the main aspects of the program.

The script `mixedFEM3D.m` reads a mesh that is stored in the file `mesh_cylinder_R0.mat`. (More well-resolved meshes are also available in the files `mesh_cylinder_R1.mat` and `mesh_cylinder_R2.mat`.) The mesh contains a discretization of a circular cylinder of radius 0.125 m and height 0.4 m, where the volume is discretized by an unstructured mesh of tetrahedrons. The file that stores the mesh also contains other variables with information about the discretization and here is a list of these variables with some descriptions:

- `no2xyz` stores the coordinates of the nodes,
- `el2no` stores the nodes of the tetrahedrons,
- `el2ma` stores the material indices of the tetrahedrons,
- `ed2no_pec` stores the nodes of the edges that are located on the surface S and should be treated as PEC,
- `ed2no_all` stores the nodes of all edges in the mesh, and
- `fa2no_all` stores the nodes of all faces (i.e. triangles) in the mesh.

In the MATLAB implementation, the mesh is also stored in a database that provides some useful functionality. In particular, the user of the database can get information about the mesh that can not be easily extracted from the variables above. Here, we summarize the functions that provide an interface to this database without giving further information on the specific implementation of the database. (The interested reader can examine the MATLAB code for further details.)

- `ElementDatabase_Init.m` initializes the database for a specific type of element. The user must provide a name for the element and, below, we use `'edges'` and `'faces'`. In addition, the user provides the size of the database. For example, the edges are indexed by two nodes for each edge and the node numbers can take all values from one to the total number of nodes in the mesh. Thus, the global number of any edge in the mesh could be stored in a matrix at the row and column indexed by the two nodes of the edge. In this manner, it would be easy to look-up the global edge number given its two global nodes. Thus, all the global edge numbers could be stored in a square matrix (i.e. a two-dimensional array) with the size equal to the number of nodes. (The actual implementation of the database is quite different for efficiency purposes but it behaves in this manner.) Similarly, faces could be stored in a three-dimensional array with the size equal to the number of nodes, where a particular face is indexed by its three nodes. The function call `ElementDatabase_Init('edges', noNum*[1 1])` initializes the edges and `ElementDatabase_Init('faces', noNum*[1 1 1])` initializes the faces, where `noNum` is the number of nodes.

- `ElementDatabase_Set.m` stores elements in the database. For example, we would store all the edges of the mesh in the database by the function call `ElementDatabase_Set('edges', ed2no_all)`. Similarly, all the faces of the mesh are stored by `ElementDatabase_Set('faces', fa2no_all)`.
- `ElementDatabase_Get.m` gets the global number of an edge or a face given its global nodes. Thus, the function call `ElementDatabase_Get('edges', [n1 n2])` returns the global edge number of the edge that connects the nodes with indices `n1` and `n2`. (The database is not sensitive to the ordering of the two nodes, i.e. `[n1 n2]` and `[n2 n1]` give the same global edge number.) An empty matrix is returned if there is no edge that connects the nodes `n1` and `n2`. This functionality is convenient since it can be used in the assembling procedure to relate local edges to global edges and, similarly, local faces to global faces. Also, it is useful for imposing the boundary conditions.
- `ElementDatabase_Cardinal.m` returns the number of elements in the database. For example, `ElementDatabase_Cardinal('faces')` gives the total number of faces that are stored in the database.

The vector `el2ma` stores the material index of the tetrahedrons in the same mannar as `el2no` stores the nodes of the tetrahedrons. Similarly, `ma2er` stores the permittivity ϵ_r associated with the different material indices and `ma2si` stores the conductivity σ associated with the different material indices. In fact, `ma2er` and `ma2si` are so-called cell arrays in MATLAB and each entry is a so-called function handle. A function handle can be used as a regular function in MATLAB and, here, such a function returns the value of the material parameter given a coordinate (x, y, z) in physical space. In the present example, the mesh stored in the file `mesh_cylinder_R0.mat` contains only one material and, consequently, we have that `el2ma` only contains the material index one. Also, `ma2er` and `ma2si` only contain one function each. However, it is possible to add more materials such that different space dependent functions are used for different regions of the computational domain. In the script `mixedFEM3D.m` presented below, the relative permittivity ϵ_r and the conductivity σ are given by the expressions

$$\epsilon_r(x, y, z) = 1 + 4\exp\left[-\frac{(x-0.125)^2 + y^2 + (z-0.3)^2}{0.1^2}\right] \tag{6.84}$$

$$\sigma(x, y, z) = \sigma_0 \exp\left[-\frac{(x+0.125)^2 + y^2 + (z-0.3)^2}{0.1^2}\right]. \tag{6.85}$$

The function `Fem_Assemble.m` assembles the global matrices **C** (referred to as `cMtx` in the program), $\mathbf{M}^{(1)}$ (referred to as `uMtx`), $\mathbf{M}^{(\epsilon_r)}$ (referred to as `eMtx`) and $\mathbf{M}^{(\sigma)}$ (referred to as `sMtx`). After imposing the boundary conditions, these matrices are used to form the matrices **A** (referred to as `aMtx` in the program) and **B** (referred to as `bMtx`). The eigenvalue problem (6.83) can be solved in two different ways: (i) a direct solver implemented by `eig` and this solver is chosen by setting `solver = 'direct'` in the beginning of the program; or (ii) an iterative solver implemented by `eigs` and this solver is chosen by setting `solver ='sparse'` in the beginning of the program. The iterative solver is suitable for large problems since it reduces the memory requirements dramatically. However, the iterative solver only solves for a few eigenvalues in the vicinity of a so-called shift point and, therefore, it is important to provide a reasonably good value for the shift. (The reader is encouraged to consult the help for `eigs` in MATLAB in order to get further information on the arguments required for `eigs`.) Here, we proceed as follows in order to determine a suitable shift. First, we solve a small eigenvalue problem (i.e. we use a relatively coarse mesh) and, for this, we exploit the direct solver that gives all the eigenvalues. Then, we identify the interesting part of the spectrum with, say, some 30 eigenvalues. Then, we compute the average of these eigenvalues and use that as a shift when we continue with more accurate computations for the same problem on more well-resolved meshes.

Once the solution is computed, we visualize the eigenvalues and the corresponding eigenvectors. Here, the vector `mVtr` in the beginning of the program contains the indices for the modes that the user wishes to visualize. The eigenvectors are visualized by the MATLAB-command `quiver3`. This command requires the vector field at a set of specific points and, here, we choose the nodes of the mesh. We exploit the following projection technique to provide that information. For the (global) electric field, we have $\boldsymbol{E} = \sum_j E_j \boldsymbol{N}_j$ and the u-component of this vector field is $E^{(u)} = \hat{\boldsymbol{u}} \cdot \boldsymbol{E} = \sum_j E_j \hat{\boldsymbol{u}} \cdot \boldsymbol{N}_j$, where u is one of the Cartesian axes. For visualization purposes, we wish to express the u-component of the electric field in terms of nodal basis functions, i.e. $E^{(u)} = \sum_j E_j^{(u)} \varphi_j$ and this can be achieved by the following FEM procedure. We multiply both sides with the weighting function φ_i and integrate over the computational domain, which gives

$$\sum_j E_j^{(u)} \int_V \varphi_i \varphi_j \, dV = \int_V \varphi_i \, (\hat{\boldsymbol{u}} \cdot \boldsymbol{E}) dV = \sum_j E_j \int_V \varphi_i \, (\hat{\boldsymbol{u}} \cdot \boldsymbol{N}_j) dV.$$

This is a system of linear equations $\mathbf{P}\mathbf{v}^{(u)} = \mathbf{Q}^{(u)}\mathbf{e}$ with $P_{ij} = \int_V \varphi_i \varphi_j \, dV$ and $Q_{ij}^{(u)} = \int_V \varphi_i \, (\hat{\boldsymbol{u}} \cdot \boldsymbol{N}_j) dV$, where the solution vector $\mathbf{v}^{(u)}$ contains the coefficients $E_j^{(u)}$ that can be used directly for plotting the u-component of the electric field at node j. The matrix **P** can be diagonalized by so-called mass lumping, where we use a quadrature scheme with quadrature points at the vertices of the tetrahedron. Such a procedure makes the solution $\mathbf{v}^{(u)} = \mathbf{P}^{-1}\mathbf{Q}^{(u)}\mathbf{e}$ of the system of linear equations cheap to compute. It should be noted that the same procedure is used to project the magnetic flux density onto the nodal elements. The reader is encouraged to examine the implementation of `ProjSol2Nodes_Assemble.m` and `ProjSol2Nodes_CmpElMtx.m` for further details.

6.6.4.4 Important Parts of the MATLAB Implementation

The main script `mixedFEM3d.m` is listed below. In the beginning of the script, the user can choose the modes to visualize and the solver that is used for the eigenvalue problem. Also, the user can specify the permittivity and conductivity. Then, the mesh is read from disk and the database is initialized. After the administration of the mesh is completed, the eigenvalue problem is assembled and solved. Finally, the eigenvalues are plotted and the eigenmodes are visualized.

```
clear all

% Select the modes that will be visualized
mVtr = 1151 + (1:20);

% Direct or sparse eigenvalue solver for small and large
% problems, respectively [solver = 'direct' or 'sparse']
solver = 'direct';

% Materials
ma2er = {@(x,y,z) 1 + 4*exp(-((x-0.125).^2+y.^2+(z-0.3).^2)/
        (0.1^2))};
ma2si = {@(x,y,z) 0.1*exp(-((x+0.125).^2+y.^2+(z-0.3).^2)/
        (0.1^2))};

% Constants
c0 = 299792458;        % speed of light in vacuum
m0 = 4*pi*1e-7;        % permeability in vacuum
e0 = 1/(m0*c0^2);      % permittivity in vacuum
z0 = sqrt(m0/e0);      % wave impedance in vacuum

% Read mesh
load mesh_cylinder_R0

% Initialize the FEM
Fem_Init(no2xyz, ed2no_all, fa2no_all)

% Find PEC edges in the database
edIdx_pec = ElementDatabase_Get('edges', ed2no_pec);
noIdx_pec = unique(ed2no_pec(:))';

% Find all edges in the database
edNum_all = ElementDatabase_Cardinal('edges');
faNum_all = ElementDatabase_Cardinal('faces');
edIdx_all = 1:edNum_all;
noIdx_all = 1:size(no2xyz,2);
```

```
% Compute the interior edges
edIdx_int = setdiff(edIdx_all, edIdx_pec);
noIdx_int = setdiff(noIdx_all, noIdx_pec);

% Assemble global matrices
[eMtx, sMtx, cMtx, uMtx] = ...
    Fem_Assemble(no2xyz, el2no, el2ma, ma2er, ma2si);

eMtx_int = eMtx(edIdx_int,edIdx_int);    % mass matrix with
                                          permittivity
sMtx_int = sMtx(edIdx_int,edIdx_int);    % mass matrix with
                                          conductivity
cMtx_int = cMtx(:,edIdx_int);            % curl matrix
uMtx_int = uMtx;                         % mass matrix with unity
                                          coefficient

edNum_dof = length(edIdx_int);
faNum_dof = faNum_all;

solIdx_bFld = 1:faNum_dof;
solIdx_eFld = faNum_dof + (1:edNum_dof);

aMtx = ...
    [sparse(faNum_dof,faNum_dof) -cMtx_int; ...
    cMtx_int.' -z0*sMtx_int];
bMtx = ...
    [uMtx_int sparse(faNum_dof,edNum_dof); ...
    sparse(edNum_dof,faNum_dof) eMtx_int];

% Solve the eigenvalue problem
if strcmp(solver, 'direct')
    aMtx = full(aMtx);
    bMtx = full(bMtx);
    [eigVtr_int, eigVal] = eig(aMtx, bMtx);
elseif strcmp(solver, 'sparse')
    [eigVtr_int, eigVal] = eigs(aMtx, 0.5*(bMtx+bMtx'), 30,
                           1i*(19.5+1i));
else
    error('unknown eigenvalue solver')
end

eigVal = diag(eigVal); % j*w/c0
[eigTmp, eigIdx_sort] = sort(real(-1i*eigVal));
eigVal     = -1i*eigVal(eigIdx_sort); % w/c0
eigVtr_int = eigVtr_int(:,eigIdx_sort);
fr = c0*eigVal/(2*pi);

% Visualize the eigenfrequencies
figure(1), clf
plot(real(fr(mVtr))/1e9, imag(fr(mVtr))/1e9, 'ks')
xlabel('Real part of eigenfrequency [GHz]')
ylabel('Imaginary part of eigenfrequency [GHz]')
grid on
```

```
% Visualize the eigenmodes
bFld_all = eigVtr_int(solIdx_bFld,:);

eFld_all = zeros(edNum_all,size(eigVtr_int,2));
eFld_all(edIdx_int,:) = eigVtr_int(solIdx_eFld,:);

[pMtx_ed2no, pMtx_fa2no] = ProjSol2Nodes_Assemble(no2xyz, el2no);

bxFld_all = (pMtx_fa2no.xc*bFld_all) / c0;
byFld_all = (pMtx_fa2no.yc*bFld_all) / c0;
bzFld_all = (pMtx_fa2no.zc*bFld_all) / c0;

exFld_all = pMtx_ed2no.xc*eFld_all;
eyFld_all = pMtx_ed2no.yc*eFld_all;
ezFld_all = pMtx_ed2no.zc*eFld_all;

for mIdx = mVtr
    figure(2), clf
    dVal = 0.4;
    for dIdx = 0:1
        for edIdx = 1:size(ed2no_pec,2)
            noTmp = ed2no_pec(:,edIdx);
            xyzTmp = no2xyz(:,noTmp);
            plot3(dIdx*dVal + xyzTmp(1,:), xyzTmp(2,:),
                xyzTmp(3,:), ...
                'Color', 0.5*[1 1 1]), hold on
        end
        if dIdx == 0
            exViz = real(exFld_all(:,mIdx).');
            eyViz = real(eyFld_all(:,mIdx).');
            ezViz = real(ezFld_all(:,mIdx).');
            quiver3(dIdx*dVal + no2xyz(1,:), no2xyz(2,:),
                no2xyz(3,:), ...
                exViz, eyViz, ezViz, 2, 'k')
        elseif dIdx == 1
            bxViz = imag(bxFld_all(:,mIdx).');
            byViz = imag(byFld_all(:,mIdx).');
            bzViz = imag(bzFld_all(:,mIdx).');
            quiver3(dIdx*dVal + no2xyz(1,:), no2xyz(2,:),
                no2xyz(3,:), ...
                bxViz, byViz, bzViz, 2, 'k')
        end
    end
    axis equal
    axis off
    view(-24,14)
    pause
end
```

The script `mixedFEM3d.m` starts with an initialization performed by the function `Fem_Init.m`. This function enumerates the local edges and faces. The local numbering in the function should be compared with the numbering of the reference tetrahedron as described in Sec. B.2.1. In addition, this function initializes the database that stores the enumeration of all the global edges and faces.

```
% -------------------------------------------------------------
% Initialize the FE-solver by numbering of local and global
% edges and faces.
% -------------------------------------------------------------
function Fem_Init(no2xyz, ed2no_all, fa2no_all)

% Arguments:
%    no2xyz = coordinates of the nodes
%    ed2no_all = nodes of all edges
%    fa2no_all = nodes of all faces
% Returns:
%    -

global ed2noLoc fa2noLoc

% Setting up the edge information.
ed2noLoc = ...
    [1 2; 2 3; 3 1; 1 4; 2 4; 3 4]';
fa2noLoc = ...
    [3 2 1; 1 2 4; 2 3 4; 3 1 4]';

% Number the edges
ElementDatabase_Init('edges', size(no2xyz,2)*[1 1])
ElementDatabase_Set('edges', ed2no_all)

% Number the faces
ElementDatabase_Init('faces', size(no2xyz,2)*[1 1 1])
ElementDatabase_Set('faces', fa2no_all)
```

The function `Fem_Assemble.m` assembles the global matrices **C** (referred to as `cMtx` in the program), $\mathbf{M}^{(1)}$ (referred to as `uMtx`), $\mathbf{M}^{(\epsilon_r)}$ (referred to as `eMtx`) and $\mathbf{M}^{(\sigma)}$ (referred to as `sMtx`). The assembling procedure does not impose any boundary conditions. Notice that the positive direction of the edges is based on the global node numbers. Similarly, the orientation of the normal of the faces is based on the global node numbers of the face.

```
% -------------------------------------------------------------
% Assemble the global matrices
% -------------------------------------------------------------
function [eMtx, sMtx, cMtx, uMtx] = ...
    Fem_Assemble(no2xyz, el2no, el2ma, ma2er, ma2si)

% Arguments:
%   no2xyz = coordinates of the nodes
%   el2no = nodes of the tetrahedrons
%   el2ma = material of the tetrahedrons
%   ma2er = relative permittivity of the materials
%   ma2si = conductivity of the materials
% Returns:
%   eMtx = mass matrix with permittivity coefficient
%   sMtx = mass matrix with conductivity coefficient
%   cMtx = curl matrix
%   uMtx = mass matrix with unity coefficient

global ed2noLoc fa2noLoc
```

```
% Global number of entities
elNumGlo = size(el2no,2);
edNumGlo = ElementDatabase_Cardinal('edges');
faNumGlo = ElementDatabase_Cardinal('faces');

% Incremental steps for each element
incRes_EE = 6*6;
incRes_FE = 4*6;
incRes_FF = 4*4;

% Initializing.
idxRes_EE = 1;
idxRes_FE = 1;
idxRes_FF = 1;

irRes_EE = zeros(incRes_EE*elNumGlo,1);
icRes_EE = zeros(incRes_EE*elNumGlo,1);

irRes_FE = zeros(incRes_FE*elNumGlo,1);
icRes_FE = zeros(incRes_FE*elNumGlo,1);

irRes_FF = zeros(incRes_FF*elNumGlo,1);
icRes_FF = zeros(incRes_FF*elNumGlo,1);

meRes_EE = zeros(incRes_EE*elNumGlo,1);
msRes_EE = zeros(incRes_EE*elNumGlo,1);
mcRes_FE = zeros(incRes_FE*elNumGlo,1);
mmRes_FF = zeros(incRes_FF*elNumGlo,1);

% Computing the contributions to the mass and
% stiffness matrices.
for elIdx = 1:elNumGlo

    no = el2no(:,elIdx);
    xyz = no2xyz(:,no);

    [eElMtx_EE, sElMtx_EE, cElMtx_FE, uElMtx_FF] = ...
        Fem_CmpElMtx(xyz, ma2er{el2ma(elIdx)}, ma2si{el2ma(elIdx)});

    noTmp = zeros(size(ed2noLoc));
    noTmp(:) = el2no(ed2noLoc(:),elIdx);
    esVtr = sign(noTmp(2,:)-noTmp(1,:));
    eiVtr = ElementDatabase_Get('edges', noTmp);

    noTmp = zeros(size(fa2noLoc));
    noTmp(:) = el2no(fa2noLoc(:),elIdx);
    fsVtr = 2*(...
        ((noTmp(1,:) < noTmp(2,:)) & (noTmp(2,:) < noTmp(3,:))) | ...
        ((noTmp(2,:) < noTmp(3,:)) & (noTmp(3,:) < noTmp(1,:))) | ...
        ((noTmp(3,:) < noTmp(1,:)) & (noTmp(1,:) < noTmp(2,:))) ...
        ) - 1;
    fiVtr = ElementDatabase_Get('faces', noTmp);

    irTmp_EE = eiVtr'*ones(size(eiVtr));
    icTmp_EE = ones(size(eiVtr'))*eiVtr;
    isTmp_EE = esVtr'*esVtr;

    irTmp_FE = fiVtr'*ones(size(eiVtr));
    icTmp_FE = ones(size(fiVtr'))*eiVtr;
    isTmp_FE = fsVtr'*esVtr;

    irTmp_FF = fiVtr'*ones(size(fiVtr));
    icTmp_FF = ones(size(fiVtr'))*fiVtr;
    isTmp_FF = fsVtr'*fsVtr;

    irRes_EE(idxRes_EE + (1:incRes_EE) - 1) = irTmp_EE(:);
    icRes_EE(idxRes_EE + (1:incRes_EE) - 1) = icTmp_EE(:);
```

```
    meRes_EE(idxRes_EE + (1:incRes_EE) - 1) = isTmp_EE(:).*eElMtx_EE(:);
    msRes_EE(idxRes_EE + (1:incRes_EE) - 1) = isTmp_EE(:).*sElMtx_EE(:);

    irRes_FE(idxRes_FE + (1:incRes_FE) - 1) = irTmp_FE(:);
    icRes_FE(idxRes_FE + (1:incRes_FE) - 1) = icTmp_FE(:);
    mcRes_FE(idxRes_FE + (1:incRes_FE) - 1) = isTmp_FE(:).*cElMtx_FE(:);

    irRes_FF(idxRes_FF + (1:incRes_FF) - 1) = irTmp_FF(:);
    icRes_FF(idxRes_FF + (1:incRes_FF) - 1) = icTmp_FF(:);
    muRes_FF(idxRes_FF + (1:incRes_FF) - 1) = isTmp_FF(:).*uElMtx_FF(:);

    idxRes_EE = idxRes_EE + incRes_EE;
    idxRes_FE = idxRes_FE + incRes_FE;
    idxRes_FF = idxRes_FF + incRes_FF;

end

eMtx = sparse(irRes_EE, icRes_EE, meRes_EE, edNumGlo, edNumGlo);
sMtx = sparse(irRes_EE, icRes_EE, msRes_EE, edNumGlo, edNumGlo);
cMtx = sparse(irRes_FE, icRes_FE, mcRes_FE, faNumGlo, edNumGlo);
uMtx = sparse(irRes_FF, icRes_FF, muRes_FF, faNumGlo, faNumGlo);
```

The element matrices are computed by the function `Fem_CmpElMtx.m`. It is probably instructive for the reader to compare this implementation in detail with the basis functions and quadrature rule given in Sect. B.2.1. Also, the reader is encouraged to compare the implementation with the discussion in Sect. 6.6.1. It should be rather straightforward to directly associated the different statements in the MATLAB function with the corresponding mathematical formulas.

```
% ---------------------------------------------------------------
% Compute element matrices for the tetrahedron by means of
% numerical integration on the reference element
% ---------------------------------------------------------------
function [eElMtx_EE, sElMtx_EE, cElMtx_FE, uElMtx_FF] = ...
    Fem_CmpElMtx(xyz, ma2er, ma2si)

% Argument:
%   xyz = the coordinates of the nodes of the element
%   ma2er = material to permittivity
%   ma2si = material to conductivity
% Returns:
%   eElMtx_EE = mass matrix with permittivity coefficient
%   sElMtx_EE = mass matrix with conductivity coefficient
%   cElMtx_FE = curl matrix
%   uElMtx_FF = mass matrix with unity coefficient

% Quadrature rule
q2u = [[5.854101966249685e-01, ...
        1.381966011250105e-01, ...
        1.381966011250105e-01]; ...
       [1.381966011250105e-01, ...
        5.854101966249685e-01, ...
        1.381966011250105e-01]; ...
       [1.381966011250105e-01, ...
        1.381966011250105e-01, ...
        5.854101966249685e-01]; ...
```

```
        [1.381966011250105e-01, ...
         1.381966011250105e-01, ...
         1.381966011250105e-01]]';
q2w = [4.166666666666666e-02; ...
       4.166666666666666e-02; ...
       4.166666666666666e-02; ...
       4.166666666666666e-02]';

% H(grad) basis functions
up{1} = 1 - q2u(1,:) - q2u(2,:) - q2u(3,:);
up{2} = q2u(1,:);
up{3} = q2u(2,:);
up{4} = q2u(3,:);

% Gradient of H(grad) basis functions
ug{1} = [-1 -1 -1]';
ug{2} = [+1 0 0]';
ug{3} = [0 +1 0]';
ug{4} = [0 0 +1]';

% H(div) basis functions
uim{1} = 2*(  cross(ug{2},ug{1})*up{3} ...
            + cross(ug{1},ug{3})*up{2} ...
            + cross(ug{3},ug{2})*up{1});
uim{2} = 2*(  cross(ug{2},ug{4})*up{1} ...
            + cross(ug{4},ug{1})*up{2} ...
            + cross(ug{1},ug{2})*up{4});
uim{3} = 2*(  cross(ug{3},ug{4})*up{2} ...
            + cross(ug{4},ug{2})*up{3} ...
            + cross(ug{2},ug{3})*up{4});
uim{4} = 2*(  cross(ug{1},ug{4})*up{3} ...
            + cross(ug{4},ug{3})*up{1} ...
            + cross(ug{3},ug{1})*up{4});

% H(curl) basis function
uin{1} = ug{2}*up{1} - ug{1}*up{2};
uin{2} = ug{3}*up{2} - ug{2}*up{3};
uin{3} = ug{1}*up{3} - ug{3}*up{1};
uin{4} = ug{4}*up{1} - ug{1}*up{4};
uin{5} = ug{4}*up{2} - ug{2}*up{4};
uin{6} = ug{4}*up{3} - ug{3}*up{4};

% Curl of H(curl) basis functions
ouTmp = ones(size(q2w));
ucn{1} = 2*cross(ug{1},ug{2})*ouTmp;
ucn{2} = 2*cross(ug{2},ug{3})*ouTmp;
ucn{3} = 2*cross(ug{3},ug{1})*ouTmp;
ucn{4} = 2*cross(ug{1},ug{4})*ouTmp;
ucn{5} = 2*cross(ug{2},ug{4})*ouTmp;
ucn{6} = 2*cross(ug{3},ug{4})*ouTmp;

% Physical coordinates
q2x = zeros(3,length(q2w));
for iIdx = 1:4
```

```
    q2x = q2x + xyz(:,iIdx)*up{iIdx};
end

% Jacobian
jac = zeros(3);
for iIdx = 1:4
    jac = jac ...
        + [xyz(1,iIdx)*ug{iIdx}, ...
           xyz(2,iIdx)*ug{iIdx}, ...
           xyz(3,iIdx)*ug{iIdx}];
end

% Mappings
det_jac = det(jac);
map_ccs = inv(jac);       % mapping for curl-conforming space
map_dcs = jac'/det_jac;   % mapping for div-conforming space
for iIdx = 1:6
    gin{iIdx} = map_ccs*uin{iIdx};
    gcn{iIdx} = map_dcs*ucn{iIdx};
end
for iIdx = 1:4
    gim{iIdx} = map_dcs*uim{iIdx};
end

% Evaluation of element matrix: epsilon Ni Nj
for iIdx = 1:6
    for jIdx = 1:6
        maTmp = ma2er(q2x(1,:),q2x(2,:),q2x(3,:));
        ipTmp = maTmp.*sum(gin{iIdx}.*gin{jIdx});
        eElMtx_EE(iIdx,jIdx) = ipTmp * q2w' * det_jac;
    end
end

% Evaluation of element matrix: sigma Ni Nj
for iIdx = 1:6
    for jIdx = 1:6
        maTmp = ma2si(q2x(1,:),q2x(2,:),q2x(3,:));
        ipTmp = maTmp.*sum(gin{iIdx}.*gin{jIdx});
        sElMtx_EE(iIdx,jIdx) = ipTmp * q2w' * det_jac;
    end
end

% Evaluation of element matrix: Mi curl_Nj
for iIdx = 1:4
    for jIdx = 1:6
        maTmp = ones(size(q2w));
        ipTmp = maTmp.*sum(gim{iIdx}.*gcn{jIdx});
        cElMtx_FE(iIdx,jIdx) = ipTmp * q2w' * det_jac;
    end
end

% Evaluation of element matrix: Mi Mj
for iIdx = 1:4
    for jIdx = 1:4
```

```
        maTmp = ones(size(q2w));
        ipTmp = maTmp.*sum(gim{iIdx}.*gim{jIdx});
        uElMtx_FF(iIdx,jIdx) = ipTmp * q2w' * det_jac;
    end
end
```

These are the most important functions in the MATLAB implementation of the eigenvalue solver. However, the reader may find it useful and interesting to also read the rest of the MATLAB functions that are available for download as described in the Preface of this book.

6.6.4.5 Numerical Results

First, we test the eigenvalue solver on a cavity with the material parameters $\epsilon = \epsilon_0$, $\mu = \mu_0$ and $\sigma = 0$. This problem can be solved analytically [5] and the solutions are categorized into so-called transverse electric (TE) and transverse magnetic (TM) modes. Each mode is indexed by three integers m, n and p (i.e. TE_{mnp} and TM_{mnp}) that describe the variation of the mode along the azimuthal, radial and axial coordinate in the cylindrical coordinate system where the z-axis coincides the the cylinder axis. A convergence study shows that the error in the resonance frequencies is proportional to h^2.

Fig. 6.30 shows the electric field (to the left) and the magnetic flux density (to the right) for the TE_{111}-mode, which has the lowest non-zero eigenfrequency in the spectrum. The electric field and the magnetic flux density are one quarter of a period out-of-phase with respect to each other and, here, the vector fields are shown when their amplitudes reach the maximum. The magnetic flux density circulates around the electric field in accordance with Maxwell's equations. The TE_{111}-mode is a degenerated mode and the eigenvalue solver correctly computes also the other TE_{111}-mode, which is rotated 90° around the cylinder axis. Similarly, Fig. 6.31 and 6.32 show the modes TM_{010} and TM_{011}, which are axisymmetric and non-degenerated. Finally, Fig. 6.33 shows one of the degenerated TE_{112} modes and Fig. 6.34 shows the axisymmetric and non-degenerated mode TM_{012}.

Finally, Fig. 6.35 shows the complex eigenfrequencies for a cavity with the inhomogeneous permittivity (6.84) and conductivity (6.85), where $0 \leq \sigma_0 \leq 0.1$. For the lossless case $\sigma_0 = 0$, the eigenfrequencies are located on the real axis. As the losses are introduced by increasing σ_0, the eigenfrequencies move into the upper half-plane, i.e. the eigenfrequencies get a positive imaginary part. This corresponds to damping with respect to time and the damping increases as the losses increase. It is interesting to notice that the eigenmodes are damped differently and, in addition, the resonance frequencies are slightly perturbed by the increased losses.

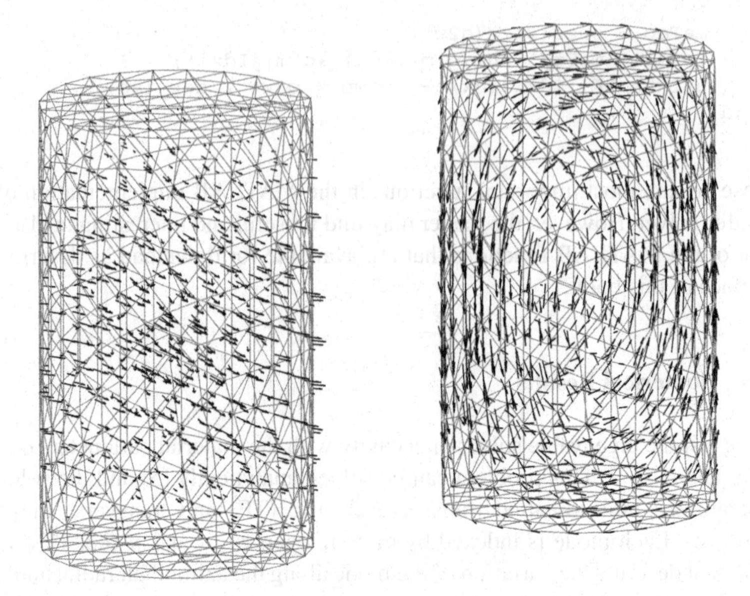

Fig. 6.30 The TE_{111} mode of a circular cylinder cavity of radius 0.125 m and height 0.4 m: left – electric field; and right – magnetic flux density. The material parameters inside the cavity are $\epsilon = \epsilon_0$, $\mu = \mu_0$ and $\sigma = 0$

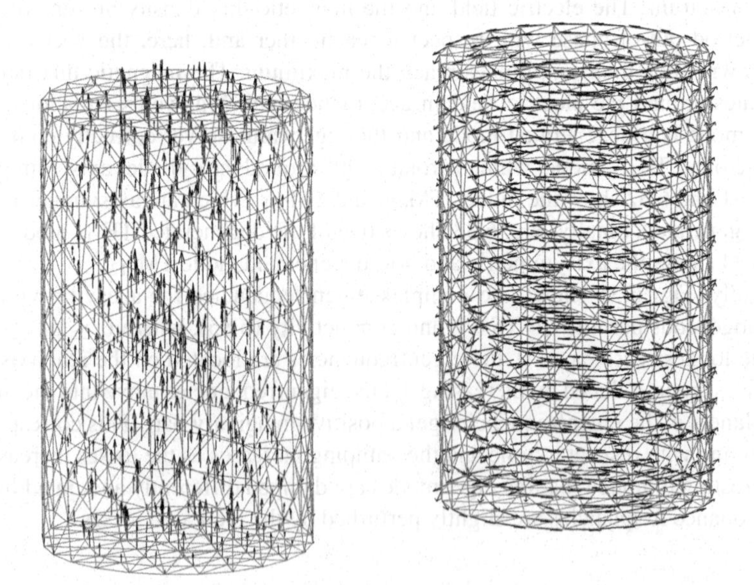

Fig. 6.31 The TE_{010} mode of a circular cylinder cavity of radius 0.125 m and height 0.4 m: left – electric field; and right – magnetic flux density. The material parameters inside the cavity are $\epsilon = \epsilon_0$, $\mu = \mu_0$ and $\sigma = 0$

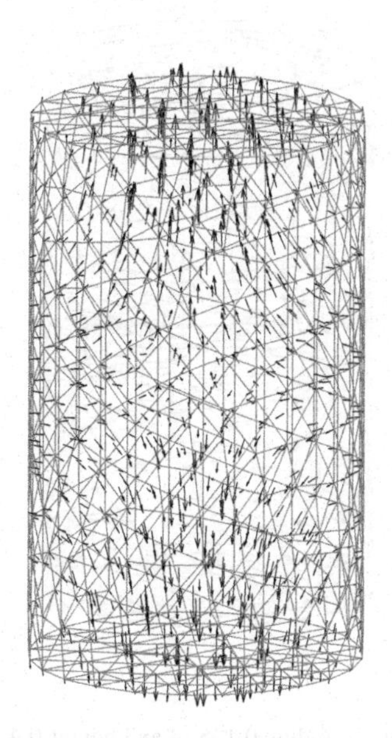

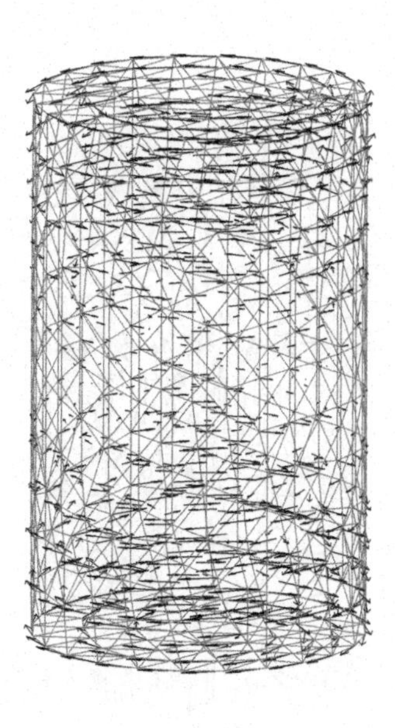

Fig. 6.32 The TE_{011} mode of a circular cylinder cavity of radius 0.125 m and height 0.4 m: left – electric field; and right – magnetic flux density. The material parameters inside the cavity are $\epsilon = \epsilon_0$, $\mu = \mu_0$ and $\sigma = 0$

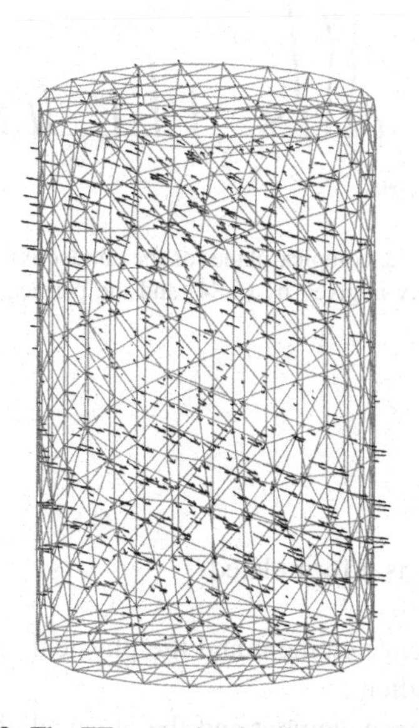

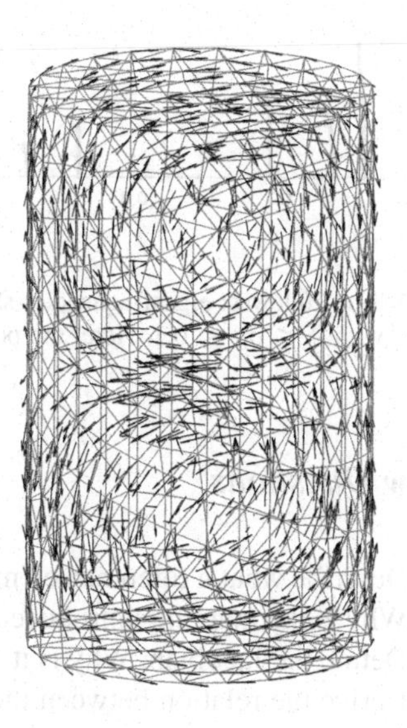

Fig. 6.33 The TE_{112} mode of a circular cylinder cavity of radius 0.125 m and height 0.4 m: left – electric field; and right – magnetic flux density. The material parameters inside the cavity are $\epsilon = \epsilon_0$, $\mu = \mu_0$ and $\sigma = 0$

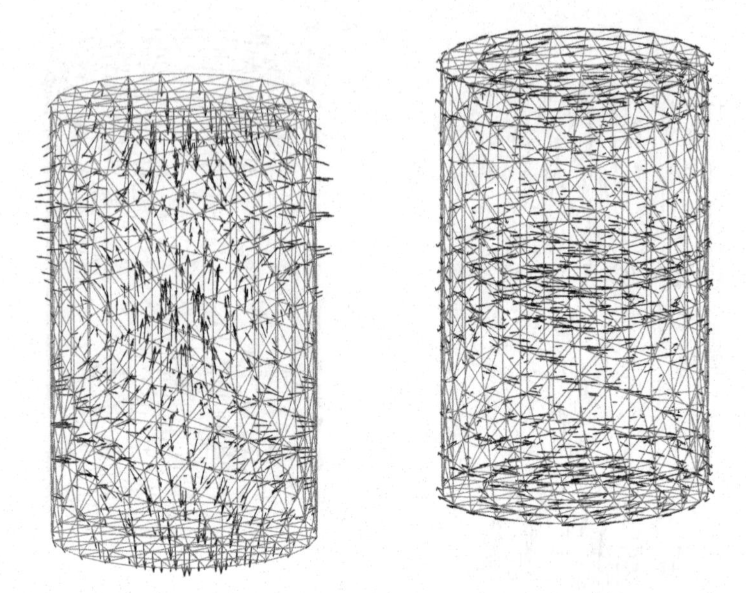

Fig. 6.34 The TM_{012} mode of a circular cylinder cavity of radius 0.125 m and height 0.4 m: left – electric field; and right – magnetic flux density. The material parameters inside the cavity are $\epsilon = \epsilon_0$, $\mu = \mu_0$ and $\sigma = 0$

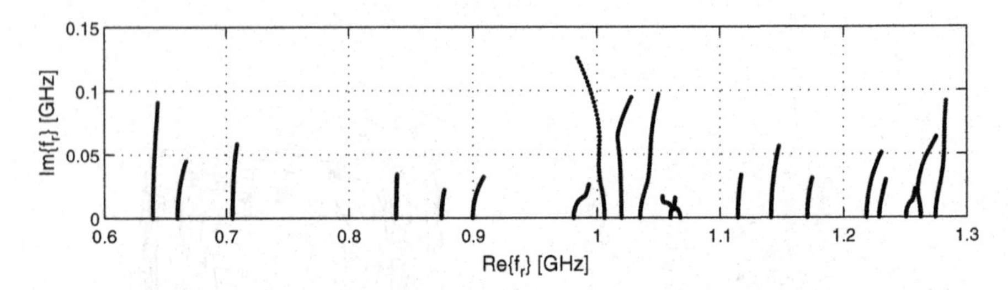

Fig. 6.35 The 20 lowest eigenfrequencies shown in the complex plane for the conductivity given by (6.85) with $0 \leq \sigma_0 \leq 0.1$. Here, the permittivity is given by (6.84) and the permeability is $\mu = \mu_0$

Review Questions

6.6-1 Describe what a reference element is.

6.6-2 Why is the mapping introduced? How is it formulated?

6.6-3 Define the Jacobian. Why is it useful?

6.6-4 Derive the relation between the reference element and the physical element for the nodal basis functions and their gradient.

6.6-5 Derive the relation between the reference element and the physical element for the edge element basis functions and their curl.

6.6-6 Describe what a quadrature rule is and how it is used.

6.7 Time-Dependent Problems

Now we consider a time evolution problem for the vector wave equation. Let us choose a simple example with a lossless region (i.e., $\sigma = 0$) and metal boundary conditions. There are no driving currents, and instead we excite the problem with a nonzero initial field. The problem can be stated as

$$\nabla \times \left(\frac{1}{\mu} \nabla \times \boldsymbol{E} \right) + \epsilon \frac{\partial^2 \boldsymbol{E}}{\partial t^2} = \boldsymbol{0} \qquad \text{in } S, \tag{6.86}$$

$$\hat{\boldsymbol{n}} \times \boldsymbol{E} = \boldsymbol{0} \qquad \text{on } L_1, \tag{6.87}$$

$$\boldsymbol{E}(\boldsymbol{r}, t = 0) = \boldsymbol{E}_0(\boldsymbol{r}) \qquad \text{in } S, \tag{6.88}$$

$$\left. \frac{\partial \boldsymbol{E}(\boldsymbol{r}, t)}{\partial t} \right|_{t=0} = \boldsymbol{0} \qquad \text{in } S. \tag{6.89}$$

Besides the boundary condition (6.87) we need two initial conditions (6.88) and (6.89), because the equation is of second order in time. The electric field is expanded in edge elements, and the coefficients E_j are now *time dependent*:

$$\boldsymbol{E}(\boldsymbol{r}, t) = \sum_{j=1}^{N_e} E_j(t) \boldsymbol{N}_j(\boldsymbol{r}). \tag{6.90}$$

Equation (6.86) is tested by taking the scalar product with the weighting function $\boldsymbol{W}_i = \boldsymbol{N}_i(\boldsymbol{r})$ and integrated (the $\nabla \times \mu^{-1} \nabla \times$-term by parts) over the computational domain.

So far, we have discretized in space but not in time. The result is a system of coupled ordinary differential equations (ODE) for the expansion coefficients

$$\mathbf{S}\mathbf{z}(t) + c_0^{-2} \mathbf{M} \frac{\partial^2 \mathbf{z}(t)}{\partial t^2} = \mathbf{0},$$

where $\mathbf{S}$ and $\mathbf{M}$ are given by (6.58)–(6.59). To solve this system of ODEs, we can use either finite differences or finite elements in time. A first attempt for time-stepping might be the centered finite difference scheme

$$\mathbf{M}\left(\mathbf{z}^{n+1} - 2\mathbf{z}^n + \mathbf{z}^{n-1}\right) = -\left(c_0 \Delta t\right)^2 \mathbf{S}\mathbf{z}^n, \tag{6.91}$$

where we need to specify $\mathbf{z}^1$ and $\mathbf{z}^2$ as initial conditions. This scheme is subject to the time-step limitation discussed in Sect. 4.4.1, $\Delta t \leq 2/\omega_{\max}$. Yet it is implicit, because the mass matrix $\mathbf{M}$ must be inverted at every time step.

$$\mathbf{z}^{n+1} = 2\mathbf{z}^n - \mathbf{z}^{n-1} - (c_0 \Delta t)^2 \mathbf{M}^{-1} \mathbf{S}\mathbf{z}^n.$$

Thus, straightforward time-stepping for FEM has two drawbacks: it is slow, because of the inversion, and the time-step is limited. There are two ways to improve on this. One can be used if the mass matrix is sufficiently close to diagonal that it can be approximated by a diagonal matrix. This is known as "mass lumping" in mechanics and leads to explicit time-stepping. Mass lumping works well for the edge elements on quadrilaterals. In fact, with some additional lumping of the stiffness matrix, time-stepped edge elements on rectangles are equivalent to the FDTD scheme. This solution gives a low number of operations per time-step, but still the time-step is limited by the CFL condition.

Mass lumping does not work for edge elements on triangles or tetrahedra, and for these elements, one must invert a system of equations on each time step. A much better method in this case is to apply a scheme that is even more implicit, so that it is stable for arbitrarily large time steps. This is achieved by averaging the stiffness term in time:

$$\mathbf{M}\left(\mathbf{z}^{n+1} - 2\mathbf{z}^{n} + \mathbf{z}^{n-1}\right) = -\,(c_0\Delta t)^2\,\mathbf{S}\left[\theta\mathbf{z}^{n+1} + (1-2\theta)\mathbf{z}^{n} + \theta\mathbf{z}^{n-1}\right]. \quad (6.92)$$

This scheme is stable for *any* time-step if $\theta \geq 1/4$. However, the scheme becomes inaccurate if the time-step is long compared with the characteristic time on which the solution evolves.

The time-stepping scheme in (6.92) was introduced in 1959 by Newmark [52], and it is often referred to as the Newmark scheme. One interesting feature of the Newmark scheme is that it reduces to the finite difference scheme (6.91) when the implicitness parameter θ is zero. In fact, the Newmark scheme can be viewed as a strict FEM scheme based on Galerkin's method and a piecewise linear expansion of the electric field in time [64]. The implicitness parameter enters through a linear combination of exact and trapezoidal integration applied to the weak form of the problem. Equation (6.92) is recovered if we use the weights $1 - 6\theta$ and 6θ for the exact and trapezoidal integration, respectively. This makes it possible to combine [63] finite difference schemes (with explicit time-stepping) with FEM (with implicit time-stepping), and moreover, it is feasible to construct relatively simple proofs of stability based on von Neumann analysis. Since the lowest term in the error expansion is of second order in Δt for the FEM with Galerkin's method, this also applies to both (6.91) and (6.92).

Review Questions

6.7-1 How and under what conditions can the implicit Newmark scheme give an explicit time-stepping schemes? What's the name of the explicit time-stepping scheme?

6.7-2 Are there any advantages of the implicit Newmark scheme compared to explicit time-stepping schemes?

6.8 Magnetostatics and Eddy Current Problems

Two-dimensional scalar calculations can be applied to problems involving magnetic materials and eddy currents. Eddy current calculations are generally made by applying the low-frequency approximation, which consists in ignoring the displacement current and setting $\epsilon_0 = 0$. Roughly speaking, the low-frequency approximation works when the geometrical dimensions of the computational domain are much smaller than a wavelength $\lambda = c/f$.

The low-frequency equations are usually solved by introducing the magnetic vector potential $\boldsymbol{A}$, such that $\boldsymbol{B} = \nabla \times \boldsymbol{A}$. The advantage of this is that the condition of solenoidal magnetic field $\nabla \cdot \boldsymbol{B} = 0$ is automatically satisfied. Note, however, that although the magnetic field is uniquely determined, the vector potential is not; any gradient of a scalar potential can be added to $\boldsymbol{A}$ without changing the magnetic field $\boldsymbol{B}$. The electric field is given by $\boldsymbol{E} = -\partial \boldsymbol{A}/\partial t - \nabla\phi$. With this representation for $\boldsymbol{B}$ and $\boldsymbol{E}$, Faraday's law is automatically satisfied. Ampère's law gives

$$\nabla \times \frac{1}{\mu} \nabla \times \boldsymbol{A} + \sigma \left(\frac{\partial \boldsymbol{A}}{\partial t} + \nabla\phi \right) = \boldsymbol{J}^s, \tag{6.93}$$

where $\boldsymbol{J}^s$ is an imposed source current, usually representing currents in coils, and $\sigma\boldsymbol{E} = -\sigma(\partial \boldsymbol{A}/\partial t + \nabla\phi)$ is the conduction current. As a consequence of the low-frequency approximation $\epsilon_0 = 0$, both sides of Poisson's equation, $\nabla \cdot (\epsilon\nabla\phi) = -\rho$, vanish, and therefore, the electrostatic potential is undetermined in the low-frequency approximation.

6.8.1 2D Formulation

For 2D problems with currents flowing in the z-direction and variations only in the x- and y-directions, the potentials can be chosen in a simple way:

$$\boldsymbol{A} = A_z(x, y)\hat{z}, \quad \phi = 0. \tag{6.94}$$

Then the magnetic field is $\boldsymbol{B} = \nabla A_z \times \hat{z}$ and the current density is

$$\nabla \times \left(\frac{1}{\mu} \nabla A_z \times \hat{z} \right) = -\hat{z} \nabla \cdot \frac{1}{\mu} \nabla A_z. \tag{6.95}$$

If the time-dependence is harmonic $\propto \exp(j\omega t)$, the z-component of Ampère's law gives

$$-\nabla \cdot \frac{1}{\mu} \nabla A_z + j\omega\sigma A_z = J_z^s, \tag{6.96}$$

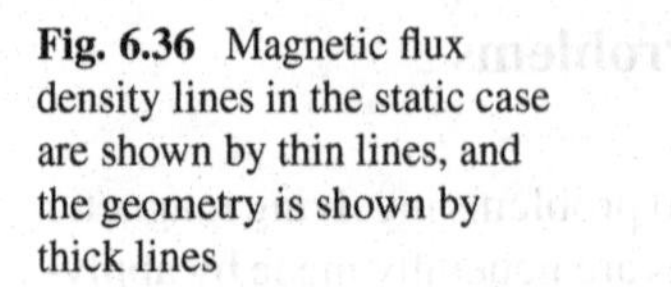

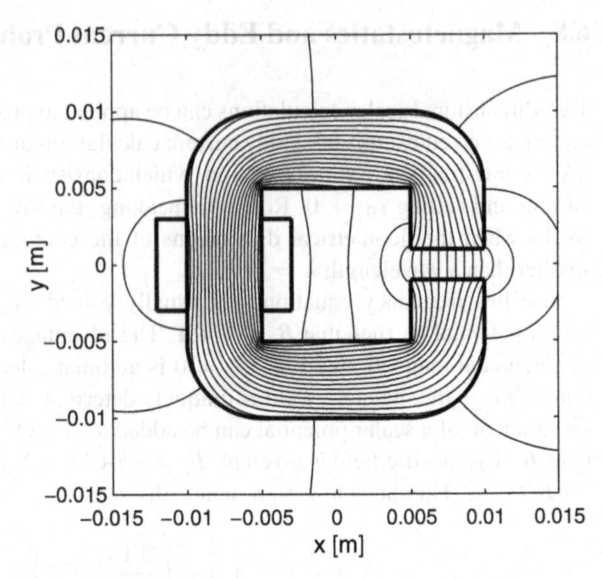

Fig. 6.36 Magnetic flux density lines in the static case are shown by thin lines, and the geometry is shown by thick lines

which is a complex, scalar Helmholtz equation. The boundary condition of the continuous normal component for $\boldsymbol{B}$ is fulfilled if A_z is continuous. The boundary condition of continuous $\hat{\boldsymbol{n}} \times \boldsymbol{H} = \hat{\boldsymbol{n}} \times \mu^{-1}(\nabla A_z \times \hat{z}) = -\hat{z}\mu^{-1}\partial A_z/\partial n$ requires continuity of $\mu^{-1}\partial A_z/\partial n$.

In microwave terminology, the 2D formulation in (6.94) and (6.96) corresponds to TM polarization. This 2D problem is readily solved using nodal elements for the vector potential A_z, and we have discussed the techniques for this in Sect. 6.3.

6.8.2 A 2D Application Problem

As a practical application, we consider the 2D electromagnet shown in Fig. 6.36. The magnetic circuit consists of an iron core ($\mu_r = 4000$) shaped like the letter C and two rectangular copper conductors. The left and right copper conductors carry source currents $+J_z^s$ and $-J_z^s$, respectively.

First we solve the *static* problem $-\nabla \cdot (\mu_r^{-1} \nabla A_z) = \mu_0 J_z^s$. We have discussed all the techniques necessary for this in Sect. 6.3, and they have been implemented in a user-friendly way in the MATLAB toolbox `pdetool`. The computed magnetic flux lines (equipotential lines for A_z) are shown in Fig. 6.36. Note the almost uniform distribution of magnetic flux lines in the core. There is also some leakage of flux, especially in the vicinity of the air gap where significant fringing occurs.

For finite frequencies, we solve (6.96). The resulting magnetic flux lines for the frequencies $f = 1.0$ Hz and $f = 10$ Hz are shown in Fig. 6.37. We have used the conductivities $\sigma_{\text{Fe}} = 10^7$ S/m for the iron core and $\sigma_{\text{Cu}} = 5.8 \cdot 10^7$ S/m for the copper conductors. The electrical conductivity reduces the penetration of the magnetic field

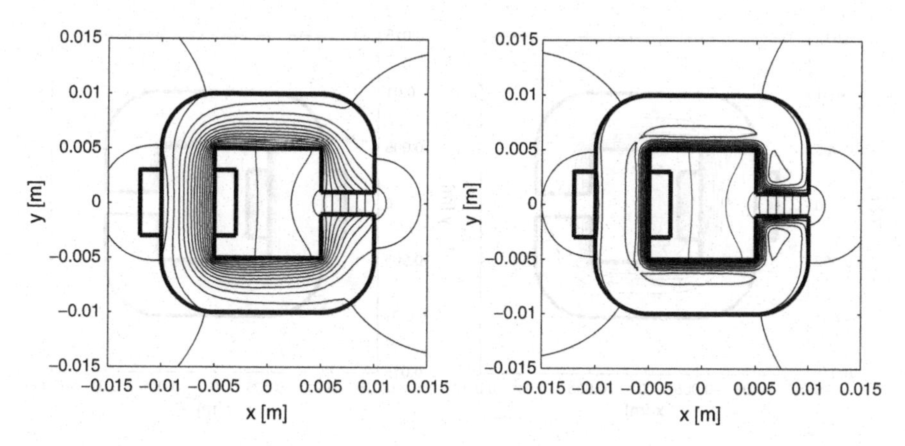

Fig. 6.37 Magnetic flux density lines at $f = 1.0$ Hz and $f = 10$ Hz are shown to the left and the right, respectively

into the iron (and to a lesser extent, into the copper) as the frequency increases. This is called the *skin effect*. The skin depth, over which the magnetic field decays by $1/e$, can be found from (6.96) as $\delta = 1/\sqrt{\pi f \mu \sigma}$. At $f = 1.0$ Hz the skin depths are 2.5 mm and 66 mm for iron and copper, respectively. For $f = 10$ Hz the skin depths are 0.8 mm (rather thin!) in iron and 21 mm in copper.

Time variation, i.e., nonzero frequency, introduces eddy currents in the conducting regions. One can see in Fig. 6.37 that the eddy currents in the iron core squeeze the magnetic flux to the inner surface of the iron core. This is where the circumference traversed by the field lines is the smallest. Note that despite the localization of the flux to one side of the iron, the field lines spread out evenly in the air gap. Here, the flux density (density of contours) is almost uniform. The reason for this is that the air gap gives the dominant contribution to the magnetic reluctance.

Contour lines for the total power dissipation density $P_t = \sigma |J_z^t|^2$ at $f = 1.0$ Hz and $f = 10$ Hz are shown in Fig. 6.38. The total current J_z^t is the sum of the source current J_z^s and the eddy current J_z^e. The source current is prescribed as a constant value in the copper region. In practice, the copper region would most likely consist of a single thin wire wound many turns around the core. This can be modeled as a uniform current distribution. The eddy currents are computed from the vector potential, $J_z^e = \sigma E_z = -j\omega\sigma A_z$.

At power frequencies, eddy currents reduce the regions where the magnetic field penetrates the iron to very thin layers. To avoid this one can use laminations that prevent eddy currents from flowing in certain directions. For the 2D electromagnet shown here, laminations in the xy-plane will inhibit the eddy currents completely. We reiterate that the 2D eddy current problem is well handled by nodal elements. This technique is extensively described in the textbook of Silvester and Ferrari [75].

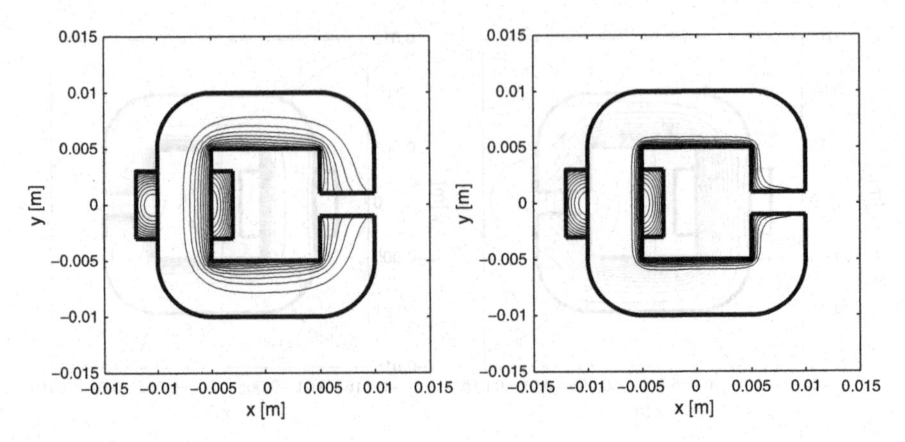

Fig. 6.38 Contour lines for the power dissipation density at f = 1.0 Hz (left) and f = 10.0 Hz (right)

6.8.3 3D Eddy Current Calculations

Here, we will give a brief introduction to eddy current calculations in three dimensions. This is a complex subject, so the discussion will be kept general, and leave out many details. Several different formulations are used for solving the low-frequency equation (6.93). Before proceeding to discuss two of these formulations, we note that the divergence of Ampère's law with $\epsilon_0 = 0$ shows that the current density has zero divergence. This must hold, both for the coil current $\boldsymbol{J}^s$ and the conduction currents $-\sigma(j\omega \boldsymbol{A} + \nabla\phi)$. Therefore, the low-frequency problem can be stated as

$$\nabla \times \frac{1}{\mu}\nabla \times \boldsymbol{A} + \sigma\left(\frac{\partial \boldsymbol{A}}{\partial t} + \nabla\phi\right) = \boldsymbol{J}^s,$$

$$\nabla \cdot \sigma\left(\frac{\partial \boldsymbol{A}}{\partial t} + \nabla\phi\right) = 0,$$

$$\nabla \cdot \boldsymbol{J}^s = 0, \tag{6.97}$$

We outline how this set of equations can be solved using nodal and edge elements.

6.8.3.1 Solution by Nodal Elements for the Components of A

The method based on nodal elements for the components of the vector potential is still used in commercial codes, despite some known difficulties. The first difficulty comes from the fact that the null-space solutions for the curl-curl operator cannot be represented by divergence-conforming elements. This problem can be cured by removing the null-space (which does not contribute to the magnetic field anyway) by adding a so-called penalty term $-\nabla\mu^{-1}\nabla\cdot\boldsymbol{A}$ to Ampère's law, so that the set of equations becomes

$$\nabla\times\frac{1}{\mu}\nabla\times\boldsymbol{A}-\nabla\frac{1}{\mu}\nabla\cdot\boldsymbol{A}+\sigma\left(\frac{\partial\boldsymbol{A}}{\partial t}+\nabla\phi\right)=\boldsymbol{J}^s, \tag{6.98}$$

$$\nabla\cdot\sigma\left(\frac{\partial\boldsymbol{A}}{\partial t}+\nabla\phi\right)=0. \tag{6.99}$$

This procedure makes the differential operator in (6.98) similar to a Laplacian and removes highly oscillatory, spurious solutions. The system (6.98)–(6.99) can be solved using Galerkin's method, where (6.98) is tested with the basis functions for $\boldsymbol{A}$ (vectorial nodal elements) and (6.99) is tested with the basis functions for ϕ (scalar nodal elements).

Taking the divergence of (6.98) and using (6.99), we get

$$\nabla^2\frac{1}{\mu}\nabla\cdot\boldsymbol{A}=0. \tag{6.100}$$

Thus, $\mu^{-1}\nabla\cdot\boldsymbol{A}$ satisfies the Laplace equation, and if $\nabla\cdot\boldsymbol{A}$ vanishes on the boundaries, this implies $\nabla\cdot\boldsymbol{A}=0$ everywhere. Therefore, the penalty term in (6.98) is numerically zero, so that it does not change Ampère's law. (Actually, it is nonzero for the spurious solutions, which are removed by adding the penalty term.) One of the advantages of the penalty term is that it changes the conditioning of the matrix by removing small eigenvalues, and therefore makes the system easier to solve by an iterative solver. Note that for this formulation, the condition of zero divergence for the conduction currents (6.99) is essential and cannot be left out. This condition is not a gauge condition, but it indirectly enforces $\nabla\cdot\boldsymbol{A}=0$, which is called the Coulomb gauge.

It turns out that this method works well, except at edges and corners where the magnetic permeability μ changes. At such edges, the magnetic field is unbounded, and the penalty formulation is not accurate. Recent work suggests that this problem can be overcome by removing the penalty term locally around such singularities.

6.8.3.2 Solution by Edge Elements for *A*

Edge elements work better for low-frequency problems, but the procedures for an efficient implementation are not simple [46]. As a first attempt, one can set the scalar potential to zero and expand the solution of (6.97) in edge elements. If the frequency is zero, one must note that the curl-curl operator has a large null-space. For the lowest-order edge elements, this null-space consists of $\boldsymbol{A}=\nabla U$, where U is a piecewise linear scalar variable. Therefore, (6.97) can be solved only if $\boldsymbol{J}^s$

has no projection on this null-space. One can ensure this, either by representing $\boldsymbol{J}^s$ as the curl of a current potential, or by subtracting the gradient of a scalar U from $\boldsymbol{J}^s$ and imposing the condition $\langle \nabla \bar{U}, \boldsymbol{J}^s - \nabla U \rangle = 0$ for all piecewise linear test functions $\bar{U}$. This procedure works excellently for static problems. It does not suffer from the accuracy problems that occur for the nodal representation at edges where μ has jumps.

If one straightforwardly extends this procedure to finite frequency, the matrix becomes ill-conditioned, and iterative solvers converge very slowly. The cure for this is a somewhat surprising procedure, which consists in introducing a scalar potential ϕ and *not* prescribing a gauge condition. Instead of a gauge condition, one requires the divergence of the conduction current to be zero, so that the system of equations is

$$\nabla \times \frac{1}{\mu} \nabla \times \boldsymbol{A} + \sigma \left(j\omega \boldsymbol{A} + \nabla \phi \right) = \boldsymbol{J}^s, \tag{6.101}$$

$$\nabla \cdot \sigma \left(j\omega \boldsymbol{A} + \nabla \phi \right) = 0. \tag{6.102}$$

Note that this system of equations is degenerate, because the second equation is the divergence of the first (assuming $\nabla \cdot \boldsymbol{J}^s = 0$). Moreover, ϕ occurs only in the combination $j\omega \boldsymbol{A} + \nabla \phi = -\boldsymbol{E}$, so that any change of $\boldsymbol{A}$ and ϕ that leaves this combination unchanged is permitted. This is precisely a gauge transformation, which does not change the physical fields. Thus, the system (6.101)–(6.102) permits any gauge, and the method is referred to as the *ungauged* formulation. Of course, the indeterminacy of the solutions implied by gauge transformations means that the matrix is singular. However, iterative methods work also for singular matrices, provided that the right-hand side is *consistent*, that is, has no projection on the null-space.

The ungauged formulation greatly reduces the number of iterations for Krylov solvers (to which an introduction is given in Appendices C and D). The ungauged formulation can be viewed as a form of preconditioner for the curl-curl equation, and it improves the complex eigenvalue spectrum of the operator. The advantage of the edge elements over the nodal element formulation with a penalty term is that the edge elements give good approximations also at corners of magnetic materials.

Eddy current calculations are more frequently carried out on hexahedral meshes than on tetrahedral ones. One reason for this is that eddy current problems often involve currents in thin layers, within the skin depth $\delta = (2/\omega\mu\sigma)^{1/2}$ of conductor surfaces. The skin depth is typically in the millimeter to centimeter range, which is small compared to the global dimensions of a motor, generator, or transformer. Therefore, high resolution is required in the direction normal to the surface of a conductor, whereas the resolution requirements in the perpendicular direction can be much less demanding. This anisotropy is easier to achieve on a hexahedral mesh than a tetrahedral one. Another anisotropy can be introduced by laminations, and these are much easier to treat on a hexahedral mesh, which can be aligned with the laminations.

Review Questions

6.8-1 What is the low-frequency approximation and when is it applicable?

6.8-2 Consider a 2D low-frequency problem in the xy-plane. Use Maxwell's equations to derive a partial differential equation for the z-component of the vector potential. How can boundary conditions for the fields be formulated in terms of the vector potential?

6.8-3 Is the vector potential uniquely defined? If not, what conditions do you need to uniquely determine the vector potential?

6.8-4 Why is the electrostatic potential undetermined in the low-frequency approximation?

6.8-5 What is the difference between the magnetostatic problem and the low-frequency eddy current problem? Give examples of how the characteristic features of the solution change. Does this influence the choice of numerical algorithms and discretizations?

6.8-6 What is a penalty term and why is it used?

6.8-7 Mention some drawbacks associated with representing the components of the vector potential in a 3D eddy current problem by nodal elements.

6.8-8 Explain what a gauge transformation is.

6.8-9 Under what conditions is it possible to solve a system of linear equations where the system matrix is singular?

6.9 Variational Methods

The FEM can also be introduced as a variational method. Variational methods are intimately related with essential conservation laws of the system, and can give valuable insights into the application problem.

As an illustration, we study an example of electrostatics in a source free region. Here $\boldsymbol{D} = \epsilon \boldsymbol{E}$ and $\boldsymbol{E} = -\nabla \phi$, where ϕ is the electric potential. The natural choice of a variational quantity is the electrostatic field energy:

$$W[\phi] = \frac{1}{2} \int_V \boldsymbol{E} \cdot \boldsymbol{D} dV = \frac{1}{2} \int_V \epsilon |\nabla \phi|^2 dV. \tag{6.103}$$

The potential for which (6.103) gives the energy does not have to be the true solution, but it must fulfill the boundary conditions. The remarkable thing is that the true potential distribution, satisfying the boundary conditions and Poisson's equation $-\nabla \cdot (\epsilon \nabla \phi) = 0$, is exactly the function that minimizes (6.103); i.e., it gives the smallest electrostatic energy of all allowed ϕ.

To show this, let ϕ_0 be the potential that minimizes (6.103). Then, change the potential slightly by adding a perturbation $\delta\phi$, and compute the electrostatic energy for the perturbed potential $\phi = \phi_0 + \delta\phi$:

$$W[\phi_0 + \delta\phi] = W[\phi_0] + \int_V \epsilon \nabla \delta\phi \cdot \nabla \phi_0 dV + O\left((\delta\phi)^2\right). \tag{6.104}$$

When $\delta\phi$ is small, the higher-order terms $O((\delta\phi)^2)$ can be dropped. When the electrostatic energy W has a minimum, the first variation $\delta W = W[\phi_0+\delta\phi]-W[\phi_0]$ must be zero. After an integration by parts, (6.104) gives the following condition for the energy to be stationary:

$$\delta W = \int_V \delta\phi \left[-\nabla \cdot (\epsilon \nabla \phi_0)\right] dV = 0. \tag{6.105}$$

If this is to hold for all perturbations $\delta\phi$, the potential ϕ_0 must satisfy $-\nabla\cdot(\epsilon\nabla\phi_0) = 0$ everywhere in V; i.e., the differential equation of electrostatics in a source-free region is satisfied.

6.9.1 Relation Between Linear Differential Equations and Quadratic Forms

In more general terms, the solution f of a self-adjoint linear differential equation $L[f] = s$ in a domain Ω corresponds to a *stationary point* for the quadratic form

$$I[f] = \frac{1}{2}\langle f, L[f]\rangle - \langle f, s\rangle. \tag{6.106}$$

We use the scalar product $\langle f, g\rangle = \int_\Omega fg \, d\Omega$, where f and g are real functions. An operator L is self-adjoint if $\langle g, L[f]\rangle = \langle f, L[g]\rangle$ for all f and g. The factor $\frac{1}{2}$ in the first term of (6.106) is needed in order to produce the correct differential equation, because the first term in I is quadratic, while the second is linear.

Now let δf be a small variation of f. We will consider variations only up to linear order in δf. We let δI denote the first-order variation of $I[f]$ when $f \rightarrow f + \delta f$ and say that $I[f]$ is stationary if

$$\delta I = 0, \ \forall \delta f. \tag{6.107}$$

Since f represents a minimum, the rate of change of I at f must be zero. Let us expand $I[f + \delta f]$ in powers of δf:

$$\begin{aligned} I[f + \delta f] &= \frac{1}{2}\langle f + \delta f, L[f + \delta f]\rangle - \langle f + \delta f, s\rangle \\ &= \frac{1}{2}\langle f, L[f]\rangle - \langle f, s\rangle \end{aligned}$$

$$+\frac{1}{2}\langle \delta f, L[f]\rangle + \frac{1}{2}\langle f, L[\delta f]\rangle - \langle \delta f, s\rangle$$
$$+\frac{1}{2}\langle \delta f, L[\delta f]\rangle$$
$$= I[f] + \delta I + O((\delta f)^2). \tag{6.108}$$

The first variation is the part that is linear in δf, that is,

$$\delta I = \frac{1}{2}\langle \delta f, L[f]\rangle + \frac{1}{2}\langle f, L[\delta f]\rangle - \langle \delta f, s\rangle.$$

In order for $I[f]$ to be stationary, the first variation must vanish:

$$\delta I = \frac{1}{2}(\langle \delta f, L[f]\rangle + \langle f, L[\delta f]\rangle) - \langle \delta f, s\rangle = 0, \quad \forall \delta f. \tag{6.109}$$

Now L is self-adjoint, i.e., $\langle f, L[\delta f]\rangle = \langle L[f], \delta f\rangle$, so the condition for I stationary becomes $\langle \delta f, L[f]\rangle - \langle \delta f, s\rangle = \langle \delta f, L[f] - s\rangle = 0$. Thus, for every admissible variation δf we have

$$\langle \delta f, L[f] - s\rangle = \int_{\Omega} \delta f(L[f] - s) d\Omega = 0. \tag{6.110}$$

Since δf is an *arbitrary* function, this requires that the residual $r = L[f] - s$ vanish *everywhere* in Ω; that is, that the differential equation $L[f] = s$ be satisfied.

The discussion above shows that we can solve the differential equation $L[f] = s$ by finding the function f that makes $I[f]$ stationary. Often, I represents the energy, and the solution of the differential equation is the one that minimizes the energy. The electrostatics problem we just discussed is an example of this.

6.9.1.1 A 1D Example

To illustrate some features of the variational method, we study a simple example in one dimension. Let $L[f] = -f''$ and $s = x^2$ with the boundary conditions $f(0) = f(1) = 0$. We make a guess for the solution f containing only two parameters a and b. The function $f(x) = ax(1-x)^3 + bx^2(1-x)$ satisfies the boundary conditions for arbitrary a and b. We seek the combination of a and b such that the differential equation is satisfied as well as possible. If it is not possible to find an exact solution, we want the "best" combination of a and b.

This can be done by computing the quadratic form I and finding its stationary point. Since the operator $L[f]$ is self-adjoint, it corresponds to a quadratic form given by $I[f] = \frac{1}{2}\langle f, L[f]\rangle - \langle f, s\rangle$, that is,

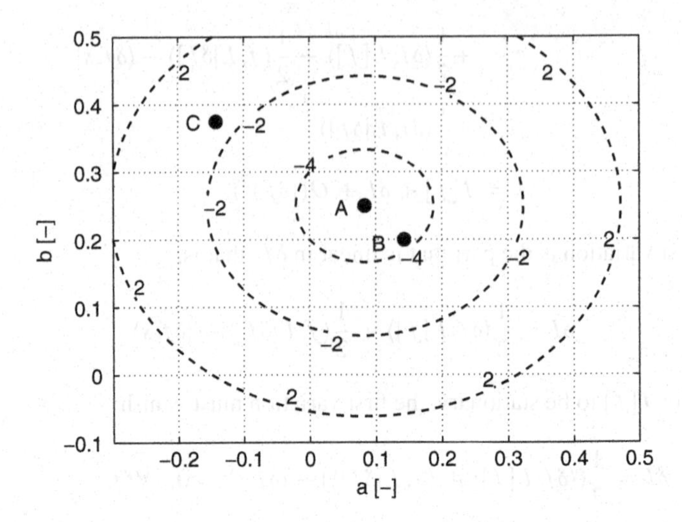

Fig. 6.39 Level contours for the quadratic form I. The stationary point is shown by the dot labeled A, and this combination of a and b solves $-f'' = x^2$. B and C are not stationary points, and they do not solve the differential equation

$$I[f] = -\frac{a}{140} - \frac{b}{30} + \frac{1}{2}\left(\frac{3a^2}{35} + \frac{2b^2}{15}\right), \tag{6.111}$$

I is a quadratic function in the parameters a and b, and Fig. 6.39 shows level contours for the quadratic form I with respect to these parameters.

There is a global minimum for I indicated by the dot labeled A in Fig. 6.39. To find the values of a and b for this minimum we set the gradient of I equal to zero:

$$\frac{\partial I}{\partial a} = -\frac{1}{140} + \frac{3\,a}{35} = 0 \tag{6.112}$$

$$\frac{\partial I}{\partial b} = -\frac{1}{30} + \frac{2\,b}{15} = 0 \tag{6.113}$$

which gives the solution $a = 1/12$ and $b = 1/4$. The corresponding solution $f(x) = x(1 - x^3)/12$ indeed solves $-f'' = x^2$, and it is shown in Fig. 6.40 by the solid curve labeled A. If the basis functions had been chosen in a less clever way, so that the true solution could not be constructed, the variational approach would have given the "best" approximation of $f(x)$.

Let us see what happens if we change the values of a and b away from the minimum A in Fig. 6.39, e.g., to the points B and C. The new combinations of a and b and their values of I are shown in Table 6.7 together with the correct solution. The functions f corresponding to B and C are also shown in Fig. 6.40.

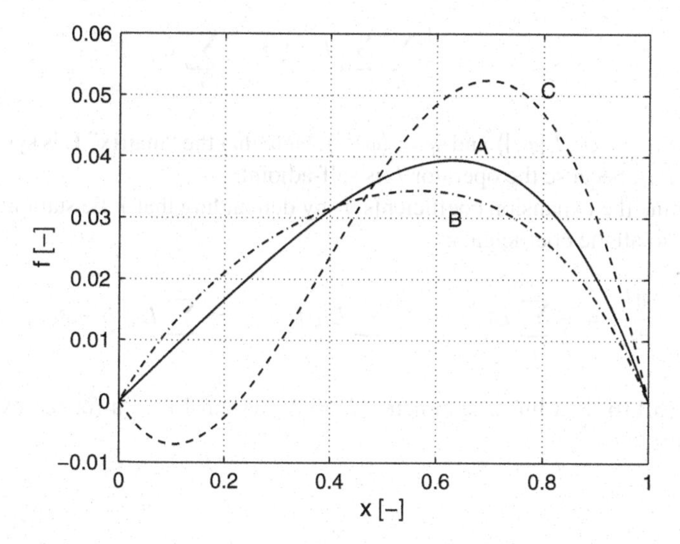

Fig. 6.40 The function solving $-f'' = x^2$ is shown by the solid curve labeled A. The two other functions labeled B and C do not satisfy the differential equation

Table 6.7 Three different combinations of the parameters. The true solution is labeled A.

Label	a	b	$I[f]$
A	1/12	1/4	−4.46
B	1/7	1/5	−4.15
C	−1/7	3/8	−1.23

6.9.2 *Rayleigh–Ritz Method*

The variational formulation gives a procedure, the *Rayleigh–Ritz* method, for finding approximate solutions of self-adjoint linear equations. It consists of the following steps:

- Approximate f by an expansion in a finite set of basis (or *trial*) functions φ_i, $i = 1, 2, \ldots, N$:

$$f(\mathbf{r}) = \sum_{i=1}^{N} f_i \varphi_i(\mathbf{r}). \tag{6.114}$$

- Evaluate the quadratic variational form I as a function of the expansion coefficients

$$\begin{aligned} I(f_1, f_2, \ldots, f_N) = I[f] &= \frac{1}{2}\langle f, L[f]\rangle - \langle f, s\rangle \\ &= \frac{1}{2}\sum_i \sum_j f_i f_j \langle \varphi_i, L[\varphi_j]\rangle - \sum_i f_i \langle \varphi_i, s\rangle \end{aligned}$$

$$= \frac{1}{2} \sum_i \sum_j L_{ij} f_i f_j - \sum_i s_i f_i, \tag{6.115}$$

where $L_{ij} = \langle \varphi_i, L[\varphi_j] \rangle$ and $s_i = \langle \varphi_i, s \rangle$. Note that the "matrix" $\mathbf{L}$ is symmetric, $L_{ij} = L_{ji}$, because the operator L is self-adjoint.

- Determine the expansion coefficients f_i by demanding that I be stationary with respect to all the coefficients:

$$0 = \frac{\partial I}{\partial f_k} = \frac{1}{2} \sum_j L_{kj} f_j + \frac{1}{2} \sum_i L_{ik} f_i - s_k = \sum_i L_{ki} f_i - s_k. \tag{6.116}$$

Equation (6.116) is a linear symmetric $N \times N$ system $\mathbf{Lf} = \mathbf{s}$ for the expansion coefficients.

6.9.3 *Galerkin's Method*

Galerkin's method is intimately connected to the variational formulation. In fact, the Rayleigh–Ritz formulation (6.116) leads to Galerkin's method for self-adjoint systems. Using the definitions of the matrix elements L_{ki} and s_k, we have from (6.116)

$$\begin{aligned} \sum_i L_{ki} f_i - s_k &= \sum_i \langle \varphi_k, L[f_i \varphi_i] \rangle - \langle \varphi_k, s \rangle \\ &= \langle \varphi_k, L[\sum_i f_i \varphi_i] - s \rangle = \langle \varphi_k, L[f] - s \rangle \\ &\Longrightarrow \int_\Omega \varphi_k (L[f] - s)\, d\Omega = 0. \end{aligned} \tag{6.117}$$

This is Galerkin's method for solving $L[f] = s$, since the weighting functions are equal to the basis functions. It is also the same as the variational condition (6.110), but the previously arbitrary weighting function δf for the residual $r = L[f] - s$ is now restricted to lie in the space of the basis functions. This shows that Galerkin's method can be derived from variational calculus.

We stress some important facts:

- For self-adjoint differential equations, the Rayleigh–Ritz and Galerkin methods are equivalent.
- The Galerkin method can be used also for non-self-adjoint problems where no variational principle can be found.
- In the more general *Petrov–Galerkin* method, the weighting functions w_i are different from the basis functions φ_i.

6.9.4 A Variational Method for Maxwell's Equations

Maxwell's equations can be put in variational form in a few different ways. One way is to apply the general prescription (6.106) to the lossless self-adjoint curl-curl equation

$$\nabla \times \mu^{-1} \nabla \times \boldsymbol{E} + \epsilon \partial^2 \boldsymbol{E} / \partial t^2 = -\partial \boldsymbol{J} / \partial t, \tag{6.118}$$

integrate both in space and time, and ignore the boundary terms. This gives the quadratic form

$$L = \iint \left(\frac{1}{2\mu} |\nabla \times \boldsymbol{E}|^2 - \frac{\epsilon}{2} \left| \frac{\partial \boldsymbol{E}}{\partial t} \right|^2 + \boldsymbol{E} \cdot \frac{\partial \boldsymbol{J}}{\partial t} \right) dV dt. \tag{6.119}$$

For a small variation of the electric field $\boldsymbol{E} \to \boldsymbol{E} + \delta \boldsymbol{E}$, the first-order change of L is

$$\delta L = \iint \left(\frac{1}{\mu} \nabla \times \boldsymbol{E} \cdot \nabla \times \delta \boldsymbol{E} - \epsilon \frac{\partial \boldsymbol{E}}{\partial t} \cdot \frac{\partial \delta \boldsymbol{E}}{\partial t} + \frac{\partial \boldsymbol{J}}{\partial t} \cdot \delta \boldsymbol{E} \right) dV dt,$$

and an integration by parts (ignoring boundary terms) gives

$$\delta L = \iint \left(\nabla \times \frac{1}{\mu} \nabla \times \boldsymbol{E} + \epsilon \frac{\partial^2 \boldsymbol{E}}{\partial t^2} + \frac{\partial \boldsymbol{J}}{\partial t} \right) \cdot \delta \boldsymbol{E} \, dV dt.$$

Thus, if $\boldsymbol{E}$ is a solution of Maxwell's equations, then $\delta L = 0$ for any $\delta \boldsymbol{E}$, which means that $L[\boldsymbol{E}]$ is stationary. Conversely, to make L stationary, i.e., $\delta L = 0$ for an arbitrary $\delta \boldsymbol{E}$, the curl-curl equation (6.118) must be satisfied.

A slight reformulation of the variational principle that is more directly related to physical quantities uses the vector and scalar potentials as independent variables. The fields are represented as

$$\boldsymbol{B} = \nabla \times \boldsymbol{A}, \quad \boldsymbol{E} = -\nabla \phi - \frac{\partial \boldsymbol{A}}{\partial t}. \tag{6.120}$$

The quadratic form is the magnetic minus the electric energy, plus terms involving the sources, integrated in space and time:

$$L = \iint \left(\frac{B^2}{2\mu} - \boldsymbol{A} \cdot \boldsymbol{J} - \frac{\epsilon E^2}{2} + \rho \phi \right) dV dt. \tag{6.121}$$

We get Maxwell's equations by setting the first variation of L with respect to ϕ and $\boldsymbol{A}$ to zero. For $\phi \to \phi + \delta \phi$, integration by parts gives the first variation

$$\delta L = \iint (\epsilon \nabla \delta\phi \cdot \boldsymbol{E} + \rho \delta\phi) \, dV dt$$

$$= \iint (\rho - \nabla \cdot \epsilon \boldsymbol{E}) \delta\phi \, dV dt = 0. \tag{6.122}$$

Therefore, $\delta L = 0$ for all $\delta\phi$ if and only if Poisson's equation $\nabla \cdot \epsilon \boldsymbol{E} = \rho$ is satisfied.

For $\boldsymbol{A} \to \boldsymbol{A} + \delta \boldsymbol{A}$ the same procedure gives

$$\delta L = \iint \left(\frac{1}{\mu} \nabla \times \delta \boldsymbol{A} \cdot \boldsymbol{B} + \frac{\partial \delta \boldsymbol{A}}{\partial t} \cdot \epsilon \boldsymbol{E} - \delta \boldsymbol{A} \cdot \boldsymbol{J} \right) dV dt$$

$$= \iint \left(\nabla \times \frac{\boldsymbol{B}}{\mu} - \frac{\partial}{\partial t} \epsilon \boldsymbol{E} - \boldsymbol{J} \right) \cdot \delta \boldsymbol{A} \, dV dt = 0. \tag{6.123}$$

Therefore, $\delta L = 0$ for all $\delta \boldsymbol{A}$ if and only if Ampère's law $\nabla \times (\boldsymbol{B}/\mu) = \partial(\epsilon \boldsymbol{E})/\partial t + \boldsymbol{J}$ holds everywhere. Faraday's law and $\nabla \cdot \boldsymbol{B} = 0$ are automatically satisfied because of the potential representation (6.120).

Review Questions

6.9-1 Motivate why variational methods are useful.
6.9-2 What are a quadratic form, functional, variation, and stationary point?
6.9-3 List and describe the steps involved in the Rayleigh–Ritz method.
6.9-4 What conditions must be fulfilled for the Rayleigh–Ritz formulation and Galerkin's method to be equivalent? Given such conditions, show that they are equivalent.
6.9-5 For Maxwell's equations, write down the quadratic form in terms of the electric field and show that a solution that makes the quadratic form stationary satisfies Maxwell's equations.
6.9-6 Repeat the previous problem when the quadratic form is expressed in terms of the potentials. Provide a physical interpretation of the constituents of the quadratic form.

Summary

- The FEM is in short:
  - To solve $L[f] = s$, divide the solution region into elements and expand the sought solution f in local basis functions $f(\boldsymbol{r}) = \sum_{i=1}^{N} f_i \phi_i(\boldsymbol{r})$.
  - Make the residual $r = L[f] - s$ orthogonal to N weighting functions w_i, $i = 1, 2, \ldots, N$ (the method of weighted residuals).

Galerkin's method $w_i = \phi_i$ is a popular choice for the weighting functions. Other choices, i.e., $w_i \neq \phi_i$, are referred to as Petrov–Galerkin, and some possibilities are collocation $w_i = \delta(\boldsymbol{r} - \boldsymbol{r}_i)$, least squares $w_i = L[\phi_i]$, and least square stabilized Galerkin $w_i = \phi_i + cL[\phi_i]$, where the parameter c is optimized.

- In one dimension with uniform meshes and f in piecewise linear elements, Galerkin's method gives

$$\frac{d^2 f}{dx^2} \to \frac{f_{i+1} - 2f_i + f_{i-1}}{h^2},$$

$$f \to \frac{f_{i+1} + 4f_i + f_{i-1}}{6},$$

where lumping (which in this case is obtained by trapezoidal integration) gives $f \to f_i$; i.e., the finite difference approximation is recovered for the Helmholtz equation in one dimension.

- For the Helmholtz scalar equation in 2D, we can use a continuous linear approximation of the solution f on a mesh of triangles. The expansion $f(\boldsymbol{r}) \approx \sum_i f_i \varphi_i(\boldsymbol{r})$ is then used to represent the solution, where φ_i is a piecewise linear basis function with $\varphi_i(\boldsymbol{r}_i) = 1$ and $\varphi_i(\boldsymbol{r}_j) = 0$ when $i \neq j$. The FEM formulation typically involves matrix entries of the type

$$-\nabla^2 \to S_{ij} = \int_S \nabla\varphi_i \cdot \nabla\varphi_j \, dS,$$

$$1 \to M_{ij} = \int_S \varphi_i \varphi_j \, dS.$$

By terminology borrowed from mechanics, **S** is referred to as the stiffness matrix and **M** as the mass matrix.

- Adaptivity can often restore nominal convergence for singular problems.
- Mixed elements for a system of coupled first-order differential equations

$$\frac{\partial E}{\partial x} = \omega\mu H, \qquad \frac{\partial H}{\partial x} = -\omega\epsilon E,$$

are treated with E expanded in piecewise linear functions (connected with integer mesh) and H expanded in piecewise constants (connected with half mesh). This gives

$$\frac{E_{i+1} - E_i}{h} = \omega\mu H_{i+\frac{1}{2}}, \qquad \frac{H_{i+\frac{1}{2}} - H_{i-\frac{1}{2}}}{h} = -\omega\epsilon\left[\frac{2}{3}E_i + \frac{1}{6}(E_{i-1} + E_{i+1})\right],$$

where the term $\omega\epsilon E$ can be lumped by the trapezoidal rule $\int_{x_i}^{x_{i+1}} f(x)dx \approx (h/2)[f(x_i) + f(x_{i+1})]$; i.e., we have $(4E_i + E_{i-1} + E_{i+1})/6 \to E_i$.

- Edge elements $\boldsymbol{N}_i$ have continuous tangential components, which makes the curl of the solution square integrable. They are often referred to as curl-conforming elements, and some distinguishing features are:
  - the basis functions $\boldsymbol{N}_i$ have unit tangential components along one edge and zero along all the other edges,
  - spurious solutions and spectral contamination are avoided,
  - the null-space of the curl operator is correctly represented.

  The formulation for the vector Helmholtz equation involves terms of the type

$$\nabla \times \nabla \times \rightarrow S_{ij} = \int_S (\nabla \times \boldsymbol{N}_i) \cdot (\nabla \times \boldsymbol{N}_j)\, dS,$$

$$1 \rightarrow M_{ij} = \int_S \boldsymbol{N}_i \cdot \boldsymbol{N}_j \, dS.$$

- Time-dependent problems use time-dependent coefficients for the spatial expansion of the field. The wave equation $\mathbf{Sz}(t) + c_0^{-2}\mathbf{M}\, \partial^2 \mathbf{z}(t)/\partial t^2 = \mathbf{0}$ can be time-stepped with the finite difference scheme

$$\mathbf{M}\left(\mathbf{z}^{n+1} - 2\mathbf{z}^n + \mathbf{z}^{n-1}\right) = -\left(c_0 \Delta t\right)^2 \mathbf{S}\mathbf{z}^n,$$

  which requires a sufficiently small time-step Δt for stability. An even more implicit scheme, derived by averaging the stiffness term in time, gives unconditional stability (provided that the implicitness parameter θ is greater than or equal to $1/4$):

$$\mathbf{M}\left(\mathbf{z}^{n+1} - 2\mathbf{z}^n + \mathbf{z}^{n-1}\right) = -\left(c_0 \Delta t\right)^2 \mathbf{S}\left[\theta \mathbf{z}^{n+1} + (1 - 2\theta)\mathbf{z}^n + \theta \mathbf{z}^{n-1}\right].$$

- The solution f of a self-adjoint linear differential equation $L[f] = s$ is a stationary point of the quadratic form

$$I[f] = \frac{1}{2}\langle f, L[f]\rangle - \langle f, s\rangle.$$

  A self-adjoint operator L satisfies $\langle f, L[g]\rangle = \langle g, L[f]\rangle$ for all f and g.
- The Rayleigh–Ritz method solves $L[f] = s$ by expanding f in global basis functions $f(\boldsymbol{r}) \approx \sum_{i=1}^{N} f_i \phi_i(\boldsymbol{r})$ and evaluating the quadratic form $I(f_1, f_2, \ldots, f_N)$. Coefficients are determined by $\partial I / \partial f_i = 0$ for all $i = 1, 2, \ldots, N$. For self-adjoint problems, the equivalent Galerkin formulation is to make the residual $r = L[f] - s$ orthogonal to all the basis functions, i.e., $\int (L[f] - s)\varphi_i \, d\Omega = 0$ for all i.

Problems

P.6-1 Derive the finite element approximation of the 1D Helmholtz equation $-(d^2/dx^2 + k^2)f = 0$ for piecewise linear elements on a nonequidistant mesh and show for the system matrix

$$A_{ij} = S_{ij} - k^2 M_{ij}$$

that the elements are

$$S_{i,i-1} = -\frac{1}{x_i - x_{i-1}}, \qquad M_{i,i-1} = \frac{x_i - x_{i-1}}{6},$$

$$S_{i,i} = \frac{1}{x_{i+1} - x_i} + \frac{1}{x_i - x_{i-1}}, \qquad M_{i,i} = \frac{x_{i+1} - x_{i-1}}{3},$$

$$S_{i,i+1} = -\frac{1}{x_{i+1} - x_i}, \qquad M_{i,i+1} = \frac{x_{i+1} - x_i}{6}.$$

Show that for a uniform mesh with cell size h this gives a discretization that is similar to the finite difference approximation, except that the mass term is weighted between adjacent nodes. Substitute a complex exponential $f = \exp(jkx)$ and show that the FEM approximation gives $k_{\text{FEM}}^2 = 24\sin^2(kh/2)/[2 + 4\cos^2(kh/2)] \approx k^2(1 + k^2h^2/12)$, so that the FEM eigenvalue converges from above. Note that the error has the same magnitude, but the opposite sign, as the FD approximation (3.18). Based on this, can you find a three-point discretized operator that gives an error $O(k^4h^4)$?

P.6-2 Consider a scattering problem where both the geometry and the sources are independent of the z-coordinate. Derive the weak formulation for the Helmholtz equation

$$-\nabla \cdot \left(\frac{1}{\mu}\nabla E_z^{\text{sc}}\right) - \omega^2 \epsilon E_z^{\text{sc}} = 0,$$

where E_z^{sc} is the scattered electric field from a metal cylinder. Impose the boundary condition $E_z^{\text{sc}} = -E_z^{\text{inc}}$ on the surface of the scatterer, where E_z^{inc} is the incident wave. The finite element mesh discretizes the region around the metal cylinder and extends some distance from the scatterer. At the exterior boundary of the mesh we apply the absorbing boundary condition $\hat{\boldsymbol{n}} \cdot \nabla E_z^{\text{sc}} = -jkE_z^{\text{sc}}$ to mimic an open region problem. What criteria must be fulfilled for this boundary condition to be accurate? To answer this question, it is useful to consider a plane wave $E_z^{\text{sc}} = E_0 \exp(-j\boldsymbol{k} \cdot \boldsymbol{r})$ that is incident on such an absorbing boundary.

P.6-3 A rectangular finite element occupies the region defined by $x_a \le x \le x_b$ and $y_a \le y \le y_b$. This element has four nodes and, also, four nodal basis functions:

$$\varphi_1^e = \frac{x_b - x}{x_b - x_a} \cdot \frac{y_b - y}{y_b - y_a}, \qquad \varphi_2^e = \frac{x - x_a}{x_b - x_a} \cdot \frac{y_b - y}{y_b - y_a},$$

$$\varphi_3^e = \frac{x - x_a}{x_b - x_a} \cdot \frac{y - y_a}{y_b - y_a}, \qquad \varphi_4^e = \frac{x_b - x}{x_b - x_a} \cdot \frac{y - y_a}{y_b - y_a}.$$

Is it feasible to apply the FEM to a mesh where such a rectangular finite element is connected to a triangular finite element so that the two share one edge? Suggest a situation in which it can be useful to discretize the solution domain with both triangles and rectangles.

P.6-4 In addition to the organization of nodes and elements, it is often necessary to include various materials and boundary conditions in the discrete representation of a FEM problem. The data structures discussed in Sect. 6.3.2 can also be extended to deal with postprocessing steps, e.g., integration along a contour. Discuss how the representation of the geometrical information relating to materials, boundary conditions, and postprocessed quantities could be implemented in a FEM computer program.

P.6-5 Consider the electrostatic problem $-\nabla\cdot(\epsilon_0\nabla\phi) = 0$. For a solution computed by the FEM with linear triangles, the potential is piecewise linear, and the corresponding electric field is piecewise constant. Given such a FEM solution, evaluate $Q_t = \oint_{L_t} \boldsymbol{D}\cdot\hat{\boldsymbol{n}}\, dl$ applied to a single triangle, where L_t is the boundary of the triangle. Evaluate also $Q_e = \oint_{L_e} \boldsymbol{D}\cdot\hat{\boldsymbol{n}}\, dl$ applied to a single edge shared by two triangles, where L_e is an integration contour enclosing the edge. Interpret the derived expressions for Q_t and Q_e. How do these quantities depend on the variation in the solution as compared to the cell size? Since the charge density is supposed to be zero, the dissatisfaction of Gauss's law could be used as a physics based indication of inaccuracy. Note that Q_t and Q_e do not give a bound on the actual error in the solution ϕ. Bounds on the error in the solution can be derived mathematically [28], but such a derivation is beyond the scope of this book.

P.6-6 In Sect. 6.8, we computed the vector potential $\boldsymbol{A} = A_z(x, y)\hat{\boldsymbol{z}}$ on an unstructured mesh of triangles. Given this solution, we used a routine that plots equipotential lines of the vector potential to visualize the flux lines of the magnetic flux density. Show that a contour where $A_z(x, y)$ is constant is also a flux line for the magnetic flux density $\boldsymbol{B}$.

P.6-7 Eliminate the magnetic field from (6.40) and (6.41). Compare this result with the FEM applied to the Helmholtz equation in one dimension,

$$\frac{d}{dx}\left(\frac{1}{\mu}\frac{dE_z}{dx}\right) + \omega^2\epsilon E_z = 0,$$

where the element matrices have been evaluated with either exact or trapezoidal integration. How do these methods relate to finite differences applied to the 1D Helmholtz equation?

P.6-8 Consider a scattering problem where both the geometry and the sources are independent of the z-coordinate. Here, we solve for the electric field $\boldsymbol{E}(x, y) = \hat{\boldsymbol{x}}\, E_x(x, y) + \hat{\boldsymbol{y}}\, E_y(x, y)$, and the computational mesh is truncated at a constant radius R from the origin. The scatterer is located at the origin. Modify the matrix entries (6.48) and the vector entries (6.50) to impose the Sommerfeld radiation condition

$$\hat{\boldsymbol{r}} \times (\nabla \times \boldsymbol{E}) + jk\hat{\boldsymbol{r}} \times (\hat{\boldsymbol{r}} \times \boldsymbol{E}) = \hat{\boldsymbol{r}} \times (\nabla \times \boldsymbol{E}^{\text{inc}}) + jk\hat{\boldsymbol{r}} \times (\hat{\boldsymbol{r}} \times \boldsymbol{E}^{\text{inc}})$$

combined with an external source that produces the prespecified incident field $\boldsymbol{E}^{\text{inc}}$. What boundary condition should be imposed on a metal scatterer? Which criteria must be fulfilled for the Sommerfeld radiation condition to be accurate?

P.6-9 A rectangular finite element occupies the region defined by $x_a^e \leq x \leq x_b^e$ and $y_a^e \leq y \leq y_b^e$. This element has four nodes and also four nodal basis functions:

$$\varphi_1^e = \frac{x_b^e - x}{x_b^e - x_a^e} \cdot \frac{y_b^e - y}{y_b^e - y_a^e}, \qquad \varphi_2^e = \frac{x - x_a^e}{x_b^e - x_a^e} \cdot \frac{y_b^e - y}{y_b^e - y_a^e},$$

$$\varphi_3^e = \frac{x - x_a^e}{x_b^e - x_a^e} \cdot \frac{y - y_a^e}{y_b^e - y_a^e}, \qquad \varphi_4^e = \frac{x_b^e - x}{x_b^e - x_a^e} \cdot \frac{y - y_a^e}{y_b^e - y_a^e}.$$

Consider an electric potential $\phi = \sum_{j=1}^{4} \phi_j \varphi_j^e$ on this rectangle. Show that the gradient of this potential falls into the space of the edge elements; i.e., the equality $\boldsymbol{E} = -\nabla\phi$ is satisfied pointwise. In other words, given arbitrary values for ϕ_j, show that there exist values for E_j such that $\sum_{j=1}^{4} E_j \boldsymbol{N}_j^e = -\sum_{j=1}^{4} \phi_j \nabla\varphi_j^e$ for every point inside the rectangle.

P.6-10 Prove that (6.92) is stable for an arbitrary time-step when $\theta \geq 1/4$ by carrying out a von Neumann stability analysis for eigenmodes of $\mathbf{Sz} = \lambda\mathbf{Mz}$, where $\lambda = \omega^2/c^2$.

P.6-11 How are solutions of the type $\boldsymbol{E} = -\nabla\phi$ treated by (6.86), (6.91), and (6.92)?

P.6-12 What is the natural choice of a variational quantity for the steady electric current problem $-\nabla\cdot(\sigma\nabla\phi) = 0$ that was treated in Sect. 6.3? How are boundary conditions treated in this case? Give a physical interpretation of the minimization of this functional and derive its first variation.

Computer Projects

C.6-1 Write a program that automatically generates a triangulation for a rectangular domain. You can use a structured mesh of rectangles and divide the rectangular elements on the diagonal to create the triangles.

C.6-2 Modify the program in Sect. 6.3.3 so that you can compute the capacitance of a capacitor with an inhomogeneous dielectric. Let the spatial dependence of the permittivity be a prespecified function of your own choice. Note that if the triangles are small compared to the variations in the permittivity, you can sample ϵ at the center of each element and assume it to be constant inside that element.

How does the error scale with the cell size given such an assumption? Will this have any impact on the order of convergence for the final algorithm? Can you improve the performance of such a method?

C.6-3 Rewrite (6.55)–(6.56) in terms of the z-component of the magnetic field. Use the program in Sect. 6.3.3 as a starting point for an implementation that solves this eigenvalue problem on a mesh of triangles. Will the static eigenvalue(s) $\omega = 0$ be reproduced by this formulation? Explain your findings.

C.6-4 Implement a FEM that solves $\nabla \times \nabla \times \boldsymbol{E} = k^2\boldsymbol{E}$ by means of rectangular finite elements. Apply your program to a 2D cavity with metal boundary and compute the eigenfrequencies. Find a test case for which the analytical result is known and perform a convergence study of the lowest eigenvalues. What order of convergence do you expect? Is this order of convergence reproduced by your program?

C.6-5 Modify the FEM function `edgeFEM2D` (in Sect. 6.5.6) so that it can treat problems where the material parameters are different in different cells. Add two extra input arguments, which are vectors with relative electric permittivity and magnetic permeability for all elements. Also modify the function such that homogeneous Dirichlet boundary conditions can be used. Can the same plot routine be used after these changes?

Chapter 7
The Method of Moments

The previous chapters of this book are devoted to the solution of Maxwell's equations on differential form, where the focus is on finite-difference schemes and the finite element method. In this chapter, Maxwell's equations are reformulated as integral equations, where the field solution is expressed in terms of superpostion integrals that involve the sources and a so-called Green's function. In this setting, we would typically have unknown sources that we wish to compute given that we have sufficient information that describes the known field. Typically, this type of formulation is useful for problems where the sources can be described by a relatively few degrees of freedom, when compared to the number of degrees of freedom that would be required for a corresponding description in terms of the fields.

In particular, we introduce the integral formulation of both electrostatics and the complete Maxwell system. The electromagnetics community normally refers to the integral formulation as the method of moments (MoM), for reasons that will be explained later. In mathematics, the MoM is often referred to as the boundary element method (BEM). We will reformulate electrostatics, for which we have previously used Poisson's and Laplace's equations, as an integral equation. In the following sections on scattering problems, we will rewrite the full Maxwell equations as an integral equation for currents on the surfaces of conductors, and apply this formulation to a scattering problem. The scattered electric field can be expressed in terms of surface currents on conductors. The condition that the tangential electric field vanishes on conductor surfaces then gives an integral equation from which we can compute the surface currents. For the interested reader, more information on the MoM can be found in, e.g., [20, 54, 87].

7.1 Integral Formulation of Electrostatics

In electrostatics, the electric potential ϕ is determined from the sources according to Poisson's equation

T. Rylander et al., *Computational Electromagnetics*, Texts in Applied Mathematics 51, DOI 10.1007/978-1-4614-5351-2_7,

$$\nabla^2 \phi = -\frac{\rho}{\epsilon_0}. \tag{7.1}$$

This is the differential equation formulation. The solution of Poisson's equation in free space can be constructed by superposing the contributions $\phi(\boldsymbol{r}) = q/4\pi\epsilon_0|\boldsymbol{r} - \boldsymbol{r}'|$ from point charges $q = \rho_v dV$ at locations $\boldsymbol{r}'$:

$$\phi(\boldsymbol{r}) = \int_V \frac{\rho(\boldsymbol{r}')dV'}{4\pi\epsilon_0|\boldsymbol{r} - \boldsymbol{r}'|}. \tag{7.2}$$

If the potential ϕ is known, (7.2) can be seen as an integral equation for the charge density ρ. The integral formulation is suited for problems such as the capacitance calculation in Chap. 3, where the potential is known on conducting boundaries and charges occur only on these boundaries. Then, the potential ϕ was given on the boundaries, $\phi = \phi_{\text{spec}} = 0$ on the outer conductor and $\phi = \phi_{\text{spec}} = 1$ on the inner one. As an alternative to solving Laplace's equation for the potential in the vacuum region, we can calculate the charges ρ_s on the conducting walls S by solving the integral equation

$$\int_S \frac{\rho_s(\boldsymbol{r}')}{4\pi\epsilon_0|\boldsymbol{r} - \boldsymbol{r}'|} dS' = \phi_{\text{spec}}(\boldsymbol{r}). \tag{7.3}$$

In the 2D capacitor problem, the surface integral reduces to a line integral, and we instead use the potential from a line charge $-(\rho_l/2\pi\epsilon_0)\ln|\boldsymbol{r} - \boldsymbol{r}'|$ as weighting, that is,

$$-\frac{1}{2\pi\epsilon_0}\int_S \rho_l(\boldsymbol{r}')\ln|\boldsymbol{r} - \boldsymbol{r}'|dl' = \phi_{\text{spec}}(\boldsymbol{r}). \tag{7.4}$$

Here, we "derived" the integral equations by referring to well-known expressions from electrostatics. However, it is useful to derive them in a more mathematical fashion, and also introduce the concept of a Green's function. The same procedures will be used to derive the electric field integral equation for the complete Maxwell system.

A characteristic property of the integral formulation is that it deals readily with open geometries. Consider the parallel plate capacitors illustrated in Fig. 7.1. In Fig. 7.1(a), the capacitor is enclosed in a conducting box, and in this case, differential equation solvers such as finite differences or finite elements work well. However, if there is no surrounding box, these methods have difficulties with truncating the open computational region, whereas the MoM works very well and has no difficulties with the open geometry; see Fig. 7.1(b). (In fact, the open geometry simplifies the MoM calculation, because it reduces the number of surfaces on which charges can reside.)

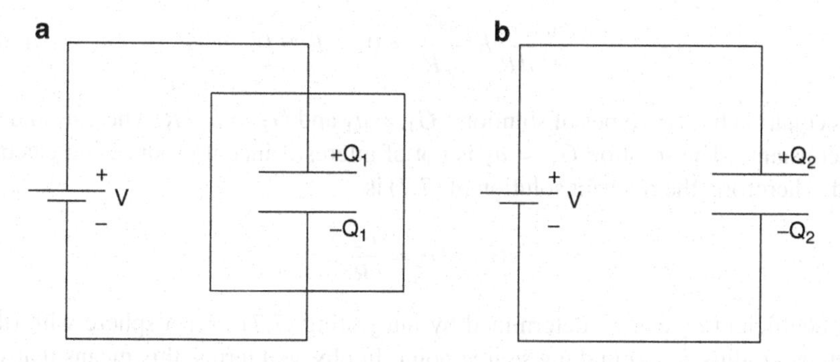

Fig. 7.1 Parallel plate capacitor in (a) closed geometry and (b) open geometry. Differential equation solvers can easily deal with the closed geometry (a), but the MoM is better adapted to deal with the open geometry (b)

7.1.1 Green's Function

Here, we introduce the concept of a Green's function $G(\boldsymbol{r}, \boldsymbol{r}')$, which represents the "field" at $\boldsymbol{r}$ produced by a point source at $\boldsymbol{r}'$. In electrostatics, the Green's function represents the electric potential at $\boldsymbol{r}$ produced by a unit charge at $\boldsymbol{r}'$. In three dimensions, this is

$$G(\boldsymbol{r}, \boldsymbol{r}') = \frac{1}{4\pi\epsilon_0|\boldsymbol{r} - \boldsymbol{r}'|}. \tag{7.5}$$

We will show how the Green's function for electrostatics can be found by solving Poisson's equation. This also serves as a preparation for the more complicated time-harmonic case, treated in Sect. 7.3.

The potential from a point charge in three dimensions satisfies Poisson's equation,

$$-\epsilon_0 \nabla^2 \phi(\boldsymbol{r}) = \delta^3(\boldsymbol{r} - \boldsymbol{r}'). \tag{7.6}$$

Here, $\delta^3(\boldsymbol{r} - \boldsymbol{r}')$ is the 3D Dirac delta function, which represents a unit point charge. It vanishes at all $\boldsymbol{r} \neq \boldsymbol{r}'$, and at $\boldsymbol{r} = \boldsymbol{r}'$ it is infinitely large, in such a way that the total charge $\int_V \delta^3(\boldsymbol{r} - \boldsymbol{r}')dV = 1$ for all volumes V where $\boldsymbol{r}'$ is an interior point. The solution $\phi(\boldsymbol{r})$ to (7.6) is the Green's function $G(\boldsymbol{r}, \boldsymbol{r}')$. Thus, for electrostatics,

$$-\epsilon_0 \nabla_r^2 G(\boldsymbol{r}, \boldsymbol{r}') = \delta^3(\boldsymbol{r} - \boldsymbol{r}'), \tag{7.7}$$

where the subscript r indicates that the differential operator acts on the $\boldsymbol{r}$ argument, the field point. By symmetry, the electrostatic potential that solves (7.7) can depend only on the distance $R = |\boldsymbol{r} - \boldsymbol{r}'|$ between the source and observation point. Therefore, except at the singularity $R = 0$, G satisfies

$$-\frac{\epsilon_0}{R^2}\frac{d}{dR}R^2\frac{dG}{dR} = 0, \quad R > 0. \tag{7.8}$$

This equation has two types of solutions, $G_1 = a_1$ and $G_2 = a_2/R$, where a_1 and a_2 are constants. The solution $G_1 = a_1$ is not of interest, since it produces no electric field. Therefore, the relevant solution of (7.7) is

$$G = G_2 = \frac{a_2}{R}$$

The coefficient a_2 can be determined by integrating (7.7) over a sphere with (the arbitrary) radius R_0 around the source point. In physical terms, this means that we equate the flux of electric displacement $\boldsymbol{D} = \epsilon_0\boldsymbol{E}$ through the surface of the sphere to the enclosed charge. By means of Gauss's theorem, the integral of the left-hand side of (7.7) is

$$\begin{aligned}-\epsilon_0\int_{R<R_0}\nabla\cdot\nabla G dV &= -\epsilon_0\oint_{R=R_0}\nabla G\cdot\hat{\boldsymbol{n}}dS \\ &= -\epsilon_0\left(-\frac{a_2}{R_0^2}\right)\cdot 4\pi R_0^2 = 4\pi\epsilon_0 a_2 \end{aligned}\tag{7.9}$$

This must be equal to the integral of the right-hand side (the enclosed charge), which is unity by definition. Therefore, $a_2 = 1/4\pi\epsilon_0$, so the Green's function for 3D electrostatics is

$$G(\boldsymbol{r},\boldsymbol{r}') = \frac{1}{4\pi\epsilon_0|\boldsymbol{r}-\boldsymbol{r}'|}. \tag{7.10}$$

To be precise, we add that the Green's function derived here is the one valid for free space, with no boundaries. The Green's function can also be defined for cases with conductors and dielectrics, but then one needs more elaborate methods to calculate it.

Assuming that all the charges reside on the surfaces of conductors, the potential can be written as

$$\phi(\boldsymbol{r}) = \int_{\text{Conductors}} G(\boldsymbol{r},\boldsymbol{r}')\rho_s(\boldsymbol{r}')dS'. \tag{7.11}$$

The two formulations, Poisson (7.1) and Coulomb (7.2) or (7.11), are equivalent. To see that, we apply the Laplace operator to (7.11), and use $\nabla_r^2 G(\boldsymbol{r},\boldsymbol{r}') = -\delta^3(\boldsymbol{r}-\boldsymbol{r}')/\epsilon_0$ to verify that the potential satisfies Poisson's equation (7.1)

$$\begin{aligned}\nabla_r^2\int G(\boldsymbol{r},\boldsymbol{r}')\rho(\boldsymbol{r}')dV' &= \int[\nabla_r^2 G(\boldsymbol{r},\boldsymbol{r}')]\rho(\boldsymbol{r}')dV' \\ &= -\frac{1}{\epsilon_0}\int\delta^3(\boldsymbol{r}-\boldsymbol{r}')\rho(\boldsymbol{r}')dV' = -\frac{\rho(\boldsymbol{r})}{\epsilon_0}.\end{aligned}$$

Therefore, the integral formulation (7.11) is equivalent to Poisson's equation.

7.1.2 General Formulation

After having formulated the electrostatic potential problem as an integral equation, we can formalize the idea to a more general problem.

Consider a differential equation

$$Df = s, \tag{7.12}$$

where D is a differential operator, f is a field, and s is the source distribution. Let $G(\boldsymbol{r}, \boldsymbol{r}')$ be the field at $\boldsymbol{r}$ produced by a point source at $\boldsymbol{r}'$, that is, G satisfies

$$D_r G(\boldsymbol{r}, \boldsymbol{r}') = \delta^3(\boldsymbol{r} - \boldsymbol{r}'). \tag{7.13}$$

By the principle of superposition, which holds for linear systems, the differential equation (7.12) can be rewritten as the integral equation

$$f(\boldsymbol{r}) = \int G(\boldsymbol{r}, \boldsymbol{r}') s(\boldsymbol{r}') dV'. \tag{7.14}$$

Direct substitution shows that (7.14) is a solution to (7.12).

The integral formulation is efficient when the sources reside on small surfaces, and it deals very easily with problems in "open" geometry, where differential equation solvers have difficulties.

7.1.3 FEM Solution

Usually some parts of finite element methodology are used for solving the integral equation. The procedures will be outlined in this section.

7.1.3.1 Basis Functions

The charge distribution is expanded in, say, N basis functions $s_k(\boldsymbol{r})$:

$$\rho_s(\boldsymbol{r}) = \sum_{k=1}^{N} a_k s_k(\boldsymbol{r}). \tag{7.15}$$

In early applications of the MoM, the basis functions were often chosen as global functions, and one tried to use as much knowledge of the solution as possible to find expansions that gave accurate results with a small number of basis functions (sometimes only 1!). Nowadays, it is more common to divide the surfaces with sources into small elements and use local basis functions. This requires less knowledge and works for much more general problems.

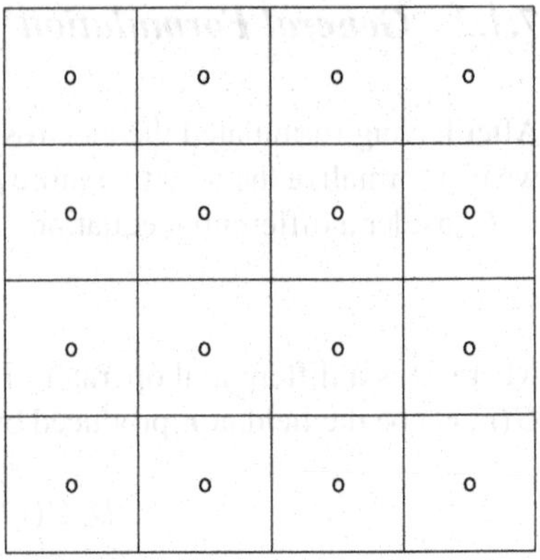

Fig. 7.2 Suitable 2D grid for MoM solution of an electrostatics problem. The charge density can be expanded in piecewise constants, and the matching points (o) can be placed at the center of each element

For convenience of notation, we introduce the potential generated by a basis function:

$$\phi_k(\boldsymbol{r}) = \int G(\boldsymbol{r}, \boldsymbol{r}')s_k(\boldsymbol{r}')dS'. \tag{7.16}$$

Then, the approximate potential becomes

$$\bar{\phi}(\boldsymbol{r}) = \sum_{k=1}^{N} a_k \phi_k(\boldsymbol{r}). \tag{7.17}$$

7.1.3.2 Testing Procedures

We want to enforce the condition $\bar{\phi} = \phi_{\text{spec}}$ on the conducting surfaces where the potential is known; that is, minimize the residual $r = \sum_k a_k \phi_k - \phi_{\text{spec}}$ on the conductors. Two methods are commonly used for minimizing the residual.

- *Point matching*, also known as *collocation* and the *Nystrom method.* Choose testing points $\boldsymbol{r}_j$, $j = 1, 2, \ldots, N$ (as many as the basis functions), and impose

$$\bar{\phi}(\boldsymbol{r}_j) = \phi_{\text{spec}}(\boldsymbol{r}_j), \quad j = 1, 2, \ldots, N. \tag{7.18}$$

  To get a well-behaved scheme, the testing points should be chosen so that each feels mainly the effects of one particular basis function. If this criterion is not fulfilled, the computed charge distribution may show a spurious oscillatory behavior, simply because the oscillating components of the charge distribution are not detected at the observation points. A good recipe for electrostatics is to choose piecewise constant basis functions and place the collocation points in the middle of each element, as shown in Fig. 7.2.
- *Weighted residuals.* Choose weighting functions w_j, $j = 1, 2, \ldots, N$ (as many as the basis functions), and impose

$$\int_{\text{Conductor}} w_j(\mathbf{r})[\bar{\phi}(\mathbf{r}) - \phi_{\text{spec}}(\mathbf{r})]dS = 0. \tag{7.19}$$

Here Galerkin's method uses $w_j(\mathbf{r}) = s_j(\mathbf{r})$. If we use global basis functions, as was common practice in the early applications of the boundary element method, Galerkin's weighting procedure (7.19) can be seen as a way of taking moments of the mismatch in the potential. This is why the electromagnetics community usually refers to the boundary element method as the MoM. Point matching corresponds to taking the test functions as delta functions $w_j(\mathbf{r}) = \delta(\mathbf{r} - \mathbf{r}_j)$.

The integrations required for (7.18) and (7.19) are generally done numerically, and the singularity of the Green's function at $\mathbf{r} = \mathbf{r}'$ needs particular attention.

Both collocation and the method of weighted residuals lead to an $N \times N$ system of equations

$$\sum_{k=1}^{N} A_{jk}a_k = b_j, \quad j = 1, 2, \ldots, N,$$

$$A_{jk} = \int w_j(\mathbf{r})\phi_k(\mathbf{r})dS = \int dS\, w_j(\mathbf{r}) \int dS'\, G(\mathbf{r}, \mathbf{r}')s_k(\mathbf{r}'),$$

$$b_j = \int w_j(\mathbf{r})\phi_{\text{spec}}(\mathbf{r})dS. \tag{7.20}$$

For the self-adjoint Poisson's equation, the Green's function is symmetric, $G(\mathbf{r}, \mathbf{r}') = G(\mathbf{r}', \mathbf{r})$, which is referred to as *reciprocity*. If one uses Galerkin's method to construct the MoM equations in (7.20), the matrix also becomes symmetric, i.e., $A_{jk} = A_{kj}$.

Review Questions

7.1-1 Compare integral formulations with differential equation formulations. Mention some pros and cons of integral formulations.

7.1-2 Give an example of suitable weighting and basis functions for (7.3).

7.1-3 What is a Green's function?

7.1-4 Derive the Green's function for Poisson's equation in 3D free space.

7.1-5 Why does the electromagnetics community refer to boundary element methods as method of moments?

7.1-6 Generalize the technique for square elements, demonstrated in Sect. 7.1.3, to a discretization that consists of triangles. Is it possible to combine squares and triangles? Could such a combination be useful?

7.1-7 What is the difference between point matching and weighted residuals?

7.2 Capacitance Problem in an Unbounded 2D Region

We will illustrate the MoM by solving a simple problem: calculate the capacitance per unit length of two equal and parallel conducting strips in free space, as illustrated in Fig. 7.3. The MoM is particularly useful for this open geometry.

To set up the equations for a 2D geometry, we note that the potential from a line charge at $\boldsymbol{r}' = (x', y')$, with line charge density density ρ_l (Coulomb/meter), is

$$\phi = -\frac{\rho_l}{2\pi\epsilon_0}\ln\frac{|\boldsymbol{r}-\boldsymbol{r}'|}{r_0},$$

where r_0 is an arbitrary constant. For the parallel plate capacitor, this gives

$$\begin{aligned}\phi(x,y) = &-\frac{1}{2\pi\epsilon_0}\int_{-w/2}^{w/2}\rho_s\left(x',\frac{a}{2}\right)\ln\sqrt{(x-x')^2+\left(y-\frac{a}{2}\right)^2}\,dx' \\ &-\frac{1}{2\pi\epsilon_0}\int_{-w/2}^{w/2}\rho_s\left(x',-\frac{a}{2}\right)\ln\sqrt{(x-x')^2+\left(y+\frac{a}{2}\right)^2}\,dx' \quad (7.21)\end{aligned}$$

This particular problem has two symmetries, both left–right symmetry,

$$\rho_s(-x', a/2) = \rho_s(x', a/2),$$

and up–down antisymmetry

$$\rho_s(x', -a/2) = -\rho_s(x', a/2).$$

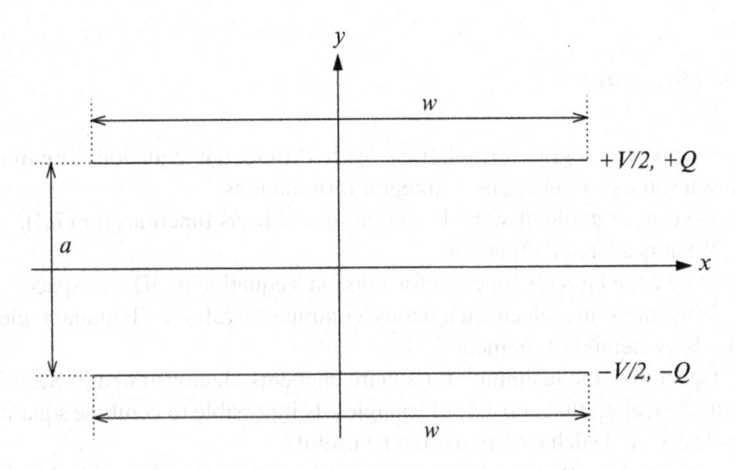

Fig. 7.3 Cross section of the capacitor

7.2.1 Integration

We divide each capacitor plate into elements $x' \in [x_i, x_{i+1}]$ and use piecewise constant basis functions to represent the charge density. The testing will be done as point matching at the midpoints of each element $x_{\text{test},i} = x_{i+\frac{1}{2}} = \frac{1}{2}(x_i + x_{i+1})$. This gives a good coupling between each basis function, which is constant on an element, and the corresponding testing point. If we chose the testing points as the nodes, they would not be able to detect the potential resulting from a charge distribution where neighboring elements have opposite charges, because contributions from two adjacent elements cancel at a node on the element boundary.

To get the potential from a piecewise constant charge distribution, we need to integrate. The singular kernel complicates the integration over the element on which the observation point is located, but the piecewise constant elements in 2D allow an exact analytical integration

$$\begin{aligned} I(x_s, x_e, d) &= -\frac{1}{2\pi\epsilon_0} \int_{x_s}^{x_e} \ln \sqrt{x^2 + d^2}\, dx \\ &= -\frac{1}{2\pi\epsilon_0} \left[\frac{1}{2} x \ln(x^2 + d^2) - x + d \arctan(x/d) \right]_{x_s}^{x_e} . \end{aligned} \tag{7.22}$$

This simplification is helpful, and we will use it. If we take into account the left–right symmetry and up–down antisymmetry, it is enough to discretize only the right half of the upper plate. We divide this into N elements with endpoints x_i, $i = 0, 1, 2, \ldots, N$. Then the potential at the point (x, y) from the assumed charge distribution can be written as

$$\begin{aligned} \phi(x, y) = \sum_{i=0}^{N-1} \rho_{i+\frac{1}{2}} \; [& I(x_i - x, x_{i+1} - x, y - a/2) \\ & + I(-x_{i+1} - x, -x_i - x, y - a/2) \\ & - I(x_i - x, x_{i+1} - x, y + a/2) \\ & - I(-x_{i-1} - x, -x_i - x, y + a/2)] . \end{aligned} \tag{7.23}$$

By choosing the testing points as $x_{i+\frac{1}{2}}$ for $i = 0, 1, \ldots, N-1$, on the upper plate we get the system of equations

$$\mathbf{A}\mathbf{r} = \mathbf{v},$$

where

$$\begin{aligned} A_{ij} = & \; I(x_j - x_{i+\frac{1}{2}}, x_{j+1} - x_{i+\frac{1}{2}}, 0) + I\left(-x_{j+1} - x_{i+\frac{1}{2}}, -x_j - x_{i+\frac{1}{2}}, 0\right) \\ & - I(x_j - x_{i+\frac{1}{2}}, x_{j+1} - x_{i+\frac{1}{2}}, a) - I\left(-x_{j+1} - x_{i+\frac{1}{2}}, -x_j - x_{i+\frac{1}{2}}, a\right), \end{aligned}$$

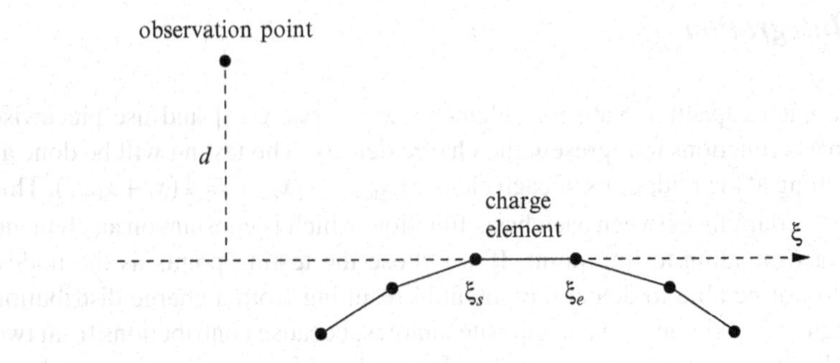

Fig. 7.4 Coordinates aligned with an element

and **v** is a column vector where all the elements are set to the potential on the upper plate $V/2$, where V is the voltage across the capacitor. Solution of this system will give the charge density on each element in the vector **r**.

7.2.2 MATLAB: MoM for General, 2D Geometries

In the introductory example, we treated a very simple geometry, with a high degree of symmetry and plane plates. However, it is easy to generalize this to a completely general 2D geometry with no symmetry and curved conductors. Fig. 7.4 shows one element and the observation point, which is assumed to lie at the normal distance d from a straight-line extension of the element. The contribution from this element to the potential at the observation point is

$$-\frac{\rho_s}{2\pi\epsilon_0}\int_{\xi_s}^{\xi_e} \ln\sqrt{x^2+d^2}\,dx = -\frac{\rho_s}{2\pi\epsilon_0}\left[\frac{1}{2}x\ln(x^2+d^2) - x + d\arctan(x/d)\right]_{\xi_s}^{\xi_e}. \tag{7.24}$$

In the following routine, we use the exact integration to generate the system matrix for point matching and general 2D geometry. Each charge-carrying element is specified by the arrays `xs` and `ys` for the starting coordinates, `xe` and `ye` for the endpoints, and `phi` for the potential. No assumption about the geometry of the plates is used.

```
% ------------------------------------------------------------------
% Compute charge distribution for 2D electrostatics by MoM
% ------------------------------------------------------------------
function [charge, sigma] = MoM2D(xs, ys, xe, ye, phi)

% Arguments:
%     xs      = x-coordinate for starting points
%     ys      = y-coordinate for starting points
%     xe      = x-coordinate for ending points
```

```
%    ye     = y-coordinate for ending points
%    phi    = the potential
% Returns:
%    sigma  = charge density for each element
%    charge = total charge on each element

xobs = 0.5*(xs + xe);                    % Observation points
yobs = 0.5*(ys + ye);
h = sqrt((xe-xs).^2 + (ye-ys).^2);   % Length of elements

% Loop over elements
for k = 1:length(xs)
  s = (   (xobs-xs(k))*(xe(k)-xs(k)) ...
       + (yobs-ys(k))*(ye(k)-ys(k)))/h(k)^2;
  d = sqrt(   (xobs-xs(k)).^2 ...
          + (yobs-ys(k)).^2 ...
          - s.^2*h(k)^2 + 1e-24);
  xis = -s*h(k);
  xie = (1-s)*h(k);
  temp =  0.5*xie.*log(xie.^2+d.^2) ...
             - xie + d.*atan(xie./d) ...
        -(0.5*xis.*log(xis.^2+d.^2) ...
             - xis + d.*atan(xis./d));
  A(:,k) = - temp(:)/(2*pi*8.854187);
end

sigma = (A\phi')';      % Charge density
charge = h.*sigma;      % Charge per element
```

[The theory behind the geometrical transformations is that a point on the straight line through $\boldsymbol{r}_s = (x_s, y_s)$ and $\boldsymbol{r}_e = (x_e, y_e)$ is $\boldsymbol{r} = \boldsymbol{r}_s + s(\boldsymbol{r}_e - \boldsymbol{r}_s)$, $-\infty < s < \infty$. The minimum distance d on this line to the observation point at $\boldsymbol{r}_o$ occurs for $s = (\boldsymbol{r}_o - \boldsymbol{r}_s)\cdot(\boldsymbol{r}_e - \boldsymbol{r}_s)/|\boldsymbol{r}_e - \boldsymbol{r}_s|^2$ and it is given by $d^2 = |\boldsymbol{r}_o - \boldsymbol{r}_s|^2 - [(\boldsymbol{r}_o - \boldsymbol{r}_s)\cdot(\boldsymbol{r}_e - \boldsymbol{r}_s)]^2/|\boldsymbol{r}_e - \boldsymbol{r}_s|^2$.]

The routine gives the charge on the elements, and this can be summed to compute the capacitance per meter. For this example, we initiate the potential to 0.5 V on the top plate and −0.5 V on the bottom one. Then the capacitance is the sum of the charges on the top plate. The computation can be called as follows (where n must be an even integer):

```
a = 1;              % Separation distance between capacitor plates
w = 1;              % Width of capacitor plates
n  = 10;            % Number of unknowns
nh = round(n/2);    % Number of elements on each plate
h  = a/nh;          % Length of the elements

% X-coordinates for starting and ending points
xs = zeros(1,n);
xe = zeros(1,n);
xs(1:nh)          = linspace(0,a-h,nh);
xs(nh+1:2*nh)     = linspace(0,a-h,nh);
xe                = xs + h;
```

Table 7.1 Capacitance for $a = w = 1$, uniform grid and analytic integration

n [-]	h [m]	C [pF/m]
10	0.20000	18.03138 50
20	0.10000	18.37294 02
30	0.06666	18.49101 21
50	0.04000	18.58699 26
70	0.02857	18.62854 17
100	0.02000	18.65986 68
140	0.01428	18.68082 79
200	0.01000	18.69658 95

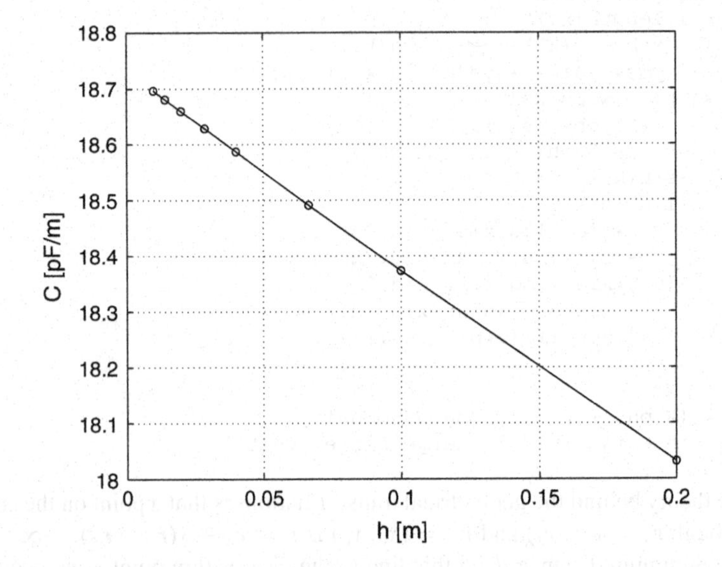

Fig. 7.5 Capacitance for $a = w = 1$, uniform grid and analytic integration, plotted versus h

```
% Y-coordinates for starting and ending points
ys = zeros(1,n);
ye = zeros(1,n);
ys(1:nh)          = 0.5*w;
ys(nh+1:2*nh)     = -0.5*w;
ye                = ys;

% Potential for the elements
V                 = zeros(1,n);
V(1:nh)           = 0.5;
V(nh+1:2*nh)      = -0.5;

% Solve the electrostatic problem
[charge, sigma] = MoM2D(xs, ys, xe, ye, V);
C = sum(charge(1:nh))
```

The results from runs with varying numbers of points are shown in Table 7.1. Fig. 7.5 shows that the convergence is linear in h.

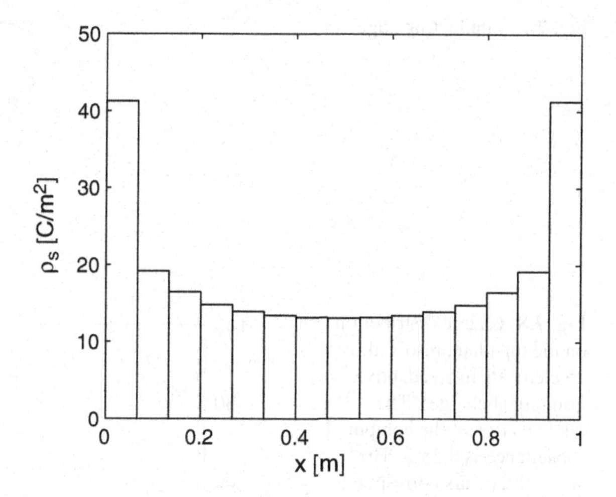

Fig. 7.6 Charge distribution on the top plate, resolved by 15 elements in a uniform grid. The relative error of the computed capacitance is 1.3%

Accurate values can be obtained from extrapolation using polynomial fits. A linear fit gives $C = 18.72858\,78$ (pF/m), quadratic 18.73349 99, cubic 18.73350 34, quartic 18.73350 27, and quintic 18.73350 27. The answer to nine digits is 18.73350 27 pF/m. For a single computation to get to within 1% of the correct answer, about 50 elements are needed.

7.2.3 *Charge Distribution*

The charge distribution on the top plate, resolved with 15 elements, is shown in Fig. 7.6.

The charge distribution for the parallel plate capacitor is singular. In this respect it is similar to the capacitance problem in Chap. 3. The nature of such singularities can be determined analytically. As an analytically solvable illustration, we consider the behavior of the electrostatic potential in the vicinity of a conductor edge in vacuum, that is, a 2D corner.

Suppose the conductor subtends an angle $\beta < 180^{\circ}$, and the vacuum region, where the potential satisfies Laplace's equation, subtends the angle $\alpha = 360^{\circ} - \beta > 180^{\circ}$; see Fig. 7.7. In cylindrical coordinates, with the edge oriented along the z-axis, the potential ϕ satisfies

$$\frac{1}{r}\frac{\partial}{\partial r}r\frac{\partial \phi}{\partial r} + \frac{1}{r^2}\frac{\partial^2 \phi}{\partial \theta^2} = 0, \quad 0 < \theta < \alpha, \tag{7.25}$$

and $\phi = 0$ for $\theta = 0, \alpha$. Relevant solutions can be found by the method of separation of variables $\phi(r, \theta) = f(r)g(\theta)$. Substituting this ansatz into Laplace's equation (7.25) and multiplying by $r^2/(f(r)g(\theta))$, we obtain

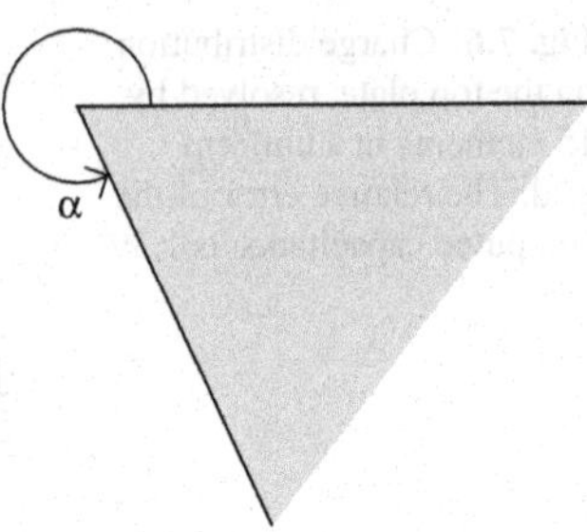

Fig. 7.7 Conducting edge

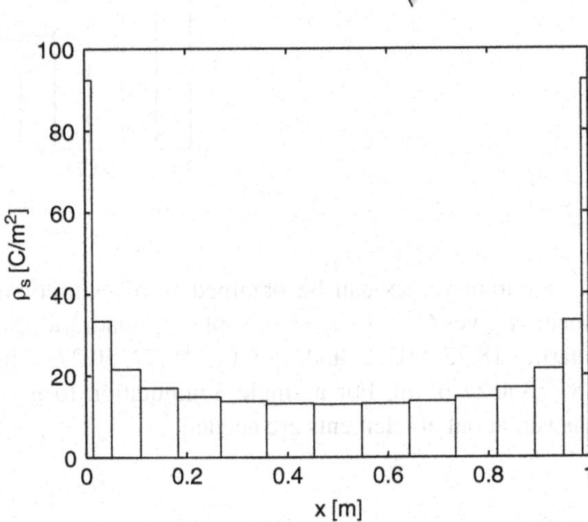

Fig. 7.8 Charge distribution on the top plate, resolved by 15 elements in an adaptive grid (equal charge). The relative error of the computed capacitance is 0.28%. The areas of the bars correspond to the charges on the corresponding elements

$$\frac{r(rf'(r))'}{f(r)} = -\frac{g''(\theta)}{g(\theta)}.$$

Since the left-hand side depends only on r and the right-hand side only on θ, both must be constant, say p^2. This gives $g(\theta) = a \sin p\theta + b \cos p\theta$ and $f(r) = cr^p + dr^{-p}$. If $p > 0$ we must choose $d = 0$ to keep the potential bounded. Thus, the acceptable solutions of separable form are

$$\phi = (a \sin p\theta + b \cos p\theta)r^p.$$

Next, we want to determine the power p. The boundary condition $\phi = 0$ at $\theta = 0$ gives $b = 0$, and $\phi = 0$ at $\theta = \alpha$ then gives $p\alpha = n\pi$, $n =$ integer. Thus, the lowest-order solution is $\phi = r^p \sin p\theta$ with $p = \pi/\alpha$. For a general opening angle α, the power p is noninteger and the smallest p is less than one if $\alpha > \pi$. For this solution, both E_r and E_θ vary as $r^{(-1+\pi/\alpha)}$. Thus, the field components tend to infinity at the corner if $\alpha > \pi$. For the edge of the capacitor plate we have $\alpha = 2\pi$, so that $E_\theta \propto r^{-1/2}$. This implies that the charge density on the plate varies as $r^{-1/2}$ near the edge.

Table 7.2 Capacitance for $a = w = 1$ and adaptive mesh

n [-]	h [m]	C [pF/m]
10	0.20000	18.32465 80
20	0.10000	18.61846 85
30	0.06666	18.68061 49
50	0.04000	18.71396 25
70	0.02857	18.72342 35
100	0.02000	18.72852 34
140	0.01428	18.73094 84
200	0.01000	18.73224 60

7.2.4 *Adaptivity*

We will use the parallel plate capacitor to illustrate the benefits of adaptive grid refinement. The elements in the middle of the strips, where the charge density is small, give small contributions to the total charge and capacitance. Some of these elements would be more efficiently used near the edges, where the charge density is high. A simple rule of thumb, which works well for adjusting the length of an element in an adaptive grid, is that the total charge on each element should be the same.

We initialize the computation with a grid where the elements have equal length to compute a first approximation. Then, the computed charge distribution can be used to generate a new grid where one seeks to distribute the charge uniformly on the elements. Such a routine is easy to implement, however, the procedure needs to be iterated several times to equalize the charge on the elements enough for a careful convergence study. The adaptively computed capacitance values are given in Table 7.2.

A plot versus h^2 shows that the adaptivity has restored the $O(h^2)$ [i.e., $O(N^{-2})$] convergence that one expects for a smooth charge distribution. Now we get 1% accuracy with fewer than 20 elements, compared to about 50 for a uniform grid. On the other hand, the calculation for each cell size had to be repeated several times to adapt the grid, so we have not really won in terms of computing time. The main use of adaptivity is in large 3D problems, where sufficient accuracy cannot be obtained without adaptivity. Another approach, which may minimize the computing time, is handmade adaptivity, where one uses knowledge about the geometry and the singularities to construct meshes that resolve the solution as well as possible with the available number of elements.

Even though the lowest order error for the adaptive grid is proportional to h^2, the extrapolations based on fitting the computed results to polynomials in h^2 are not very accurate. The reason for this is that the power series for the adaptive results also contains odd powers of h, such as h^3 and h^5. If we fit the results versus polynomials in h, quadratic extrapolation gives 18.73732 85, quartic 18.73351 51, and sixth-order 18.73350 26. The adaptive grid strongly improves the accuracy for a given number of elements, but in fact, the extrapolated results are somewhat less accurate than for a uniform grid.

7.2.5 Numerical Integration

As an alternative to exact analytical integration, one can use numerical integration. Then, the logarithmic singularity causes difficulties, and there are several possibilities to deal with this. Letting x represent points at the middle of an element, we could choose:

- Midpoint integration: $\int_{x-h/2}^{x+h/2} f(x')dx' \approx hf(x)$. This diverges for the "self contribution" where the observation point is the midpoint of the element $x_{obs} = x$.
- Trapezoidal rule: $\int_{x-h/2}^{x+h/2} f(x')dx' \approx \frac{1}{2}h[f(x-h/2)+f(x+h/2)]$ (relative error $O(h^2)$ for regular functions). However, this gives a large error for $f(x) = \ln x$.
- Gaussian integration: $\int_{x-h/2}^{x+h/2} f(x')dx' \approx \frac{1}{2}h[f(x_1)+f(x_2)]$, where $x_{1,2} = x \pm (h/2)/\sqrt{3}$, error $O(h^4)$ for regular functions. This, too, gives a large error if $f(x) = \ln x$.
- Special integration for a logarithmic singularity

$$\int_{x-h/2}^{x+h/2} f(x')dx' \approx \frac{1}{2}h[f(x_1)+f(x_2)], \quad x_{1,2} = x \pm (h/2)/e.$$

  The error is $O(h^2)$ for regular functions, and the formula is exact for $f(x) = \ln x$.

  To test these integration schemes, we compare results for the approximations

$$\int_{x-h/2}^{x+h/2} f(x)dx \approx \frac{h}{2}[f(x-\eta h/2)+f(x+\eta h/2)] \tag{7.26}$$

with different values of the parameter η. Tests show that $\eta \approx 1/e$ gives the most accurate results. Results for numerical integration and the two-strip capacitor, with $\eta = 1/e$ with and without adaptivity, are shown in Table 7.3.

For $\eta = 1/e$, the convergence on a uniform grid is close to linear in h. Polynomial fits to the results for a uniform grid in Table 7.3 gives the following extrapolations: for a linear fit 18.781, a quadratic fit 18.757, and a cubic fit 18.747. This is less accurate than for the exact integration because the integration scheme does not properly account for the contributions from neighboring cells, which are also affected by the singularity of the Green's function.

Fig. 7.9 shows the results for the analytic and numerical integration with adaptive grid refinement. Evidently, errors can come from the integration as well as from the

Table 7.3 Capacitance for $a = w = 1$ with numerical integration (7.26), $\eta = 1/e$, and with uniform and adaptive mesh

n [-]	h [m]	C [pF/m] uniform	C [pF/m] adaptive
10	0.20000	18.14722	18.48546 67
20	0.10000	18.44493	18.71508 74
30	0.06666	18.54435	18.74847 85
50	0.04000	18.62297	18.75628 29
70	0.02857	18.65609	18.75413 22
100	0.02000	18.68052	18.75026 85
140	0.01428	18.69650	18.74659 82
200	0.01000	18.70824	18.74326 22

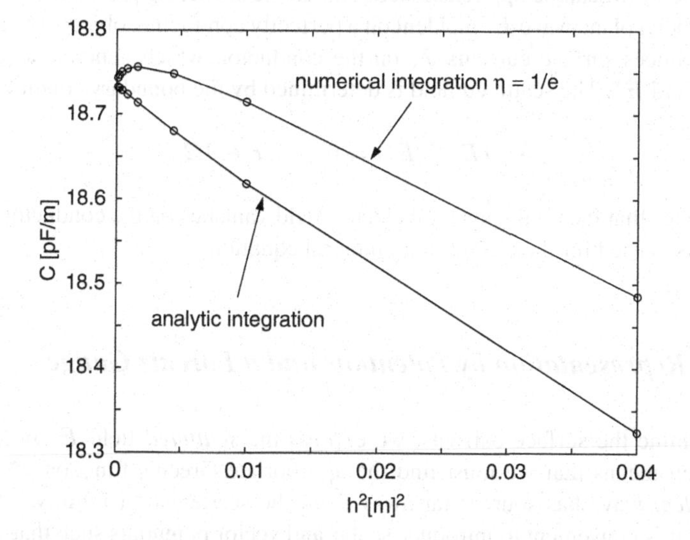

Fig. 7.9 Results for numerical and analytic integration and adaptive mesh versus h^2

expansion in finite elements, but the difference between the exact and numerical integration is rather small, about 1% on the coarsest grid.

Review Questions

7.2-1 Why is point-matching attractive for a charge distribution that is expanded in piecewise constant basis functions?

7.2-2 Derive, in two dimensions, the asymptotic behavior for the electrostatic potential and field in the vicinity of a metal corner with an opening angle α.

7.2-3 Adaptivity typically involves solving the same problem several times, which implies some additional work. Still, adaptivity is often very useful. Why?

7.2-4 Describe a simple adaptive scheme for a parallel plate capacitor problem.

7.2-5 List some integration rules that can be used for (7.21).

7.2-6 Mention an example in which numerical integration can be useful.

7.3 Electromagnetic Scattering

The MoM is frequently applied to scattering problems in the frequency domain. Electromagnetic scattering can be used for many detection applications, such as detecting aircraft by radar. A more demanding goal is to determine the properties of the scattering object from the scattered field. This is called inverse scattering, which is an important method for nondestructive testing. The MoM is also used for magnetostatics [75] and eddy current problems, for example to handle currents induced on thin conducting shells. The book of Peterson [54] gives a good account of how the MoM can be applied to electromagnetic scattering problems.

Consider a plane wave $\boldsymbol{E}^i$ incident on a perfectly conducting object. The incident wave produces surface currents $\boldsymbol{J}_s$ on the conductor, which generate a scattered electric field $\boldsymbol{E}^s$. The scattered field is determined by the boundary condition

$$\hat{\boldsymbol{n}} \times (\boldsymbol{E}^i + \boldsymbol{E}^s) = 0, \qquad \boldsymbol{r} \in \partial\Omega_c, \tag{7.27}$$

which states that the total tangential electric field vanishes on the conductor surface $\partial\Omega_c$. This is used for the electric field integral equation.

7.3.1 *Representation by Potentials and a Lorentz Gauge*

To determine the surface currents, we express the *scattered* field $\boldsymbol{E}^s$ in terms of $\boldsymbol{J}_s$, which means that we must find the appropriate Green's function. (Note that the *incident* wave has sources far away from the scatterer, "at infinity.") For this purpose, it is convenient to introduce scalar and vector potentials such that

$$\boldsymbol{E} = -\nabla\phi - \frac{\partial \boldsymbol{A}}{\partial t}, \qquad \boldsymbol{B} = \nabla \times \boldsymbol{A}. \tag{7.28}$$

With this representation, Faraday's law $\partial\boldsymbol{B}/\partial t = -\nabla\times\boldsymbol{E}$ is automatically satisfied. We substitute the potential representation (7.28) into Ampère's law

$$\nabla \times \boldsymbol{B} = \mu_0\boldsymbol{J} + \epsilon_0\mu_0\frac{\partial \boldsymbol{E}}{\partial t}. \tag{7.29}$$

Using $\nabla \times \boldsymbol{B} = \nabla \times \nabla \times \boldsymbol{A} = \nabla(\nabla \cdot \boldsymbol{A}) - \nabla^2\boldsymbol{A}$ (and assuming $\exp(j\omega t)$ time dependence), this gives

$$\nabla(\nabla \cdot \boldsymbol{A}) - \nabla^2\boldsymbol{A} = \mu_0\boldsymbol{J} - j\omega\epsilon_0\mu_0(\nabla\phi + j\omega\boldsymbol{A}). \tag{7.30}$$

As pointed out previously for eddy current problems, the potentials $\boldsymbol{A}$ and ϕ are not uniquely determined; one can always make a "gauge transformation" $\boldsymbol{A}' = \boldsymbol{A} + \nabla U$ and $\phi' = \phi - \partial U/\partial t$ without changing the physical fields $\boldsymbol{E}$ and $\boldsymbol{B}$.

To solve for the potentials uniquely, we have to specify a condition that determines the gauging potential U. This is called the *gauge condition*. One choice that makes (7.30) particularly easy to solve is the Lorentz gauge, which makes the two gradient terms in (7.30) cancel:

$$\nabla \cdot \boldsymbol{A} = -j\omega\epsilon_0\mu_0\phi. \tag{7.31}$$

Equation (7.30) with the Lorentz gauge condition (7.31) reduces to the vector Helmholtz equation

$$-\left(\nabla^2 + \frac{\omega^2}{c^2}\right)\boldsymbol{A} = \mu_0\boldsymbol{J}.$$

7.3.2 Green's Function for the Vector Potential

The Cartesian components of $\boldsymbol{A}$ satisfy *scalar* Helmholtz equations

$$-(\nabla^2 + k^2)A_i = \mu_0 J_i, \quad k = \omega/c, \tag{7.32}$$

which can be solved component by component. Here, the subindex i is x, y, or z. The Helmholtz equation (7.32) is similar to Poisson's equation, for which we derived the integral representation in Sect. 7.1. We proceed in similar ways here.

We define the Green's function for the vector potential $G(\boldsymbol{r},\boldsymbol{r}')$ as the ith component of the vector potential produced by a "point current" in the ith direction $\boldsymbol{J} = \hat{\boldsymbol{x}}_i\delta^3(\boldsymbol{r}-\boldsymbol{r}')$. Then, G satisfies

$$-\frac{1}{\mu_0}\left(\nabla_r^2 + k^2\right)G(\boldsymbol{r},\boldsymbol{r}') = \delta^3(\boldsymbol{r}-\boldsymbol{r}'). \tag{7.33}$$

The vector potential constructed by superposition

$$A_i(\boldsymbol{r}) = \int G(\boldsymbol{r},\boldsymbol{r}')J_i(\boldsymbol{r}')dV' \tag{7.34}$$

then satisfies the Helmholtz equation (7.32).

The derivation of the Green's function closely parallels that in electrostatics. We start by noting that $G(\boldsymbol{r},\boldsymbol{r}')$ can depend only on the distance between the source and observation points $R = |\boldsymbol{r}-\boldsymbol{r}'|$. Therefore, in three dimensions, (7.33) gives

$$-\frac{1}{\mu_0}\left(\frac{1}{R^2}\frac{d}{dR}R^2\frac{dG}{dR} + k^2 G\right) = 0, \quad R > 0.$$

It is easy to verify that two independent solutions of this equation are $G_1 = \exp(jkR)/R$ and $G_2 = \exp(-jkR)/R$. When these are combined with the assumed $\exp(j\omega t)$ time dependence, G_1 produces constant phase surfaces such that $kR+\omega t$ is constant, or $dR/dt = -\omega/k = -c$. That is, the constant phase surfaces move *towards* the source with the speed of light. Thus G_1 represents incoming waves, which are absorbed by the "source" currents, and these waves are called *advanced* solutions. Although they are indeed solutions of Maxwell's equations, they do not respect the principle of causality, and are not of physical interest. For $G_2 \propto \exp(-jkR)/R$, on the other hand, the constant phase surfaces satisfy $dR/dt = \omega/k = c$, so G_2 represents waves radiated *away* from the source. These solutions respect causality and are called *retarded*. They are the relevant solutions to (7.33). Thus, we pick $G(\boldsymbol{r}, \boldsymbol{r}') = a\exp(-jkR)/R$. To determine the normalization constant a, we proceed as in Sect. 7.1.1. Integrate (7.33) over a sphere of radius R_0, and to simplify the evaluation, we let R_0 tend to zero. The integral of the left-hand side becomes

$$-\frac{1}{\mu_0}\int_{R<R_0} (\nabla_r^2 + k^2)\, G\, dV = -\frac{1}{\mu_0}\left[\int_{R<R_0} \nabla_r \cdot \nabla_r G\, dV + O(k^2 R_0^2)\right].$$

Only the first term remains nonzero in the limit $R_0 \to 0$. By Gauss's theorem, this piece can be rewritten as a surface integral

$$\begin{aligned}
-\frac{1}{\mu_0}\oint_{R=R_0} \nabla G \cdot \hat{\boldsymbol{n}}\, dS &= -\frac{1}{\mu_0}\frac{dG}{dR}\bigg|_{R=R_0} 4\pi R_0^2 \\
&= -\frac{a}{\mu_0}\left(-\frac{1}{R_0^2} - \frac{jk}{R_0}\right)\exp(-jkR_0)4\pi R_0^2 \\
&\to \frac{4\pi a}{\mu_0}, \quad \text{as} \quad R_0 \to 0.
\end{aligned}$$

This must be equal to the integral over the right-hand side in (7.33), which is 1 by definition. Therefore, the normalizing coefficient is $a = \mu_0/4\pi$, and the Green's function for the vector potential is

$$G = \frac{\mu_0}{4\pi}\frac{\exp(-jkR)}{R}, \quad R = |\boldsymbol{r} - \boldsymbol{r}'|. \tag{7.35}$$

Using superposition and the fact that all currents occur on the surfaces of conductors, we can write the solution of (7.32) for each component of the vector potential as

$$A_i(\boldsymbol{r}) = \int_{\partial\Omega_c} G(\boldsymbol{r}, \boldsymbol{r}')\, \hat{\boldsymbol{x}}_i \cdot \boldsymbol{J}_s(\boldsymbol{r}')\, dS'$$

with the Green's function (7.35), where $\hat{\boldsymbol{x}}_i \cdot \boldsymbol{J}_s$ is component i of the surface current $\boldsymbol{J}_s$. Therefore, the full vector potential is

$$\boldsymbol{A}(\boldsymbol{r}) = \frac{\mu_0}{4\pi}\int_{\partial\Omega_c} \frac{\exp(-jkR)}{R}\boldsymbol{J}_s(\boldsymbol{r}')dS'. \tag{7.36}$$

We can find an equation for the scalar potential ϕ by taking the divergence of Ampère's law (7.29), substituting the potential representation for $\boldsymbol{E}$, and using the Lorentz gauge condition

$$-(\nabla^2 + k^2)\phi = \frac{j}{\omega\epsilon_0}\nabla \cdot \boldsymbol{J} = \frac{\rho}{\epsilon_0}. \tag{7.37}$$

Here, we used the equation of continuity for charge

$$j\omega\rho + \nabla \cdot \boldsymbol{J} = 0.$$

Equation (7.37) is again a scalar Helmholtz equation with the solution

$$\phi(\boldsymbol{r}) = \frac{1}{4\pi\epsilon_0}\int_{\partial\Omega_c} \frac{\exp(-jkR)}{R}\rho(\boldsymbol{r}')dS'. \tag{7.38}$$

7.3.3 The Electric Field Integral Equation

We now have expressions for the potentials in terms of the surface currents. The scattered electric field is given by

$$\begin{aligned} \boldsymbol{E}^s &= -j\omega\boldsymbol{A} - \nabla\phi \\ &= -\frac{j\omega\mu_0}{4\pi}\int_{\partial\Omega_c} \frac{\exp(-jkR)}{R}\boldsymbol{J}_s(\boldsymbol{r}')dS' \\ &\quad -\frac{j}{4\pi\epsilon_0\omega}\nabla\int_{\partial\Omega_c} \frac{\exp(-jkR)}{R}\nabla' \cdot \boldsymbol{J}_s(\boldsymbol{r}')dS'. \end{aligned} \tag{7.39}$$

The condition the surface currents have to satisfy is that the tangential component of the total field, which is the sum of the incident field and the scattered field generated by the surface currents, vanish on the surface of the conductor:

$$\boldsymbol{E}^s_{\tan} + \boldsymbol{E}^i_{\tan} = \boldsymbol{0}. \tag{7.40}$$

Combining this with (7.39), we obtain the electric field integral equation (EFIE)

$$\begin{aligned} \boldsymbol{E}^i_{\tan} &= \frac{j\omega\mu_0}{4\pi}\int_{\partial\Omega_c} \frac{\exp(-jkR)}{R}\boldsymbol{J}_s(\boldsymbol{r}')dS'\bigg|_{\tan} \\ &\quad +\frac{j}{4\pi\epsilon_0\omega}\nabla\int_{\partial\Omega_c} \frac{\exp(-jkR)}{R}\nabla' \cdot \boldsymbol{J}_s(\boldsymbol{r}')dS'\bigg|_{\tan}. \end{aligned} \tag{7.41}$$

Unfortunately, integral equations such as the EFIE are somewhat difficult to solve numerically. First of all, as will be discussed in Sect. 7.4, it is necessary to take proper account of the singularity in the Green's function to get a scheme that converges to the correct answer when the resolution is increased. The presence of this singularity causes difficulties for the integration. Numerical integration schemes that work well for smooth integrands can give very inaccurate results, and in practice, the singularity needs special treatment. One successful approach is to pull out some simplified part of the Green's function that contains the singularity and use an analytic integration for this part. The remaining, nonsingular, part of the Green's function can then be integrated by a standard numerical integration formula.

A more physics-related difficulty with the EFIE is the presence of "internal resonances." Consider a scattering problem in which the scatterer consists of a closed PEC surface, e.g., a conducting sphere. If we solve this problem using the EFIE, the integral equation has no information to tell it that the interior of the sphere is conducting. Therefore, the EFIE allows cavity eigenmodes that are internal to the sphere. At the resonance frequencies for these modes, they can be part of the solution without excitation by external sources, and the system matrix becomes singular. There is a cure for the problem of internal resonances, which consists in adding the magnetic field integral equation (MFIE) to the EFIE. The MFIE has different internal resonances than the EFIE, and with a suitable weighting of the two integral equations, all internal resonances are eliminated [54]. The summed equation is called the combined field integral equation (CFIE).

7.3.3.1 FEM Solution

To solve the EFIE (7.41) for a 3D problem using finite elements and Galerkin's method, we first need a suitable base for expanding $\boldsymbol{J}_s$:

$$\boldsymbol{J}_s(\boldsymbol{r}) = \sum_{i=1}^{N} a_i \boldsymbol{s}_i(\boldsymbol{r}). \tag{7.42}$$

To see what kind of elements are required, we work out the form of the matrix elements, which are obtained by multiplying the EFIE by a test (= basis) function $\boldsymbol{s}_i(\boldsymbol{r})$ and integrating over the PEC surfaces. We integrate the second term in (7.41) by parts and assume that no current can leave or enter the conductor, so that the edge term vanishes:

$$\begin{aligned}\int_{\partial\Omega_c} \boldsymbol{s}_i \cdot \nabla\phi \, dS &= \int_{\partial\Omega_c} [\nabla\cdot(\boldsymbol{s}_i\phi) - \phi\nabla\cdot\boldsymbol{s}_i] \, dS \\ &= \oint_{\partial\partial\Omega_c} \hat{\boldsymbol{n}}\cdot\boldsymbol{s}_i\phi \, dl - \int_{\partial\Omega_c} \phi\nabla\cdot\boldsymbol{s}_i \, dS = -\int_{\partial\Omega_c} \phi\nabla\cdot\boldsymbol{s}_i \, dS.\end{aligned}$$

This then gives the system of equations

Fig. 7.10 Rao–Wilton–Glisson basis function extending over two triangular elements

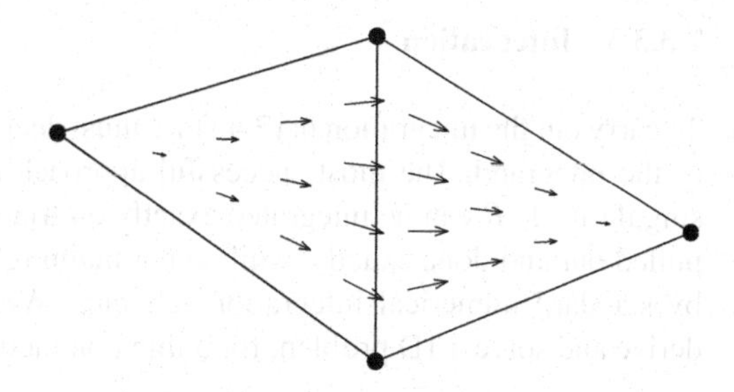

$$-\int_{\partial\Omega_c} \boldsymbol{s}_i \cdot \boldsymbol{E}^i_{\tan}\, dS = \sum_{j=1}^{N} A_{ij} a_j, \tag{7.43}$$

where the matrix elements are given by

$$A_{ij} = -\frac{j\omega\mu_0}{4\pi}\int_{\partial\Omega_c} \boldsymbol{s}_i(\boldsymbol{r}) \cdot \int_{\partial\Omega_c} \boldsymbol{s}_j(\boldsymbol{r}')\frac{\exp(-jkR)}{R} dS' dS$$
$$+ \frac{j}{4\pi\epsilon_0\omega}\int_{\partial\Omega_c} \nabla\cdot \boldsymbol{s}_i(\boldsymbol{r}) \int_{\partial\Omega_c} \nabla'\cdot \boldsymbol{s}_j(\boldsymbol{r}')\frac{\exp(-jkR)}{R} dS' dS. \tag{7.44}$$

7.3.3.2 Choice of Elements

Equation (7.44) indicates that we need basis functions for which $\nabla\cdot\boldsymbol{s}$ is nonsingular. This requirement is different from that for the differential form of Maxwell's equations, where $\nabla\times\boldsymbol{E}$ has to be square integrable. For the differential formulation of Maxwell's equations, the successful choice is curl-conforming edge elements, whose tangential component is continuous at cell interfaces. The integral formulation requires *divergence-conforming* elements, whose normal component is continuous across cell boundaries. For a 2D problem with a 1D boundary, say $\boldsymbol{J} = J_z(z)\hat{z}$, this can be achieved using piecewise linear elements. In 3D domains, with 2D boundaries, divergence-conforming elements can be constructed as the cross product of the edge elements on a surface and the surface normal $\hat{\boldsymbol{n}}$:

$$\boldsymbol{s}_{\mathrm{RWG}}(\boldsymbol{r}) = \hat{\boldsymbol{n}} \times \boldsymbol{N}(\boldsymbol{r}). \tag{7.45}$$

These are called Rao–Wilton–Glisson (RWG) elements after their inventors [58]. In polar coordinates with respect to the corner opposing the edge with which each basis function is associated, $\boldsymbol{s}_{\mathrm{RWG}}(\boldsymbol{r}) \propto r\hat{\boldsymbol{r}}$. A complete basis function extending over two triangles is shown in Fig. 7.10.

Appendix B contains information on divergence-conforming elements for triangles (see Sect. B.1.1) and quadrilaterals (see Sect. B.1.2), where both these elements can be used for the MoM formulated on surfaces.

7.3.3.3 Integration

To carry out the integration in (7.44) we must decide how to deal with the singularity of the integrand. The most successful approach [31] exploits the fact that the 3D singularity $1/R$ can be integrated exactly on triangles. Therefore, this piece can be pulled out and done exactly, while the remaining, bounded terms can be integrated by standard numerical integration schemes. We will use these considerations to derive and solve a 1D problem for a thin conducting wire in Sect. 7.4.

7.3.4 The Magnetic Field Integral Equation

The technical details of the derivation of the MFIE are somewhat subtle and lengthy as compared to the EFIE. For a complete derivation of the MFIE, the reader is referred to the literature [56, 87]. Here, we settle for stating the result for smooth PEC scatterers (that do not have sharp corners or edges)

$$-\boldsymbol{H}^{i}_{\tan} = \frac{1}{2}\hat{\boldsymbol{n}} \times \boldsymbol{J}_s(\boldsymbol{r}) + \frac{1}{4\pi} \fint_{\partial\Omega_c} \nabla\left(\frac{\exp(-jkR)}{R}\right) \times \boldsymbol{J}_s(\boldsymbol{r}')dS'\bigg|_{\tan} . \quad (7.46)$$

Here, the integral (with the bar) is evaluated in the principal-value sense [56], and it is interpreted in the following way. The domain of integration excludes an infinitesimal area around the observation point, and the contribution from the excluded area is accounted for by the term $\frac{1}{2}\hat{\boldsymbol{n}} \times \boldsymbol{J}_s(\boldsymbol{r})$. As previously mentioned, the MFIE also allows cavity eigenmodes that are internal to a conducting body. However, the MFIE has different internal resonances than the EFIE. It should be noted that the MFIE is valid only for closed surfaces, while the EFIE can be applied to both closed and open surfaces.

7.3.4.1 FEM Solution, Choice of Elements, and Integration

A FEM solution that parallels the one for the EFIE would use triangular elements. It is then useful to consider the current that flows on a single flat triangle K. We note that for the case when both the observation point $\boldsymbol{r}$ and the source point $\boldsymbol{r}'$ are located on K, both the gradient of the Green's function and the surface current density are in the plane of K, and therefore, their cross product is perpendicular to K. Since only the tangential component is included in the MFIE, the contribution from element K to the integral in (7.46) is zero when the observation point $\boldsymbol{r}$ is located on K. This is the case when the MFIE is tested, and therefore, the singularity of the integrand in the MFIE does not feature in the same way as for the EFIE. In fact, it has already been integrated analytically during the derivation of the MFIE, and it is included in the term $\frac{1}{2}\hat{\boldsymbol{n}} \times \boldsymbol{J}_s(\boldsymbol{r})$.

To solve the MFIE for a 3D problem using finite elements and Galerkin's method, we use the same basis for the current as we employed for the EFIE; i.e., the current is expanded in the RWG basis functions as shown in (7.42), and we test with $\hat{\boldsymbol{n}} \times \boldsymbol{s}_i$. This gives the system of linear equations

$$-\int_{\partial\Omega_c} (\hat{\boldsymbol{n}} \times \boldsymbol{s}_i) \cdot \boldsymbol{H}^i_{\tan} dS = \sum_{j=1}^{N} B_{ij} a_j, \tag{7.47}$$

where the matrix elements are given by

$$\begin{aligned} B_{ij} &= \frac{1}{2} \int_{\partial\Omega_c} (\hat{\boldsymbol{n}} \times \boldsymbol{s}_i) \cdot (\hat{\boldsymbol{n}} \times \boldsymbol{s}_j) dS \\ &+ \frac{1}{4\pi} \int_{\partial\Omega_c} (\hat{\boldsymbol{n}} \times \boldsymbol{s}_i) \cdot \fint_{\partial\Omega_c} \nabla \left(\frac{\exp(-jkR)}{R} \right) \times \boldsymbol{s}_j \, dS' dS. \end{aligned} \tag{7.48}$$

7.3.5 The Combined Field Integral Equation

With a suitable linear combination of (7.43) and (7.47), often referred to as the combined field integral equation (CFIE), the problems associated with internal resonances can be avoided [54]. This gives the system of linear equations

$$-\alpha \int_{\partial\Omega_c} \boldsymbol{s}_i \cdot \boldsymbol{E}^i_{\tan} dS + (1-\alpha) Z_0 \int_{\partial\Omega_c} (\hat{\boldsymbol{n}} \times \boldsymbol{s}_i) \cdot \boldsymbol{H}^i_{\tan} dS = \sum_{j=1}^{N} C_{ij} a_j, \tag{7.49}$$

where the matrix elements are given by $C_{ij} = \alpha A_{ij} - (1-\alpha) Z_0 B_{ij}$, and $0 < \alpha < 1$ is a weighting parameter.

Review Questions

7.3-1 What boundary conditions are used in the derivation of the EFIE?

7.3-2 What relation between the scalar and vector potential is used to define the Lorentz gauge? What are the consequences of this particular gauge?

7.3-3 Derive the Green's function for the scalar and vector potential for the 3D free-space case combined with the Lorentz gauge.

7.3-4 List some difficulties and useful techniques concerning the evaluation of the integrals that occur in the EFIE.

7.3-5 Describe, in words, the problems with internal resonance and mention a remedy.

7.3-6 Use the FEM to write down a system of linear equations that correspond to the EFIE. List the steps of the assembling procedure needed for this problem.
7.3-7 What basis function should be used for a PEC body treated with the EFIE and why? How does this relate to the MFIE and the CFIE?
7.3-8 What boundary conditions are satisfied by (7.36) and (7.38)?
7.3-9 Show that the matrix associated with the EFIE derived by FEM techniques and Galerkin's method is symmetric.
7.3-10 Relate the divergence-conforming and curl-conforming basis functions on triangles.
7.3-11 Why is the CFIE useful?

7.4 Scattering on Thin Wires

Here we consider scattering of electromagnetic waves by thin conducting wires. The analysis can be extended to study dipole antennas of finite length and thickness. We consider a plane wave incident on a wire of length L and radius a, aligned with the z-axis; see Fig. 7.11.

For simplicity we assume normal incidence

$$E_z^i = E_0 \exp(-jkx), \quad k = \omega/c.$$

If the wire is very thin compared with a wavelength, $ka \ll 1$, the incident wave is nearly constant, $E_z^i \approx E_0$, on the surface of the wire, and the surface current

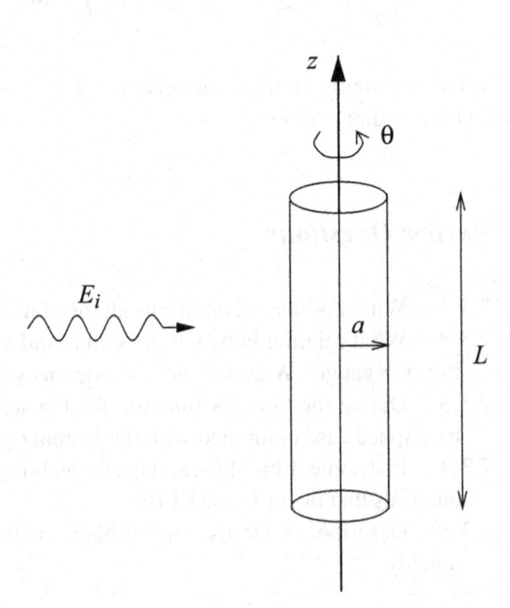

Fig. 7.11 Electromagnetic wave incident on thin wire

must be approximately $\boldsymbol{J}_s \approx J_z(z)\hat{\mathbf{z}}$. Then, (7.41) gives for the z-component of the scattered field

$$E_0 = -E_z^s = \frac{j\omega\mu_0}{4\pi} \int_{-L/2}^{L/2} \int_0^{2\pi} \frac{\exp(-jkR)}{R} J_z(z')\, a\, d\theta' dz'$$
$$+ \frac{j\omega\mu_0}{4\pi k^2} \frac{\partial}{\partial z} \int_{-L/2}^{L/2} \int_0^{2\pi} \frac{\exp(-jkR)}{R} \frac{\partial J_z(z')}{\partial z'} a\, d\theta' dz'. \quad (7.50)$$

In the integration over the wire surface, J_z is independent of θ', so the only θ'-dependence comes from R. According to the cosine theorem, the distance between two points on the wire surface satisfies

$$R^2 = (z - z')^2 + a^2 + a^2 - 2a^2 \cos(\theta - \theta') = (z - z')^2 + 4a^2 \sin^2 \frac{\theta - \theta'}{2}.$$

Carrying out the θ'-integration in (7.50), we obtain, for $|z| \le L/2$,

$$\frac{4\pi E_0}{j\omega\mu_0} = \int_{-L/2}^{L/2} G(z - z') I(z') dz' + \frac{1}{k^2} \frac{d}{dz} \int_{-L/2}^{L/2} G(z - z') \frac{dI}{dz'}(z') dz'. \quad (7.51)$$

Here $I = 2\pi a J_z$ is the total current on the surface of the wire, and the kernel of the resulting 1D integral equation is

$$G(z - z') = \frac{1}{2\pi} \int_0^{2\pi} \frac{\exp(-jkR)}{R} d\theta'. \quad (7.52)$$

7.4.1 Hallén's Equation

The 1D version of the EFIE in (7.51) can be simplified by means of a reformulation found by Hallén. Integrating the second term in (7.51) by parts and using $I(\pm L/2) = 0$ and $(d/dz')G(z - z') = -(d/dz)G(z - z')$, the equation can be written as

$$\frac{4\pi E_0}{j\omega\mu_0} = \left(1 + \frac{1}{k^2} \frac{d^2}{dz^2}\right) H, \quad (7.53)$$

$$H(z) = \int_{-L/2}^{L/2} G(z - z') I(z') dz'. \quad (7.54)$$

Equation (7.53) can be regarded as a differential equation for H, and this equation is easy to solve. Its general solution is an arbitrary homogeneous solution, for instance $4\pi E_0 / j\omega\mu_0$, added to the general solution of the homogeneous equation

$C \cos kz + D \sin kz$. When the incident wave has no z-dependence, the solution must be symmetric with respect to the midpoint of the wire. Therefore, $D = 0$, and the solution is

$$H(z) + C \cos kz = \frac{4\pi E_0}{j\omega\mu_0}, \quad |z| \leq L/2.$$

Combining this with (7.54), we obtain

$$\int_{-L/2}^{L/2} G(z-z')I(z')dz' + C \cos kz = \frac{4\pi E_0}{j\omega\mu_0}, \tag{7.55}$$

which is known as Hallén's equation. The differential order of the integral equation (7.51) has been reduced at the expense of introducing an extra constant of integration.

7.4.2 *Valid Approximation for the 1D Kernel*

As mentioned earlier, it is important to evaluate the $1/R$ singularity of the EFIE correctly, and this should be respected when we seek an expression for the 1D Green's function G. We isolate the singularity by writing

$$\frac{\exp(-jkR)}{R} = \frac{1}{R} + \frac{\exp(-jkR) - 1}{R},$$

where only the first part is singular. This gives

$$G = G_0 + G_1,$$

$$G_0 = \frac{1}{2\pi}\int_0^{2\pi} \frac{d\theta'}{R}, \quad G_1 = \frac{1}{2\pi}\int_0^{2\pi} \frac{\exp(-jkR) - 1}{R} d\theta'. \tag{7.56}$$

The advantage of the splitting is that the singular part G_0 can be evaluated exactly:

$$G_0(\zeta) = \frac{2}{\pi\sqrt{\zeta^2 + 4a^2}} K\left(\frac{4a^2}{\zeta^2 + 4a^2}\right), \quad \zeta = z - z',$$

where

$$K(m) = \int_0^{\pi/2} \frac{d\phi}{\sqrt{1 - m \sin^2 \phi}}$$

is the complete elliptic integral of the first kind. The function $G_0(\zeta)$, which contains the singular part of the 3D Green's function, is logarithmically singular when $\zeta \to 0$.

For the nonsingular part G_1, we can use less accurate approximations suitable for thin wires, such as replacing the current on the wire surface with the total current placed at the center of the wire. This means that we approximate $R \approx \sqrt{(z-z')^2 + a^2}$ in G_1, which is then straightforward to calculate. Thus, the total kernel is approximated as

$$G(\zeta) \approx \frac{2}{\pi\sqrt{\zeta^2 + 4a^2}} K\left(\frac{4a^2}{\zeta^2 + 4a^2}\right) + \frac{\exp(-jk\sqrt{\zeta^2 + a^2}) - 1}{\sqrt{\zeta^2 + a^2}}, \quad \zeta = z - z'. \tag{7.57}$$

7.4.2.1 Nonsingular Kernel Gives Spurious Solutions

If we used the approximation $R \approx \sqrt{(z-z')^2 + a^2}$ also in G_0 (that is, approximate the current on the wire surface by the same total current on its axis), the 1D kernel would lose its singularity. It can be shown that Hallén's equation (7.55) with such a smoothed kernel does not have regular solutions. If one tries to solve Hallén's equation with a nonsingular approximation for $G(z)$, the solution does not converge, but instead develops more and more short-wavelength oscillations when the resolution is increased. The reason for this is that a smooth Green's function $G(z)$ underestimates the fields created by short-wavelength currents. To create the short-wavelength components of the electric field that occur near the endpoints of the wire (for $|z| > L/2$), the smooth approximation of G requires too strong short-wavelength components in the current. As a consequence, the current density does not converge as the resolution increases.

Nevertheless, such approximations of the Green's function have been used in the past, for instance, in old versions of the NEC code, which is popular for work on thin wires. It produces acceptable results as long as the resolution in the z-direction is coarse compared to the radius of the wire, $\Delta z \gg a$. However, when the resolution is increased so that $\Delta z < a$, the current develops oscillations and the computation diverges rather than converge as the mesh is refined. This is yet another example of spurious solutions.

7.4.3 *Numerical Solution*

To evaluate the integrals in (7.54), we can either do numerical integration adapted to a logarithmic singularity, as discussed in Sect. 7.2.5, or attempt a more rigorous treatment, where the logarithmic singularity is separated out and integrated exactly. To avoid excessive work on a problem that is already an approximation, we settle for numerical integration. The elliptic integral can be accurately evaluated by using a series expansion such as given by Abramowitz and Stegun [2].

We divide the wire into elements, and expand the current in piecewise linear functions and use point matching. For piecewise linear current, the point matching of E_z should be made at the nodes, since this is where the piecewise linear basis function has its main influence. The boundary condition $I(\pm L/2) = 0$ eliminates the unknowns for I on the endpoints. To determine the constant C in (7.55) we use the condition that the equation is satisfied also at the endpoint $z = L/2$. This gives us as many conditions as we have unknowns.

7.4.4 MATLAB: Hallén's Equation

In the following routine, we use the techniques described above to solve Hallén's equation. Each current-carrying element is specified by the arrays `zs` for the starting coordinates, `ze` for the endpoints, and `E0` for the electric field.

```
% ------------------------------------------------------------------
% Compute current distribution for Hallen's equation by MoM
% ------------------------------------------------------------------
function [Iz, C, Imi] = EFIE(zs, ze, E0, a, k0)

% Arguments:
%    zs    = z-coordinate for starting points
%    ze    = z-coordinate for ending points
%    E0    = the incoming Ez and Iz the total current on
%            each element
%    a     = the radius of the wire
%    k0    = the wavenumber
% Returns:
%    Iz    = the current density along the wire
%    C     = the constant for the homogeneous solution 'cos(k0*z)'
%    Imi   = the current density on the midpoint of the wire

xi    = 0.5 - sqrt(0.25-exp(-2)); % an integration parameter
n     = length(zs) - 1;           % number of unknowns equals
                                  % the number of interior nodes


zobs  = ze;
z1    = zs + xi*(ze-zs);   % Integration points
z2    = ze + xi*(zs-ze);   % Integration points
hh    = (zs-ze)/2;

as4   = 4*a^2;             % Precomputation of constant
A     = zeros(n+1);        % System matrix

% Loop over elements
for idx = 1:n+1

  z      = zobs - z1(idx);
  zsq    = z.^2;
  za     = sqrt(zsq+a^2);
```

```
  EIK    = eval_EIK(as4./(as4 + zsq));
  temp1 = 2*EIK./(pi*sqrt(as4 + zsq)) + (exp(j*k0*za)-1)./za;

  z      = zobs - z2(idx);
  zsq    = z.^2;
  za     = sqrt(zsq+a^2);
  EIK    = eval_EIK(as4./(as4 + zsq));
  temp2 = 2*EIK./(pi*sqrt(as4 + zsq)) + (exp(j*k0*za)-1)./za;

  if (idx > 1)
    A(:,idx-1) = A(:,idx-1) ...
        + hh(idx)*((1-xi)*temp1(:) + xi*temp2(:));
  end

  if (idx < n+1)
    A(:,idx)   = A(:,idx)   ...
        + hh(idx)*(xi*temp1(:) + (1-xi)*temp2(:));
  end

end

lastrow = A(n+1,1:n);

for i = 1:n
  A(n+1,i) = 0.5*(lastrow(i)+lastrow(n+1-i));
end

A(n+1,n+1) = cos(k0*zs(1));

for i = 1:n
  A(i,n+1) = cos(k0*ze(i));
end

I   = (A\E0')';
Iz  = I(1:n);
Imi = I(round((n+1)/2));
C   = I(n+1);



% ---------------------------------------------------------------
% Evaluate the complete elliptic integral of the first kind
% by means of a polynomial approximation [M Abramowitz and
% I A Stegun, Handbook of Mathematical Functions, National
% Bureau of Standards, 1965]
% ---------------------------------------------------------------
function EIK = eval_EIK(x)

% Arguments:
%     x    = argument for K(x) in the interval 0 <= x < 1
% Returns:
%     EIK  = the value of the complete elliptic integral of
%            the first kind (with an error less than 2e-8)
```

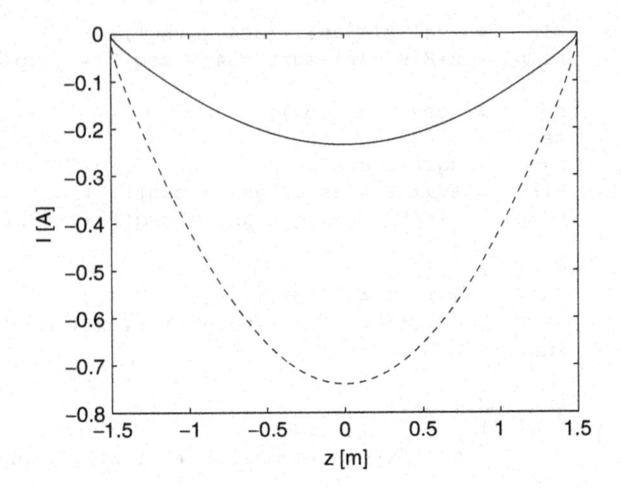

Fig. 7.12 Induced current on a wire with $kL = 3$, $ka = 0.02$: solid curve - real part and dashed curve - imaginary part

```
a   = [0.01451196212; 0.03742563713; 0.03590092383; ...
       0.09666344259; 1.38629436112];
b   = [0.00441787012; 0.03328355346; 0.06880248576; ...
       0.12498593597; 0.50000000000];

m1  = 1 - x;
EIK = polyval(a,m1)  - polyval(b,m1).*log(m1);
```

The routine computes the current distribution $I(z)$ and the constant C in Hallén's equation (7.55). Next, we present some numerical results, where, for example, Fig. 7.12 can be generated by the following script.

```
n          = 200;  % Number of cells
k0         = 1;    % Wavenumber
a          = 0.02; % Radius
L          = 3.0;  % Length
h          = L/n;  % Cell size

% Z-coordinate for starting and ending points of the segments
zs         = zeros(1,n);
ze         = zeros(1,n);
zs(1:n)    = linspace(-L/2, L/2-h, n);
ze         = zs + h;
E0         = ones(1,n);

% Solve Hallen's equation
[Iz, C, Imi] = EFIE(zs, ze, E0, a, k0);

% Plot the results
figure(1), clf
plot([zs(1) zs(1:end-1)+h/2 ze(end)], ...
     [0 real(Iz) 0], 'k-'), hold on
```

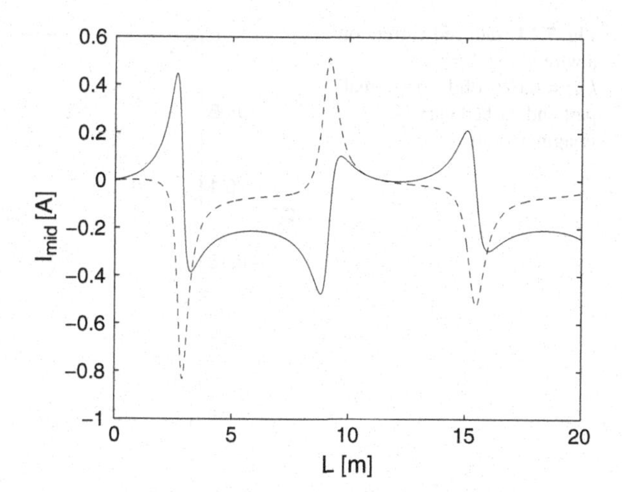

Fig. 7.13 Induced current at the midpoint of a wire as a function of L for $k = 1$, and $a = 0.02$: solid curve - real part and dashed curve - imaginary part

```
plot([zs(1) zs(1:end-1)+h/2 ze(end)], ...
     [0 imag(Iz) 0], 'k--')
xlabel('z [m]'), ylabel('I [A]')
```

7.4.5 Numerical Results

Fig. 7.12 shows the current distribution on a dipole with $kL = 3$ and $ka = 0.02$ when the dipole is resonantly excited. The calculation used the approximation (7.57) for G, which has the correct singularity. The current has steep gradients near the endpoints of the dipole, and here the charge density $\propto dI/dz$ is singular. This is similar to the singular charge distribution we found for electrostatics near the edge of the parallel plate capacitor.

Fig. 7.13 shows the induced current at the midpoint of the wire as a function of L for $k = 1$ and $a = 0.02$. Note the resonances around $kL = n\pi$, where n is an odd integer.

One may wonder why there are no resonances when kL/π is an even integer. Fig. 7.14 shows the current distribution on a dipole with $kL = 5.9$ and $ka = 0.02$ when the dipole is not strongly excited. Nevertheless, the dipole has a natural oscillation mode near this frequency. However, this mode has a full wavelength oscillation over the wire and is odd around the center point. Therefore, it does not get excited by the incident plane wave. The current induced on the wire for $kL = 5.9$ is even around the midpoint of the wire, and this is not a resonant mode of the wire at this frequency.

Fig. 7.15 shows the current distribution on a dipole with $kL = 9.2$ and $ka = 0.02$ when the dipole is resonant. The natural oscillation mode of the dipole at this frequency has a 1.5 wavelength, and this mode has a net coupling to the incident plane wave.

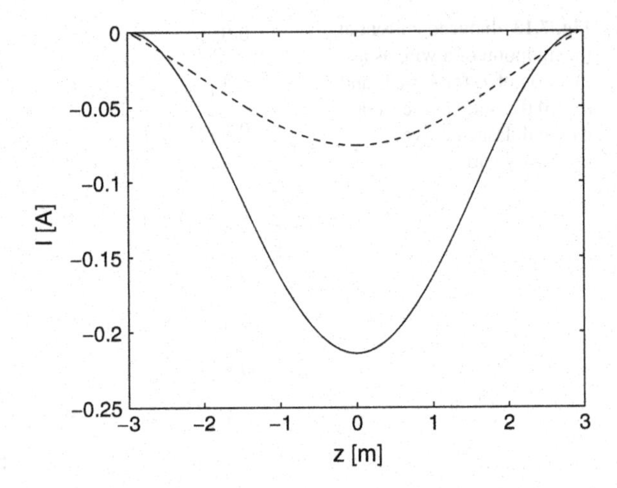

Fig. 7.14 Induced current on a wire with $kL = 5.9$, $ka = 0.02$: solid curve - real part and dashed curve - imaginary part

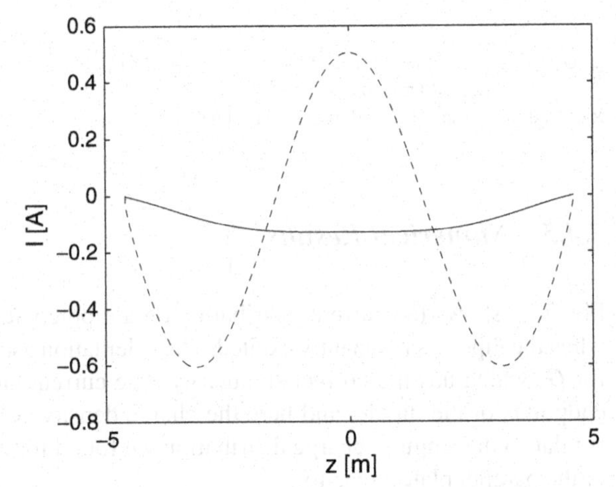

Fig. 7.15 Induced current on a wire with $kL = 9.2$, $ka = 0.02$: solid curve - real part and dashed curve - imaginary part

Review Questions

7.4-1 Derive (7.50) from (7.41). What assumptions did you use?

7.4-2 Perform the derivations required to arrive at Hallén's equation.

7.4-3 Write the Green's function for 3D free space as a sum of a singular part and a regular part. Show that the regular part is bounded as $R \to 0$.

7.4-4 Describe the steps and assumptions required to arrive at (7.57).

7.4-5 What can happen if the Green's function is too smooth; i.e., its singularity is neglected?

7.4-6 Give an example of weighting and basis functions that can be used for Hallén's equation. Write down the corresponding system of linear equations.

7.4-7 Why does Fig. 7.13 show resonances at $kL = n\pi$ only for odd integers n and not even integers?

Summary

- Consider a problem modeled by the differential equation $Df = s$, where D is a differential operator, f is a field and s is the source. The Green's function $G(\boldsymbol{r}, \boldsymbol{r}')$ satisfies $D_r G(\boldsymbol{r}, \boldsymbol{r}') = \delta(\boldsymbol{r} - \boldsymbol{r}')$, where D_r takes derivatives with respect to the unprimed coordinates. Given the Green's function, the differential equation can be written as an *integral equation*

$$f(\boldsymbol{r}) = \int G(\boldsymbol{r}, \boldsymbol{r}')s(\boldsymbol{r}')dV'.$$

  For Poisson's equation $-\epsilon_0 \nabla^2 \phi = \rho$, we have
  - the 3D Green's function $G(\boldsymbol{r}, \boldsymbol{r}') = |\boldsymbol{r} - \boldsymbol{r}'|/(4\pi\epsilon_0)$ and
  - the 2D Green's function $G(\boldsymbol{r}, \boldsymbol{r}') = -1/(2\pi\epsilon_0) \ln |\boldsymbol{r} - \boldsymbol{r}'|$.

- The method of moments (MoM) solves an integral equation by a finite element expansion; i.e., the sources $s(\boldsymbol{r}) \approx \sum_k \alpha_k s_k(\boldsymbol{r})$ are expanded in terms of basis functions $s_k(\boldsymbol{r})$. Choose as many weighting functions $w_k(\boldsymbol{r})$ as there are basis functions. Determine the coefficients α_k by multiplying the Green's function expression for $f - f_{\text{prescribed}}$ and integrating in space. Two usual choices for the weighting functions are
  - collocation with $w_k(\boldsymbol{r}) = \delta^3(\boldsymbol{r} - \boldsymbol{r}_k)$, which evaluates the field at the point $\boldsymbol{r} = \boldsymbol{r}_k$, and
  - Galerkin's method $w_k(\boldsymbol{r}) = s_k(\boldsymbol{r})$.

  The integrand of the integrals in a MoM formulation are often decomposed into a singular part and a regular part. Preferably, the singular part is treated analytically and the nonsingular part by numerical integration.
- Scattering from conducting bodies is often treated by MoM. In the Lorentz gauge, the scattered electric field can be expressed as

$$\boldsymbol{E}^s = \frac{1}{j\omega\epsilon_0\mu_0}[\nabla(\nabla \cdot \boldsymbol{A}) + k^2 \boldsymbol{A}],$$

  where $\boldsymbol{A}$ is expressed in the induced surface current as

$$\boldsymbol{A} = \frac{\mu_0}{4\pi} \int \frac{e^{-jkR}}{R} \boldsymbol{J}_s(\boldsymbol{r}')dS',$$

  where $R = |\boldsymbol{r} - \boldsymbol{r}'|$. Equivalently, $\boldsymbol{E}^s = -j\omega \boldsymbol{A} - \nabla\phi$ with $\boldsymbol{A}$ as above and

$$\phi = \frac{1}{4\pi\epsilon_0} \int \frac{e^{-jkR}}{R} \rho_s(\boldsymbol{r}')dS',$$

where $j\omega\rho_s + \nabla\cdot\boldsymbol{J}_s = 0$. The equation $\hat{\boldsymbol{n}}\times(\boldsymbol{E}^s + \boldsymbol{E}^{inc}) = 0$ is solved by the MoM on the surfaces of the conductors. The current $\boldsymbol{J}_s$ should be expanded in Rao–Wilton–Glisson basis functions, since they have continuous normal components at cell edges.

- The EFIE suffers from "internal resonances". At these resonance frequencies, the solution is wrong and the system matrix may become singular. The MFIE has different internal resonances than the EFIE, and with a suitable linear combination of the two integral equations, the internal resonances (and the problems associated with them) can be avoided. The summed equation is called the CFIE.
- Scattering from thin wires is often treated by the MoM combined with certain approximations. If the surface current is replaced by a line current $I(z)$ on axis, the MoM does not converge as the resolution is increased, but increasing wiggles appear. Short-wavelength oscillations are screened by the distance from the center to the surface of the wire. Convergence is achieved by using a more accurate Green's function that keeps the correct singularity at $\boldsymbol{r} = \boldsymbol{r}'$. Then, the electric field produced by fine-scale variations in the current is better represented.

Problems

P.7-1 Green's functions are normally constructed so that the boundary conditions are accounted for. Given the free-space Green's function in (7.10), derive the corresponding Green's function that can be used for a problem with a PEC ground plane at $z = 0$. Such a Green's function allows for an algorithm that avoids an explicit discretization of the ground plane.

P.7-2 Show that the MoM matrix in (7.20) is symmetric and positive definite if Galerkin's method is used. How does this relate to the corresponding matrices derived by the FEM?

P.7-3 In Sect. 7.2, compare the capacitor problem with and without the exploitation of symmetries. How much computational resources, in terms of memory requirements and floating point operations, can be saved by the use of symmetry?

P.7-4 Can the algorithm in Sect. 7.2 be generalized to also include dielectric materials? Discuss how the formulation would change.

P.7-5 Apart from sharp metal corners, are there other situations in which a reduced order of convergence can be expected?

P.7-6 Does the distance between two similar sharp metal corners influence the order of convergence? What order of convergence do you expect from a problem with two different sharp metal corners?

P.7-7 Use (7.36) to derive a Green's function for the vector potential that includes a PEC ground plane at $z = 0$. Perform the same derivation for the scalar potential in (7.38). Given the fields

$$\mathbf{E} = -\nabla\phi - j\omega\mathbf{A}, \quad \mathbf{B} = \nabla \times \mathbf{A},$$

verify that the boundary conditions are satisfied at the ground plane.

P.7-8 Derive the EFIE directly from Maxwell's equations.

P.7-9 For a 2D problem where a PEC cylinder is aligned with the z-axis, choose basis functions for the current $\mathbf{J}_s(x, y)$, and charge distributions $\rho_s(x, y)$. Does the choice depend on whether the TE or TM case is considered?

P.7-10 Show that (7.39) can also be written as $\mathbf{E}^s = c^2(\nabla\nabla \cdot \mathbf{A} + k^2\mathbf{A})/j\omega$ with $\mathbf{A}$ given by (7.36).

P.7-11 Is the matrix (7.48) associated with the MFIE symmetric?

P.7-12 Write down the RWG basis functions for a rectangle.

P.7-13 Try to solve (7.55) with only one basis function. What type of basis function do you choose?

Computer Projects

C.7-1 Use the algorithm in Sect. 7.2 to compute the capacitance for two parallel circular cylinders of radius a and a separation distance d, which also can be solved for analytically. What order of convergence do you expect, and do your expectations agree with the numerical experiment?

C.7-2 Evaluate (7.56) by brute force and compare the result to the approximation in (7.57). Conclusions?

C.7-3 Use the approach described in Sect. 7.4 to implement the MoM for scattering from thin wires. Reproduce some of the results presented in Sect. 7.4 for validation. Can you generalize your formulation and program so that you can solve problems where three wires or more are connected at the same point? What type of basis function do you need at such a junction and how do you test the integral equation?

C.7-4 Use the approach described in Sect. 7.3 to implement the MoM for scattering from metal surfaces. Discretize a PEC sphere by triangles and solve the scattering problem. How does the solution compare with analytical results [5]? Try to reproduce the problem with internal resonances.

Chapter 8
Summary and Overview

The goal of any analysis or optimization is to achieve sufficient accuracy with minimum effort, where effort usually is interpreted as computational cost in terms of computational time and memory requirements. However, there may also be a considerable effort associated with other issues such as the programming of the numerical algorithm or the construction of geometrical descriptions suitable for the the computations at hand.

Faced with an electromagnetic problem, say an antenna in the vicinity of a human body, we need to find a numerical algorithm that can yield sufficiently accurate results without an excessive effort. Naturally, there are a number of aspects that will guide the choice of computational method. For example, the electromagnetic problem at hand may involve boundary conditions that are necessary for a realistic model but difficult to treat for some computational methods. Complex materials with nonlinearities, anisotropies, or dispersive characteristics can also eliminate some numerical algorithms. The typical length scales of the problem is another important aspect that should be considered. In linear problems, the wavelengths present are determined by the frequency contents of the excitation and the materials. Other length scales that should be considered are the skin-depth and the size of the geometrical features present. Each of these length scales typically covers a certain range, and the combination of them can yield a significant interval (which can require certain approximations if a direct analysis is not feasible). In a typical low-frequency application, for example, the wavelength is on the order of thousands of kilometers, and the geometrical size on the order of meters (possibly down to millimeters for laminations and thin wires) while the skin-depth is typically in the range from millimeters to centimeters.

In some situations, one method is competitive for a part of the problem while another algorithm is better suited for the remaining parts. It is then attractive to combine the different algorithms to form a so-called hybrid method. Such methods can be challenging to construct, and many attempts have failed to preserve important properties of Maxwell's equations. However, successful hybrid methods offer possibilities to treat significantly larger classes of problems.

T. Rylander et al., *Computational Electromagnetics*, Texts in Applied Mathematics 51, DOI 10.1007/978-1-4614-5351-2_8,

Table 8.1 Scalings for the number of operations with frequency f and the number of iterations N_{it}

	FEM/FDTD	MoM-matrix	MoM-MLFMA
2D	f^3	$N_{it} f^2$	$N_{it} f \log f$
3D	f^4	$N_{it} f^4$	$N_{it} f^2 \log f$

One of the major challenges in CEM is to model systems that are electrically large, that is, for which the spatial extent D is many wavelengths λ in three dimensions. In this setting, it is useful to compare how the number of floating-point operations and the memory requirements for the different methods scale with the wave frequency f for a system of fixed spatial extent (where we will consider objects with geometrical features that are on the order of the wavelength or larger). Table 8.1 summarizes the scalings with frequency for the methods treated in this book and the MLFMA extension of the MoM.

It should be pointed out that there are multipliers in front of the scalings in Table 8.1, and that these coefficients can be quite significant. For instance, the multiplier is large for the MLFMA (which is a version of the MoM), so that the application problems need to be quite large before this method is competitive. However, the MLFMA is the most competitive full-wave method for very large scale scattering problems, e.g., to compute the radar cross section for an entire aircraft. In this chapter, we present a more detailed discussion of these scalings. Also, we briefly discuss a selection of other methods. There is a large number of numerical algorithms in CEM, and it is beyond the scope of this book to give a complete account.

8.1 Differential Equation Solvers

Differential equation solvers are used for both frequency- and time-domain computations. They can be applied to both driven problems and eigenvalue problems.

For differential equation solvers in frequency domain, one often uses iterative solvers (especially in three dimensions), and brief introductions to this subject are given in Appendices C and D. Generally, the number of iterations needed for convergence scales as the square root of the condition number κ of the matrix, where the condition number is the ratio of largest to smallest eigenvalues of the matrix. The smallest eigenvalues of the curl-curl or Laplace operator are independent of the resolution. The largest eigenvalues of these second-order operators scale as $1/h^2$; see, e.g., (3.17). Given that the frequency f dictates the cell size $h \propto 1/f$, the largest eigenvalue of a second-order operator scales as f^2. Therefore, the condition number $\kappa = \lambda_{\max}/\lambda_{\min} \propto f^2$, and generally, the number of iterations scales as $\sqrt{\kappa} = f$. The matrix generated by a differential equation formulation is sparse, so the number of operations per iteration is proportional to the number of unknowns, i.e., $\propto f^2$ and $\propto f^3$ for 2D and 3D problems, respectively.

Therefore, for frequency-domain FEM (or finite difference methods) the total number of operations scales as f^3 in 2D and f^4 in 3D (for a *single* frequency).

For the differential equation solvers in time domain, the time-step varies as $h \propto 1/f$, and for a fixed time interval the number of time-steps scales as $1/\Delta t \propto f$. Therefore, the number of operations for time-domain methods (such as the FDTD) and the frequency-domain methods (e.g., FEM) scales as $f \times f^2 = f^3$ in 2D and $f \times f^3 = f^4$ in 3D. But the time-domain method gives a *complete* frequency spectrum, as compared to a standard frequency-domain method that requires one computation for a single frequency.

In the following, unless stated otherwise, we focus on the scalings for 3D methods.

8.1.1 Finite-Difference Time-Domain

To keep a certain relative phase error, the FDTD needs a certain number of points per wavelength λ/h; 1% phase error requires about 18 cells per wavelength. To keep this accuracy, the number of cells in any direction, D/h, scales as f, while the maximum time-step scales as $\Delta t \propto h \propto f^{-1}$. Consequently, the total number of operations scales as f^4. If one asks for a fixed *absolute* phase error across the whole system, the number space steps scales as $f^{3/2}$, and the number of operations becomes $O(f^6)$. In this case, higher-order methods are more advantageous. So far, higher-order methods are not used very much for electromagnetic problems, but work in this area is underway.

Time-domain methods generate time sequences that can be Fourier transformed to give a full frequency spectrum in $O(f^4)$ operations. This, plus the simplicity of the FDTD, are the main reasons for its popularity. The major drawback of the FDTD is that it is tied to structured grids, which force oblique boundaries to appear as "staircases."

8.1.2 Finite-Volume Time-Domain

Finite volume time-domain (FVTD) methods generate discrete equations by integrating the Ampère and Faraday laws over each grid cell [60, 94]:

$$\int_{V_e} \epsilon \frac{\boldsymbol{E}^{n+1} - \boldsymbol{E}^n}{\Delta t} dV = \oint_{A_e} \hat{\boldsymbol{n}} \times \boldsymbol{H}^{n+\frac{1}{2}} dS - \int_{V_e} \boldsymbol{J}^{n+\frac{1}{2}} dV,$$

$$\int_{V_h} \mu \frac{\boldsymbol{H}^{n+\frac{1}{2}} - \boldsymbol{H}^{n-\frac{1}{2}}}{\Delta t} dV = -\oint_{A_h} \hat{\boldsymbol{n}} \times \boldsymbol{E}^n dS,$$

where superscripts indicate time. Two grids are used: the "primary" and "dual" grids. The electric field is defined on the vertices of the primary grid (cells V_h), and the magnetic field is defined on those of the dual grid (cells V_e), the vertices of which are the centers of the primary cells. Unlike the FDTD, the FVTD does not conserve electric and magnetic charges. Madsen and Ziolkowski [47,60] constructed an "FDTD correction" to accomplish this.

The FVTD is explicit and therefore efficient, as long as the cells are of reasonably uniform size; otherwise, very small time-steps are required, and they degrade the performance of the method. The primary grid can be made of tetrahedra, which gives the method good ability to model complex geometry. A drawback of the FVTD is the appearance of a weak "late time" instability [47, 60, 95]. This can be prevented by adding dissipation, which, however, may decrease the accuracy of the algorithm. The operation count scales the same way as for the FDTD.

8.1.3 Finite Element Method

The finite element method easily handles complex geometry, and FEM is used both in frequency- and time-domain analyses. Together with standard iterative solvers, a frequency-domain calculation requires $O(f^4)$ operations per frequency. The scaling in time-domain calculations is the same as for the FDTD, but time-domain FEM typically involves at least a factor of 10 more operations.

A valuable property of the finite element method, in comparison to the FVTD, is that both the mass matrix and the stiffness matrix are symmetric and real, which guarantees that the eigenvalues ω^2 of $\nabla \times \mu^{-1} \nabla \times \boldsymbol{E} = \omega^2 \epsilon \boldsymbol{E}$ are real. Combined with a suitable time-stepping scheme, this leads to a stable algorithm. The symmetric, or reciprocal, property of the FEM appears not to hold for finite volume discretizations. In fact, lack of symmetry is a likely explanation of the late-time instability observed for many schemes.

8.1.4 Transmission Line Method

Transmission line methods (TLM) work with combinations of electric and magnetic fields, represented as pulses propagating on a 3D grid of transmission lines. At the intersections, the *nodes*, the pulses are scattered according to scattering matrices **S**. By imposing the condition that **S** be unitary, energy conservation can be enforced, and hence stability achieved.

TLM based on so-called expanded nodes was described by Hoefer [37]. An improved, symmetrical condensed node was introduced by Johns [43]. Celuch-Marcysiak and Gwarek [17] proved the equivalence of a transmission line network with a circuit model for a nonuniform grid in 2D. An equivalence with an FDTD formulation was established on a *uniform* 3D grid by Chen et al. [18].

8.1.5 Finite Integration Technique

The finite integration technique [90] (FIT) is based on the integral representation of Maxwell's equations. The FIT reduces to the FDTD scheme on grids consisting of cubes, and for that case, the derivation of the FIT is very similar to the integral representation in Sect. 5.2.4. The fields are represented in terms of electric and magnetic voltages (organized in the vectors $\bar{\mathbf{e}}$ and $\bar{\mathbf{h}}$, respectively). These are related to the electric and magnetic fluxes (organized in the vectors $\bar{\bar{\mathbf{d}}}$ and $\bar{\bar{\mathbf{b}}}$, respectively) by the constitutive relations (expressed as $\bar{\bar{\mathbf{d}}} = \mathbf{M}_\epsilon \bar{\mathbf{e}}$ and $\bar{\mathbf{h}} = \mathbf{M}_{\mu^{-1}} \bar{\bar{\mathbf{b}}}$). Maxwell's equations (in source-free space) can then be written in the form

$$\mathbf{C}\bar{\mathbf{e}} = -\frac{d}{dt}\bar{\bar{\mathbf{b}}},$$

$$\widetilde{\mathbf{C}}\bar{\mathbf{h}} = \frac{d}{dt}\bar{\bar{\mathbf{d}}}.$$

For wave problems, the time derivatives are discretized in the leap-frog sense. Here, $\mathbf{C}$ and $\widetilde{\mathbf{C}}$ are the curl operators (matrices with elements 0 or ± 1) on the primary and dual meshes, respectively. Similarly, Gauss's law can be stated $\widetilde{\mathbf{D}}\bar{\bar{\mathbf{d}}} = \mathbf{q}$, and the condition of solenoidal magnetic flux density as $\mathbf{D}\bar{\bar{\mathbf{b}}} = \mathbf{0}$, where $\mathbf{D}$ and $\widetilde{\mathbf{D}}$ represent the divergence on the primary and dual meshes, respectively. The matrix corresponding to the gradient operator is then the transpose of the divergence matrix. The matrix operators correctly reproduce well-known properties; for example, the zero divergence of the curl is $\mathbf{DC} = \mathbf{0}$ and the zero curl of the gradient is $\mathbf{C}\widetilde{\mathbf{D}}^{\mathrm{T}} = \mathbf{0}$. This allows for various manipulations; for example, the vector wave equation can be written as $\widetilde{\mathbf{C}}\mathbf{M}_{\mu^{-1}}\mathbf{C}\bar{\mathbf{e}} + \mathbf{M}_\epsilon \partial^2 \bar{\mathbf{e}}/\partial t^2 = 0$.

Weiland and coworkers [71, 83] have investigated stable local refinement and nonorthogonal meshes for the FDTD scheme. The property $\mathbf{C} = \widetilde{\mathbf{C}}^{\mathrm{T}}$ is important for stability, and the (typically) diagonal matrices $\mathbf{M}_\epsilon$ and $\mathbf{M}_{\mu^{-1}}$ allow for explicit time-stepping. Thus, the FIT has the same scalings as the FDTD, but it allows for curved meshes and local refinement combined with stable time-stepping.

8.2 Integral Equation Solvers

For integral equations, the number of unknowns is much smaller than for volume discretizations such as FDTD or FEM, but the matrix is dense. The integral formulation is nevertheless superior for large problems because of a rather recent development called the fast multipole method (FMM). The hierarchical version of this method is called the MLFMA, multilevel fast multipole algorithm [20]. The operation count then becomes $\propto N_{it} f \log f$ in 2D and $N_{it} f^2 \log f$ in 3D. This is superior to the differential equation solvers if $N_{it} < O(f^2)$, which is generally the case. The drawback of the MLFMA is that it is quite complicated to program, and in particular, to parallelize. Integral equation methods, or the method of moments, solve either the EFIE, MFIE, or the CFIE [54] on surfaces of conductors and dielectrics.

8.2.1 Frequency-Domain Integral Equations

In frequency-domain formulations, both the EFIE and the MFIE may suffer from internal resonance; this can be avoided by using a suitable linear combination of the two equations: the CFIE.

A main advantage of the MoM is the low number of unknowns, which scale with frequency as $O(f^2)$. The drawback is that the matrix is dense. Therefore, if one attempts direct solution by LU decomposition, the operation count has a very unfavorable $O(f^6)$ scaling. In geometries that are only partly 3D, this can be improved on by Fourier transformation in the main direction of symmetry [45] or by using the Toepliz property of the MoM matrix to apply CG-FFT techniques [54]. However, for truly 3D problems other methods for solving the linear system are needed.

Iterative solvers, such as the conjugate gradient (CG) method or Krylov methods, improve the scaling. The iterative algorithms are based on matrix–vector multiplications, and with a dense matrix a conventional multiplication takes $O(f^4)$ operations. The total operation count then becomes $O(N_{it} f^4)$, where the number of iterations N_{it} can be hard to predict. Song and Chew [77] report $N_{it} \propto f^{1/2}$ for problems with only closed surfaces. Thus, the scaling becomes $f^{4.5}$ for each frequency, which is not competitive with differential equation solvers. However, recently several methods have been developed to reduce the number of operations for a matrix–vector multiplication, that is, in computing the field from given sources.

8.2.1.1 Fast Multipole Methods

A very successful scheme to replace the matrix multiplication is the fast multipole method (FMM) introduced by Rokhlin [61, 62] and developed into the multilevel fast multipole algorithm (MLFMA) by a group at the University of Illinois [78].

The FMM is described in an accessible way in [21]. The first step is to divide the simulation region into boxes, each containing a moderate number of grid cells. Fields from grid cells in the same, or an adjacent, box are computed in the standard way. The fields produced by sources farther away are computed by first generating a multipole expansion for the sources, then projecting this onto a set of plane waves in the observation box, from which one obtains the fields at each observation point. The savings come from the fact that only a moderate number L of terms are needed in the multipole expansion. A semi-empirical formula for the number of terms needed to achieve double precision accuracy is $L = kD + 10\ln(\pi + kD)$, and the required number of plane waves scales as L^2. Minimizing the total number of operations, one finds that the optimum number of elements per box scales as the square root of the total number of elements N and that the total operation count scales as $N^{3/2}$. The MLFMA repeats this algorithm in a hierarchical way on all scales and achieves a scaling $O(N \log N)$. This algorithm has been implemented in the FISC code [79].

A nice analogy of the FMM is a telephone network. If every one of N customers is connected by a direct line to every other customer, the number of connections scales as N^2. However, by introducing "hubs," the number of connections can be reduced. To make a telephone call, a customer (the source point) calls the local hub (the multipole expansion), which calls another hub (the plane waves), which finally calls the recipient of the call (the observation point).

We can conclude that for 3D problems the FMM gives an $O(f^3)$ and the MLFMA an $O(f^2 \log f)$ scaling for the operation count per iteration. These represent significant reductions from the $O(f^4)$ scaling for straightforward matrix–vector multiplication. If the number of iterations scales as $f^{1/2}$, the frequency-domain MoM is clearly competitive with time-domain differential equation solvers for large problems. However, it takes a problem of significant size for the FMM or MLFMA to be competitive, with at least several thousand unknowns. The FMM and MLFMA also imply large savings in storage because the full matrix is never stored.

8.2.1.2 Other Fast Methods

The impedance matrix localization technique (IML) [15, 16] is a matrix algebra routine that transforms to a basis for the source distribution that radiates into narrow beams. This makes the MoM matrix sparse. The method can be incorporated in existing MoM programs to sparsify an already computed matrix.

Also, wavelet transforms have been used in MoM calculations [30, 86]. Wavelet transforms work excellently in static problems where the integral kernel is nonoscillatory, and reduce the operation count to $O(N \log N)$. For electrically large systems ($D \gg \lambda$) with oscillatory kernels, Wagner and Chew [86] found that the standard wavelet transform reduces the number of operations only to βN^2, with $\beta \approx 0.1$. More recently, Golik [30] tested discrete wavelet packet similarity transformations together with thresholding of the matrix elements. As the system size was increased, with a fixed number of cells per wavelength, the number of nonzero matrix elements scaled more slowly than N^2; the numerical results suggested an $O(N^{4/3})$ scaling.

8.2.2 Time-Domain Integral Equations

Time-domain integral equations (TDIE) is a relatively new area of research. The first approaches straightforwardly discretized the time-domain form of the EFIE [57,66] and the MFIE [76] in space and time. The time-domain MFIE can be written as

$$2\pi \boldsymbol{J}(\boldsymbol{r},t) = 2\pi \hat{\boldsymbol{n}} \times \boldsymbol{H}_i(\boldsymbol{r},t) + \hat{\boldsymbol{n}} \times \int_S \left(\boldsymbol{J}(\boldsymbol{r}',\tau) + \frac{R}{c} \frac{\partial \boldsymbol{J}(\boldsymbol{r}',\tau)}{\partial t} \right) \times \frac{\hat{\boldsymbol{R}}}{R^2} dS', \quad (8.1)$$

where $\tau = t - R/c$ is the retarded time and $R = |\boldsymbol{r} - \boldsymbol{r}'|$. In the discretized version, the solution has to be saved over the time that it takes a light wave to traverse the entire simulation region, so the storage requirement for the solution scales as f^3 (as for a volume discretization). The matrix storage scales as f^4, so that for very large problems the matrix may have to be recomputed, or some fast scheme is needed for the field calculation. The operation count scales as f^5, which is worse than for differential equation solvers.

The early TDIE algorithms were unstable and required dissipation for stability [76, 85]. This problem appears to have been overcome recently for the EFIE by a variational formulation together with strict FEM techniques both in space and time [1].

Another TDIE solver has been developed by Walker and coworkers [11, 25] for the MFIE. Applying finite element techniques to (8.1), Bluck and Walker [11] derived an algorithm that is somewhat implicit. The algorithm needs to be implicit, because on every new time-level, "new," or unknown, currents enter into the surface integral in (8.1) within regions of radius $c\Delta t$ around each observation point. The resulting implicit algorithm was found to be stable if the time-step *exceeds* the time it takes a light wave to traverse the largest spatial element. (The degree of implicitness increases with the time-step.) This code has been used to compute scattering data when the scatterer is illuminated by a short pulse of duration $\propto f^{-1}$; see [25]. In this mode of operation, the operation count scales very favorably with frequency. This is because the number of elements, both in the region where one needs to integrate (illuminated source points) and in the region where the resulting field is significant (illuminated observation points), scales only as f. It is superfluous to calculate near-vanishing fields in the nonilluminated regions, and this strongly reduces the operation count if the incoming pulse is short (and the scattering surface is convex so that there are no multiple reflections).

8.3 Hybrid Methods

The different basic techniques used in CEM all have their strengths and limitations. One way to achieve performance that is better than two individual methods is to combine them into a so-called hybrid method. This can be difficult but very useful once a good and reliable formulation is found. There is a vast number of hybrid methods, and here we mention only a few of them in order to introduce the concept of hybridization.

The FDTD is efficient, but has difficulties with complex geometry. Therefore, hybrid methods have been formulated to combine efficiency with the ability to treat complex geometry. The hybrid schemes combine the FDTD with either an FVTD [60, 94, 95] or time-domain FEM [50, 92]. These methods typically experience late-time instabilities [50, 95]. Rylander and Bondeson formulated a *stable* hybrid scheme [63] that combines the FDTD with FEM on unstructured meshes. Where the structured and unstructured grids join, the mass and stiffness matrices are constructed in a special way to preserve symmetry. This makes it straightforward to achieve stability without dissipation. The scheme uses an implicit solver on the unstructured grid. It has been verified that the algorithm is stable for time-steps up to the stability limit of the FDTD. The advanced TLM, FIT, and hybrid FEM-FDTD are efficient and stable solvers that can handle complex geometry. The FEM-FDTD combination may have an advantage in being more easily coupled to standard grid generators and is more adequate for adaptive mesh refinement.

When differential equation solvers are applied to problems in unbounded geometries, the computational region must be truncated. Several methods for radiative boundary conditions have been formulated for differential equations solvers, where the perfectly matched layers [9, 55] is the preferred choice in most cases. For electrically very large problems, the volume discretizing solvers find competition from recently developed integral equation methods, which are well suited to analyze objects in free-space. For open-region problems that involve objects with complicated materials, it can be useful to use a FEM for the object and its immediate surrounding, combined with a MoM for the remaining free-space environment. It is feasible to construct frequency-domain formulations that combine the MoM and FEM. These are often referred to as finite element–boundary integral formulations, or FE-BI for short. The FE-BI formulation by Botha and Jin [13] is based on variational principles for the continuous quantities, and it yields symmetric matrices that preserve reciprocity explicitly, which reflects important properties of Maxwell's equations.

Appendix A
Projects

This appendix features five computer projects that cover the most important aspects of the theoretical material in this book: (1) convergence and extrapolation, (2) finite differences in the frequency domain for a 1D electromagnetic problem, (3) finite-difference time-domain for a 3D electrodynamic problem, (4) finite elements for a 2D eigenvalue problem, and (5) method of moments for a 2D electrostatic problem. Each computer project is presented in terms of a continuum formulation of the physical situation, and for some of the problems a part of its numerical treatment is also included. Next, a number of assignments are listed such as the derivation of the complete numerical scheme, the computer implementation, and a sequence of numerical tests. The assignments can be assessed by, for example, (1) a written report that is reviewed, (2) oral examination with or without access to a computer, or (3) a presentation in front of a larger group of students and teachers. Access to a computer allows for an exploratory and interactive testing of the computer implementation that may be difficult to achieve otherwise.

In the written or oral assessment, it is important that students attempt to provide mathematical and logical arguments to support the choices, derivations, implementations, results, and conclusions that constitute the computer project work. Such a presentation could consist of the following parts:

- Description of the continuum problem.
- Description of the numerical algorithm by means of suitable derivations together with their computer implementation. Note that it is essential that the computer implementation be well documented: (1) input variables for each function, (2) output variables for each function, (3) purpose and usage for each function, and (4) some sort of overall description of the program and its functions and scripts.
- Presentation of the numerical tests together with interpretations of the results that relate to the theoretical foundation in terms of, for example, the order of convergence, extrapolation, and estimation of the numerical error.
- Conclusion of the computer project work, which may include (1) brief summary of the work, (2) the main results, and (3) important implications of these results.

T. Rylander et al., *Computational Electromagnetics*, Texts in Applied Mathematics 51, DOI 10.1007/978-1-4614-5351-2,

A.1 Convergence and Extrapolation

A.1.1 Problem Description

In electrostatic problems, the surface charge density is singular close to sharp metal edges or corners. For a computation of the capacitance, this is problematic when the total charge on such a conducting body must be computed to determine the capacitance. Consider a situation with a given surface charge density that can be described without errors by means of an analytical formula. For example, the surface charge density close to the edge of a parallel plate capacitor varies as $x^{-(\pi-\beta)/(2\pi-\beta)}$, where x is the distance to the edge and β is the angle subtended by the metal edge as described in Sect. 7.2.3. This leads to integrals of the type

$$\int_a^b x^{\xi}\,dx, \tag{A.1}$$

where $0 \leq a < b$. Here, the case $a > 0$ yields a regular integrand and, from a mathematical viewpoint, this type of integrand features the general behavior of the surface charge distribution on a smooth metal surface. Furthermore, $a = 0$, in combination with $\xi < 0$, yields a singular integrand, which corresponds to the surface charge distribution in the vicinity of a sharp metal edge. (Note that according to the derviations in Sect. 7.2.3, an infinitely thin metal plate with $\beta = 0$ yields $\xi = -1/2$, which is the smallest possible value for ξ for such an electrostatic situation. However, we also investigate other cases, such as $\xi = -3/2$, subsequently since it provides additional understanding of some mathematical difficulties that are associated with singularities.)

A.1.2 Assignments

Implementation

Write a program that evaluates the integral (A.1) by means of midpoint integration. The program should allow for midpoint integration with n subintervals of length $h = (b-a)/n$. In what follows, the value of the numerically evaluated integral is denoted by $I_{\text{midp}}(h)$. The analytically calculated value of the integral (A.1) is denoted I_0.

Numerical Test #1

Execute your program with $a = 1$ and $b = 2$ for $\xi = -3/2, -1/2, 1/2, 3/2$.

- Evaluate the absolute error $e(h) = |I_{\text{midp}}(h) - I_0|$ as a function of the resolution controlled by h, where you should use the expression for I_0 that you get from analytical integration. Does the computed result converge to the analytical answer? What is the order of convergence? Does the order of convergence agree with what you expect from an analysis of the problem?

- Extrapolate the numerically computed result to zero cell size: fit the function $I_{\text{model}}(h) = c_0 + c_\alpha h^\alpha$ to the computed data under the assumption that the constants c_0, c_α, and α are unknown. How well does the model $I_{\text{model}}(h)$ compare with the computed data $I_{\text{midp}}(h)$? Compare the extrapolated value c_0 and the estimated order of convergence α with the analytical results. Test your extrapolation method on different sets of computed data. How do the data influence the possibilities for accurate extrapolation?

Numerical Test #2

Execute your program with $a = 0$ and $b = 2$ for $\xi = -3/2, -1/2, 1/2, 3/2$.

- Evaluate the absolute error $e(h) = |I_{\text{midp}}(h) - I_0|$ as a function of the resolution controlled by h, where you should use the expression for I_0 that you get from analytical integration. Does the computed result converge to the analytical answer? What is the order of convergence? Does the order of convergence agree with what you expect from an analysis of the problem?
- Extrapolate the numerically computed result to zero cell size: fit the function $I_{\text{model}}(h) = c_0 + c_\alpha h^\alpha$ to the computed data under the assumption that the constants c_0, c_α, and α are unknown. How well does the model $I_{\text{model}}(h)$ compare with the computed data $I_{\text{midp}}(h)$? Compare the extrapolated value c_0 and the estimated order of convergence α with the analytical results. Test your extrapolation method on different sets of computed data. How do the data influence the possibilities for accurate extrapolation?

A.2 Finite Differences in the Frequency Domain

A.2.1 Problem Description

Consider an electromagnetic plane wave that propagates toward a large flat window of glass, as shown in Fig. A.1. We wish to compute the reflected and transmitted wave. The glass window has a thickness of $2a$.

The material parameters are $\epsilon(x)$, $\mu(x) = \mu_0$, and $\sigma(x)$ in the glass, where $-a \le x \le a$. The medium outside the window is air with $\epsilon(x) = \epsilon_0$, $\mu(x) = \mu_0$, and $\sigma(x) = 0$ for $|x| > a$. The total electric field satisfies the differential equation

$$-\frac{d^2 E_z(x)}{dx^2} + \mu_0 \left[j\omega\sigma(x) - \omega^2 \epsilon(x) \right] E_z(x) = 0. \tag{A.2}$$

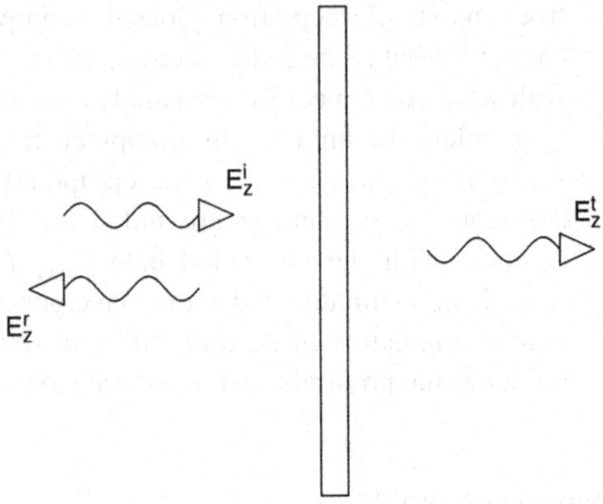

Fig. A.1 Glass window of thickness $2a$ with an incident field E_z^i, reflected field E_z^r, and transmitted field E_z^t

A.2.2 Assignments

Here we introduce some of the techniques used in electromagnetic scattering problems. One such technique is to impose the incoming wave by matching an expression for the incoming wave to the numerical solution in the vacuum region. The matching is done in a vacuum (outside the scatterer) where we know analytically how the incoming field behaves. In a similar manner, the reflected and transmitted waves can be described in the vacuum region. Following this procedure for our 1D problem we can do the matching of the discretized interior region to the fields outside at two points $x = \pm b$. We need to have $b > a$, so that the matching points are in the vacuum.

Formulate Boundary Condition

The first task is to derive the appropriate boundary conditions at $x = -b$ and $x = b$ analytically. Here is some guidance.

Let the incident field be $E_z^{\mathrm{i}}(x) = E_0^{\mathrm{i}} \exp(-jk_0 x)$. Introduce the reflected field as $E_z^{\mathrm{r}}(x) = E_0^{\mathrm{r}} \exp(+jk_0 x)$ and the transmitted field as $E_z^{\mathrm{t}}(x) = E_0^{\mathrm{t}} \exp(-jk_0 x)$. According to these expressions, the total field is (1) the superposition of the incident and reflected field for the region $x < -a$ and (2) equal to the transmitted field for the region $x > a$. Equation (A.2) yields the dispersion relation $k_0 = \omega/c_0$ for the vacuum region, where $c_0 = 1/\sqrt{\epsilon_0 \mu_0}$.

What is a priori known and unknown in the expressions above? How can this information be used to formulate the appropriate boundary conditions at $x = -b$ and $x = b$, respectively? The boundary conditions should only involve already known information (such as the incident wave) and the desired solution E_z and its first derivative, where E_z is the total electric field. (Consequently, we wish to find boundary conditions that do not explicitly involve the scattered electric field since this is unknown before the scattering problem is solved.) Note that b is quite arbitrary as long as it is larger than a.

Generate Grid and System Matrix

We use two different grids:

G1 The first grid is chosen such that the material interfaces $x = \pm a$ fall in between grid points. We use the grid points $x_n = (n+\frac{1}{2})\Delta x$ with $\Delta x = a/N$, an integer $N \geq 3$, and $n = -2N, -2N+1, \ldots, 2N-1$.

G2 The second grid is chosen such that the material interfaces $x = \pm a$ fall on grid points. We use the grid points $x_n = n\Delta x$ with $\Delta x = a/N$, an integer $N \geq 3$, and $n = -2N, -2N+1, \ldots, 2N$.

These grids discretize a region of (roughly) length $4a$ with at least three grid points in each vacuum region outside the window.

Denote the unknowns at the grid points by ζ_n, i.e., $E_z(x_n) = \zeta_n$. Discretize the differential equation (A.2) and your boundary conditions using finite differences, both with an error that is proportional to h^2. The boundary condition involving no higher than first derivatives is best centered on the half-grid. Alternatively, a numerical boundary condition can be formulated on the integer grid if more than two grid points are used.

Using the boundary conditions and the differential equation, we have a system of linear equations $\mathbf{Az} = \mathbf{b}$ to solve, where $\mathbf{z} = [\zeta_1, \zeta_2, \ldots, \zeta_{N_{\text{gp}}}]^T$ and N_{gp} is the number of grid points. Write down the matrix $\mathbf{A}$ and the right-hand-side vector $\mathbf{b}$ for the special case where $N = 3$ for the discretization G1. What are the similarities and differences when you follow the same procedure for the discretization G2? What may the implications be and how can you handle this?

Find a way to compute the reflection coefficient R and transmission coefficient T from the numerical solution, where the following definitions are used:

$$R = \frac{E_0^{\text{r}}}{E_0^{\text{i}}}, \tag{A.3}$$

$$T = \frac{E_0^{\text{t}}}{E_0^{\text{i}}}. \tag{A.4}$$

Implementation

Implement your numerical algorithm for an arbitrary $N \geq 3$ and both discretizations G1 and G2. Given input variables that describe the physical situation and its discretization, the implementation should yield the reflection and transmission coefficients as output variables.

Numerical Test #1

Test your implementation on the case where the glass window has constant relative permittivity ϵ_r and conductivity σ. The reflection and transmission coefficients can be calculated analytically in this case and are given by

$$R = \frac{(k_0^2 - k_1^2)}{\Delta} e^{j2ak_0}(e^{j4ak_1} - 1),$$
$$T = \frac{k_0 k_1}{\Delta} 4 e^{j2a(k_0+k_1)}, \tag{A.5}$$

where $\Delta = (k_0 + k_1)^2 e^{j4ak_1} - (k_0 - k_1)^2$, $k_0 = \omega / c_0$, and $k_1 = \sqrt{\epsilon_r k_0^2 - j\omega\mu_0\sigma}$.

Use the thickness $a = 2\,\text{cm}$ in combination with the constant material parameters $\epsilon_r = 2.5$ and $\sigma = 0.02$ S/m.

For the frequency $\omega = 3 \cdot 10^9$ rad/s, compute R and T numerically on the discretization `G1` for a set of appropriately chosen values of N. Do the numerically computed values of R and T converge toward the analytical values? Which order of convergence do you find?

Numerical Test #2

Now repeat the preceding test for the discretization `G2`. Did this change the convergence properties? If so, why? Incidentally, how do you choose the permittivity at $x = \pm a$?

Numerical Test #3

Also, compute R and T as functions of frequency between 0.1 and 10 GHz. You can use your favorite discretization (`G1` or `G2`) with a fixed value of N. How does the error change with respect to the frequency? Explain your findings.

Numerical Test #4

Compute R and T as a function of frequency between 0.1 and 10 GHz for an inhomogeneous window in the region $|x| \leq a$ with the following material parameters:

$$\sigma(x) = 0.02 p(x),$$
$$\epsilon(x) = \epsilon_0 [1 + 5p(x)],$$

where

$$p(x) = 1 - \left(\frac{x}{a}\right)^2$$

is a parabolic profile with $p(\pm a) = 0$.

A.3 Finite-Difference Time-Domain Scheme

A.3.1 Problem Description

Waveguides and filters are important components of many complex microwave systems. Here we consider the characteristic features for some relatively simple structures that provide a filtering functionality in waveguides. The quantities of interest are the reflection and transmission coefficients as a function of frequency. In what follows, we will limit the discussion to waveguide structures with rectangular cross sections and a finite-difference time-domain (FDTD) scheme [80, 93].

The FDTD model deals with the part of the waveguide that contains the filtering structure. At each of the two ends of the filter, a shorter section of a rectangular waveguide is attached and truncated by a port for computational modeling purposes. (The physical waveguide would normally extend beyond the ports, but that part is not included in the computational model considered here.)

Modal Representation for a Rectangular Waveguide

In an air-filled rectangular waveguide with the transverse dimensions L_x and L_y, we can decompose the electric and magnetic fields into *transverse electric* (TE) and *transverse magnetic* (TM) modes; see [19] for a detailed discussion. Each mode has its own propagation constant

$$k_z = \sqrt{\left(\frac{\omega}{c}\right)^2 - k_t^2}, \tag{A.6}$$

where k_t^2 are the eigenvalues of the transverse problem for H_z (TE case) or E_z (TM case), i.e.,

$$k_t^2 = \left(\frac{n_x \pi}{L_x}\right)^2 + \left(\frac{n_y \pi}{L_y}\right)^2. \tag{A.7}$$

For TE modes, n_x and n_y are nonnegative integers that satisfy $n_x + n_y > 0$. For TM modes, both n_x and n_y are positive integers.

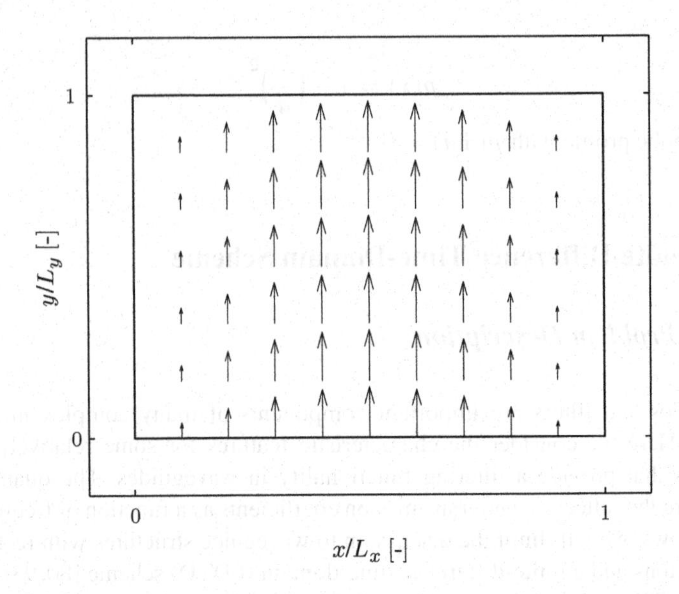

Fig. A.2 Modal field of TE_{10} mode

Numbering the modes from 1 to ∞, we can express the electric field in the waveguide as a superposition of both TE and TM modes by

$$\boldsymbol{E}(x, y, z, t) = \sum_{m=1}^{\infty} V_m(z, t)\, \boldsymbol{e}_m(x, y), \tag{A.8}$$

where $V_m(z, t)$ is the modal amplitude, or *voltage*, of mode m (which can be either a TE or a TM mode) and $\boldsymbol{e}_m(x, y)$ is its modal field. The modal field, $\boldsymbol{e}_m(x, y)$, for the TE_{10} mode is shown in Fig. A.2.

In what follows, we will consider situations where the frequency range of interest and the dimensions (L_x, L_y) of the waveguide are chosen such that k_z is real for the TE_{10} mode and imaginary for all other modes. Thus, the only mode that propagates is the TE_{10} mode. All the other modes are evanescent and decay exponentially along the waveguide axis. Consequently, the TE_{10} mode is the only mode present at a sufficiently large distance from any source or irregularity in the waveguide that may excite higher-order modes.

Computation of Scattering Parameters

A filter can be characterized in terms of its reflection and transmission coefficient, and in a more general setting, these are often referred to as the scattering parameters, or simply the S-parameters. Fig. A.3 shows a rectangular waveguide (without the filtering structure) that is truncated at two ports for computational modeling

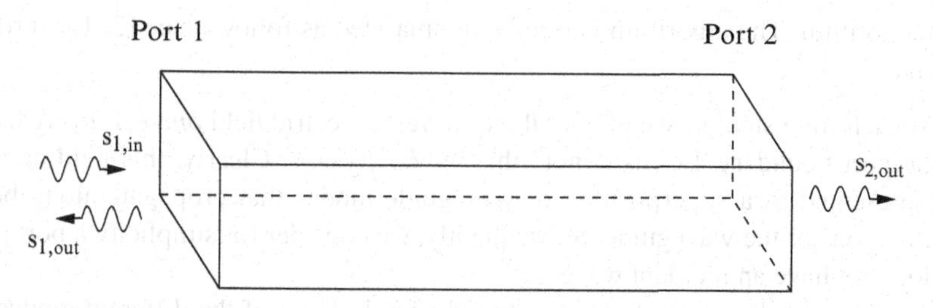

Fig. A.3 Illustration of incoming and outgoing waves

purposes. The S-parameters can be computed given the relation between the amplitudes of the TE_{10} mode at the ports: (1) an incident wave is launched at one port, (2) the reflected wave is recorded at the same port, and (3) the transmitted wave is recorded at the other port.

Let $s_{1,\text{in}}(t)$ be the amplitude of the incoming TE_{10} wave at port 1, and let $s_{1,\text{out}}(t)$ and $s_{2,\text{out}}(t)$ be the amplitudes of the outgoing TE_{10} waves at ports 1 and 2, respectively. The Fourier transform of these signals gives $S_{1,\text{in}}(\omega)$, $S_{1,\text{out}}(\omega)$, and $S_{2,\text{out}}(\omega)$. The relation between the amplitudes at the two ports is usually described by the so-called S-parameters:

$$S_{11}(\omega) = \frac{S_{1,\text{out}}(\omega)}{S_{1,\text{in}}(\omega)}, \tag{A.9}$$

$$S_{12}(\omega) = \frac{S_{2,\text{out}}(\omega)}{S_{1,\text{in}}(\omega)}. \tag{A.10}$$

The scattering parameter S_{11} is recognized as the reflection coefficient and S_{12} as the transmission coefficient. A further extension to an n-port network is rather straightforward and for such cases it is convenient to represent the S-parameters in matrix form, an $n \times n$ scattering matrix $\mathbf{S}$ with the elements S_{ij}.

Numerical Modeling

The interior of the waveguide is discretized by an FDTD grid [80, 93]. A wave can be launched at one of the ports and then propagated through the waveguide by means of Maxwell's equations represented by the FDTD scheme applied to the grid in between the ports. Consequently, a filtering structure can be modeled in detail by the FDTD scheme and its reflection and the transmission coefficient computed from the fields at the ports.

The special type of boundary condition that is required at the ports is already implemented in the MATLAB program provided as a starting point for the tasks below. However, it is useful and interesting to have some understanding of the port algorithm. The algorithm is briefly summarized as follows (see [3] for further details):

- At each time step, n, we extract the transverse electric field *one cell away* from the port boundary. Let us denote this by $\boldsymbol{E}_t|^n_{p,q,N_z-1}$. Clearly, this field can be represented as a superposition of waveguide modes that propagate along both directions of the waveguide. Subsequently, we consider for simplicity a port that does not have an incident wave.
- With this result we can compute the voltages $V_m|^n_{N_z-1}$ of the different modes m one cell away from the boundary:

$$V_m|^n_{N_z-1} = \sum_{p,q} \Delta x \, \Delta y \, \boldsymbol{E}_t|^n_{p,q,N_z-1} \cdot \boldsymbol{e}_m|_{p,q}. \tag{A.11}$$

  For a waveguide port without an incident field, the decomposed field only consists of waveguide modes that are propagating away from the interior of the computational domain. For a port with an incident field, we could easily compute the incident field at the plane one cell away from the port boundary. Then we subtract the incident field from the total field to get the field associated with modes that are propagating away from the computational domain.
- Each mode can be modeled by a 1D wave equation:

$$\frac{\partial^2 V_m}{\partial z^2} - \frac{1}{c_0^2}\frac{\partial^2 V_m}{\partial t^2} - h_m^2 V_m = 0, \tag{A.12}$$

  which can be discretized as

$$V_m|^{n+1}_r = A V_m|^n_r + B(V_m|^n_{r+1} + V_m|^n_{r-1}) - V_m|^{n-1}_r, \tag{A.13}$$

  where $V_m|^n_r \equiv V_m(r\Delta z, n\Delta t)$ and

$$B = \left(\frac{c_0 \Delta t}{\Delta z}\right)^2,$$

$$A = 2 - 2B - (c_0 \Delta t \, h_m)^2.$$

  The impulse response $I_m|^n = V_m|^n_1$ for this 1D wave equation can be computed from Eq. (A.13). Given this impulse response, we can use a 1D convolution to compute the voltages *on* the boundary that coincides with the port:

$$V_m|^n_{N_z} = V_m|^n_{N_z-1} * I_m|^n = \sum_{j=1}^{n} V_m|^{n-j}_{N_z-1} \cdot I_m|^j. \tag{A.14}$$

- Now, we know the modal voltages on the port boundary. The total electric field on the boundary is a linear combination of the modal fields

$$\boldsymbol{E}|^n_{p,q,N_z} = \sum_{m=1}^{M} V_m|^n_{N_z}\, \boldsymbol{e}_m|_{p,q}, \tag{A.15}$$

and this solution is explicitly written into the FDTD grid before the next update of the interior grid points that are located inside the computational domain.

Implementation: A Point of Departure

The following supporting MATLAB script and functions are set up such that they can be used directly as a basis for the solution of the assignments that follow. Here is a brief description of each MATLAB file:

- `Main.m`: Setup of the problem with allocation of memory for variables that store the electromagnetic fields, port information, excitation pulse, etc. Given the setup of the computational problem, the routine contains a loop that should include the update expressions of the FDTD scheme in the bulk of the computational domain. Further, extraction of the scattering parameters and addition of possible metal objects are included in this routine.
- `ComputeTEModes.m` : Computes the transverse electric field for the TE modes and the corresponding cutoff frequencies associated with the FDTD discretization.
- `ComputeTMModes.m` : Computes the transverse electric field for the TM modes and the corresponding cutoff frequencies associated with the FDTD discretization.
- `ComputeIR.m`: Computes the impulse response for all the modes included in the analysis.

If you prefer to use another programming language, these MATLAB files could also be used as a point of departure. In that case, you will have to implement this functionality in the language of your choice. It deserves to be emphasized that if you prefer to work with MATLAB, it would be useful to read and study programs written by other programmers since that would give you the opportunity to develop your programming style and find new solutions to programming problems that you may encounter. In what follows, it is assumed that the functionality provided by `Main.m`, `ComputeTEModes.m`, `ComputeTMModes.m`, and `ComputeIR.m` is in place.

A.3.2 Assignments

Consider an empty waveguide with the dimensions:

$$L_x = 40.0 \text{ mm},\; L_y = 22.5 \text{ mm, and } L_z = 160.0 \text{ mm}. \tag{A.16}$$

The waveguide ports are located at $z = 0$ mm and $z = 160.0$ mm. The waveguide is excited by a TE_{10} mode at $z = 0$ by a Gaussian-modulated sinusoidal pulse, which contains energy in the frequency interval from 4 to 7 GHz. The port located at $z = 0$ is transparent to the reflected field, which essentially corresponds to the rectangular waveguide's continuing indefinitely for the region $z < 0$. Similarly, the port located at $z = L_z$ is transparent to the transmitted field, which continues to propagation in the positive z-direction as if the waveguide continued indefinitely for the region $z > L_z$. The MATLAB implementation (which may be used as a point of departure) is set up for these particular dimensions and this excitation.

Implementation

Implement the update loops for Faraday's and Ampère's law according to the FDTD scheme in three dimensions for an empty rectangular waveguide. It is sufficient to use a cell size of $h = 2.5$ mm for the following tests, which makes the computational domain consist of $16 \times 9 \times 64$ cells. However, it is often useful to make an implementation such that it is possible to change the cell size to allow for convergence studies.

Numerical Test #1

What is the expected reflected $s_{1,\text{out}}(t)$ and transmitted $s_{2,\text{out}}(t)$ solution for an empty waveguide given the Gaussian excitation pulse? Test your code and see if the result is what you expected. What is the cutoff frequency of the TE_{10} mode? Which mode has the second lowest cutoff frequency, and what frequency is that?

Implementation

Implement a postprocessing step that transforms the time-domain scattering amplitudes $s_{1,\text{out}}(t)$ and $s_{2,\text{out}}(t)$ to their corresponding frequency-domain quantities, and provide code that evaluates the scattering parameters (A.9) and (A.10).

Numerical Test #2

Verify that your implementation of the postprocessing step works as expected for the empty waveguide. You can calculate the analytical scattering parameters (A.9) and (A.10), which makes a careful comparison feasible. Comment on your findings by an interpretation of the numerical errors.

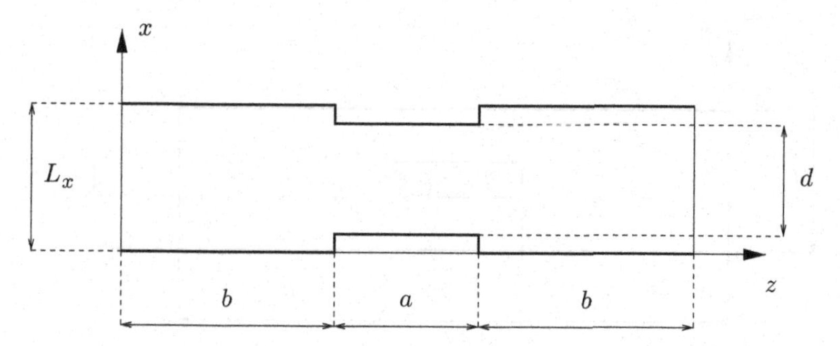

Fig. A.4 Waveguide with a narrow midsection

Implementation

Change the program so that you can analyze a waveguide that has a somewhat more narrow midsection, as shown in Fig. A.4. The dimensions are a = 4 cm, b = 6 cm, and d = 3 cm. The geometry is independent of the y-coordinate and symmetric with respect to the plane $x = L_x/2$. (The extra walls of the waveguide are also perfectly conducting.) It may be noted that this problem can also be solved by a 2D code, should such a code be available, for comparative purposes.

Numerical Test #3

Compute $|S_{11}(\omega)|$ and $|S_{12}(\omega)|$ for this modified geometry. Comment on your findings and provide an explanation of the transmission (and reflection) as a function of frequency.

Implementation

A bit more challenging problem is a metallic block placed "on the floor" of the waveguide, as shown in Fig. A.5, which contains a top and side view of the metallic block indicated by the shaded region. The dimensions in Fig. A.5 are a = 2 cm, b = 7 cm, d = 0.5 cm, and h = 1.75 cm. The geometry is again symmetric with respect to the plane $x = L_x/2$, but this geometry yields a full 3D problem.

Numerical Test #4

Compute $|S_{11}(\omega)|$ and $|S_{12}(\omega)|$ for this modified geometry and comment on your findings. In fact, the added metal block yields a filtering structure with pass band characteristics. Can you explain the physics that causes this resonance? The field computation may give some information about the resonance.

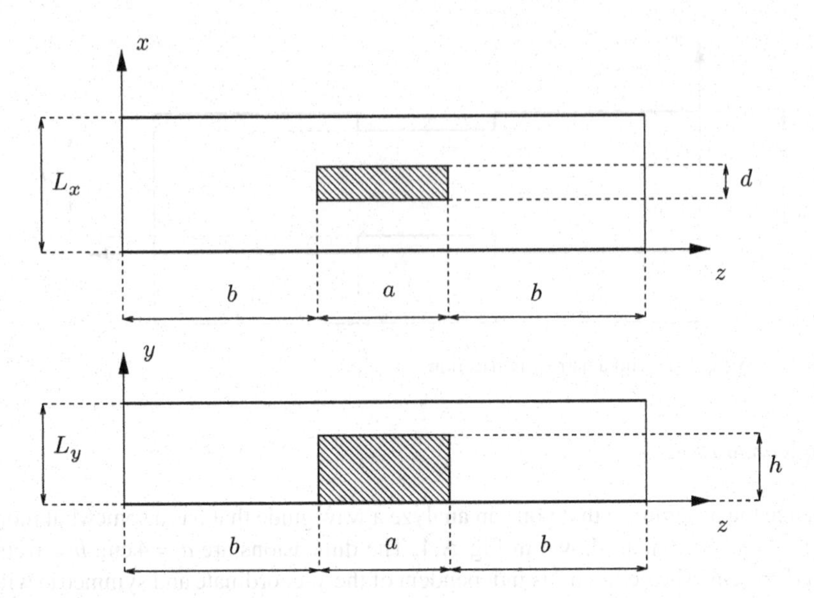

Fig. A.5 Waveguide with metal block placed "on the floor"

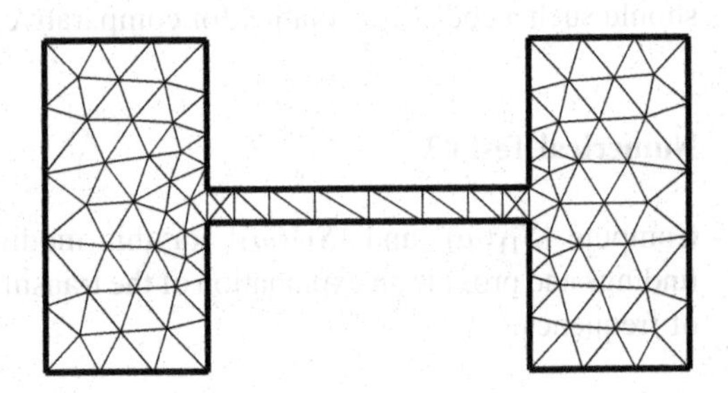

Fig. A.6 Cross section of a ridge waveguide discretized by *triangular finite elements*

A.4 Finite Element Method

A.4.1 Problem Description

A ridge waveguide has a cross section designed to allow for a single mode of propagation over larger bandwidths than a rectangular waveguide. A typical cross section of a ridge waveguide is shown in Fig. A.6. In what follows, the cross section of the waveguide is denoted by S and its closed boundary by L.

To compute the cutoff frequencies, we solve the eigenvalue problem

$$-\nabla^2 H_z = k_t^2 H_z \qquad \text{in } S, \tag{A.17}$$

$$\hat{\boldsymbol{n}} \cdot \nabla H_z = 0 \qquad \text{on } L \tag{A.18}$$

for the TE modes. For the TM modes we solve

$$-\nabla^2 E_z = k_t^2 E_z \qquad \text{in } S, \tag{A.19}$$

$$E_z = 0 \qquad \text{on } L, \tag{A.20}$$

where S is the interior of the waveguide's cross section and L its boundary. The transverse wave number is denoted by k_t and the longitudinal wave number by k_z, i.e., $k^2 = (\omega/c_0)^2 = k_t^2 + k_z^2$. More information on the theory of waveguides can be found in the literature, cf. [19].

Implementation: A Point of Departure

The following supporting MATLAB script and functions are set up such that they can be used directly as a basis for the solution of the assignments that follow. Here is a brief description of each MATLAB file:

- `Main.m` : Reads the grid, assembles the matrices, and solves the eigenvalue problem.
- `CmpElMtx.m` : Empty function where you can implement the computation of the element matrices.
- `ReadGrid.m` : Empty function where you can implement the reading of the meshes.
- `VisualizeMode.m`: Empty function where you can implement the visualization of the eigenmodes. This function should visualize two fields: (1) the field component parallel to the axis of the waveguide by means of colors and (2) its curl by vectors.

If you prefer to use another programming language, these MATLAB files could also be used as a point of departure. In that case, you will have to implement this functionality in the language of your choice. It deserves to be emphasized that if you prefer to work with MATLAB, it would be useful to read and study programs written by other programmers as that would give you the opportunity to develop your programming style and find new solutions to programming problems that you may encounter. In what follows, it is assumed that the functionality provided by `Main.m`, `CmpElMtx.m`, `ReadGrid.m`, and `Visualize.m` is in place.

Meshes

The MATLAB files come with meshes, which are stored in plain text files. The meshes discretize both a rectangular waveguide and a ridge waveguide. For each geometry, there are three discretizations that can be used for convergence studies.

- Meshes for rectangular waveguides of width $L_x = 2$ cm and height $L_y = 1$ cm.

 - `grid_rectangular_res1.txt`—coarse mesh
 - `grid_rectangular_res2.txt`—once hierarchically refined mesh
 - `grid_rectangular_res3.txt`—twice hierarchically refined mesh

- Mesh for a ridge waveguide of outer dimensions 2 and 1 cm. The spacing between the teeth is 0.1 cm, and their width is 1 cm. The mesh is shown in Fig. A.6.
 - `grid_ridge_res1.txt` – coarse mesh
 - `grid_ridge_res2.txt` – once hierarchically refined mesh
 - `grid_ridge_res3.txt` – twice hierarchically refined mesh

If you prefer to create your own meshes, there is a number of choices: (1) use a commercial tool, (2) use a freely available tool such as Triangle [73, 74], or (3) write your own mesh generator.

A.4.2 Assignments

Weak Form

Use the two eigenvalue problems as a point of departure to show that the corresponding weak forms can be written as

$$\int_S \nabla w_i \cdot \nabla H_z \, dS = k_t^2 \int_S w_i \, H_z \, dS,$$

$$\int_S \nabla w_i \cdot \nabla E_z \, dS = k_t^2 \int_S w_i \, E_z \, dS.$$

Here, the z-component of the electric and magnetic fields are expanded in and tested by nodal basis functions ϕ_i. The testing must be done in accordance with the boundary conditions. Observe that for the TE modes we have a Neumann boundary condition, while for the TM modes we have a Dirichlet boundary condition. What are the implications of this? How do the two problems differ from each other?

Element Matrices

The two eigenvalue problems can be expressed in the form $\mathbf{Ax} = k_t^2\mathbf{Bx}$. Here, $\mathbf{A}$ and $\mathbf{B}$ are square matrices having different dimensions for the two different eigenvalue problems. To assemble the matrices $\mathbf{A}$ and $\mathbf{B}$, we sum contributions from each triangle, i.e., we need to compute the following integrals:

$$A_{ij}^e = \int_{S^e} \nabla \phi_i \cdot \nabla \phi_j \, dS,$$

$$B_{ij}^e = \int_{S^e} \phi_i \phi_j \, dS.$$

To evaluate these integrals, you may find the following formula useful:

$$\int_{S^e} (\phi_1^e)^\alpha (\phi_2^e)^\beta (\phi_3^e)^\gamma \, dS = 2S^e \frac{\alpha!\beta!\gamma!}{(\alpha + \beta + \gamma + 2)!}, \tag{A.21}$$

where S^e is the area of element e. Here, the constants α, β, and γ are nonnegative integers.

Implementation

Implement the functionality of `ReadGrid.m`. This function takes the name of the file that stores the mesh as the argument. Then it reads the data in this file and returns the mesh in a format that is ready to use for the remaining program.

Numerical Test #1

Compute the 20 lowest k_t and their corresponding cutoff frequencies for the rectangular waveguide and compare to the analytical expression

$$k_t = \frac{\omega}{c_0} = \sqrt{\left(\frac{\pi n_x}{L_x}\right)^2 + \left(\frac{\pi n_y}{L_y}\right)^2}, \tag{A.22}$$

where $n_x = 0, 1, \ldots$ and $n_y = 0, 1, \ldots$, excluding $n_x = n_y = 0$.

- Perform a convergence test for the lowest eigenmode. Does this cutoff frequency converge to the analytical value? What is the order of convergence?
- Visualize the five lowest eigenmodes. Do the eigenmodes compare well with their analytical counterparts?

Numerical Test #2

Compute the 20 lowest k_t and their corresponding frequencies for the ridge waveguide. How do the two lowest cutoff frequencies change as compared to the rectangular waveguide? Compare the ratio f_2/f_1 between the two lowest frequencies ($f_2 > f_1$) for the ridge and rectangular waveguides. Can you explain the reason for this difference? How could you influence the ratio f_2/f_1 by changing the geometry?

Numerical Test #3

Perform a convergence test for the lowest eigenmode of the ridge waveguide. What is the extrapolated cutoff frequency? What is the order of convergence? Do you achieve an optimal order of convergence? If not, why is the order of convergence reduced?

Numerical Test #4

For the ridge waveguide, visualize both the longitudinal field component E_z together with $\nabla \times (\hat{z}E_z)$ for the five lowest eigenmodes. Also, visualize both the longitudinal field component H_z together with $\nabla \times (\hat{z}H_z)$ for the five lowest eigenmodes. Comment on how your visualizations relate to the cutoff frequencies. How do the eigenmodes compare with the rectangular waveguide modes?

A.5 Method of Moments

A.5.1 Problem Description

We seek the capacitance per unit length of a circular conducting wire placed over the middle of a conducting strip. The wire radius is a, and the center of the wire is located at a height h over the strip. The strip has width w and thickness b. The geometry is assumed to be two-dimensional and the structure is located in free space. For the computation of the capacitance per unit, it is therefore appropriate to use the method of moments.

For a capacitor where the two plates have different shape, the two plates must have opposite total charges. However, they will not in general have opposite potentials. Since Poisson's equation is linear, the charges and potentials must satisfy the linear relation

$$q_1 = C_{11}V_1 + C_{12}V_2,$$
$$q_2 = C_{21}V_1 + C_{22}V_2,$$

where q_i are the charges on the plates and V_i their potentials. The matrix elements C_{ij} are referred to as capacitance coefficients. Note that the capacitance coefficients in themselves are not physically very relevant and that they depend on how the normalizing distance is chosen in the expression for the potential from a line charge. The coefficients can be computed by considering two cases where V_1 and V_2 take linearly independent values, e.g.,

1. $V_1 = 1$ and $V_2 = 0$ yield the values for C_{11} and C_{21}.
2. $V_1 = 0$ and $V_2 = 1$ yield the values for C_{12} and C_{22}.

To define the capacitance in cylindrical geometry, we consider cases where the net charge is zero. This implies that $q_1 = -q_2 = q$. Then the capacitance is defined as

$$C = \frac{q}{V_1 - V_2}. \tag{A.23}$$

A.5.2 *Assignments*

First express the net capacitance C analytically in terms of the capacitance coefficients C_{11}, C_{12}, C_{21}, and C_{22}.

Implementation

Write a program to evaluate the capacitance of a general 2D, two-conductor capacitor and geometry part to generate the geometry described previously. For the 2D method of moments, you can use the MATLAB function `2DMoM.m` described in Sect. 7.2.2.

Numerical Test #1

Test the program by running the case $a = 1\,\text{mm}$, $b = 0.5\,\text{mm}$, $d = 2\,\text{mm}$, and $w = 20\,\text{mm}$. Determine the order of convergence and make an extrapolation to zero grid size within 1 % accuracy.

Numerical Test #2

Verify that your result for C is reasonably close to the analytical result for an infinite plate, i.e., $w = \infty$. (The analytical result can be derived by means of image theory for a circular cylinder next to an infinitely large ground plane, where details can be found in [19].) For this part, give the extrapolated coefficients C_{ij} as intermediate results.

Numerical Test #3

Keeping all the other parameters the same, set $w = 0.5\,\text{mm}$. Check the order of convergence and extrapolate to zero grid size, with at most 1 % error.

Numerical Test #4

Now you have some idea of what resolution is required. Based on this experience, compute and plot the capacitance as a function of width for the interval $0.2\,\text{mm} < w < 5\,\text{mm}$.

Appendix B
A Collection of the Lowest-Order Finite Elements

This book describes a number of different types of approximations that exploit representations in terms of finite elements. Useful information concerning these approximations is collected in this appendix for a range of typical element shapes: (1) the triangle, (2) the quadrilateral, (3) the tetrahedron, (4) the prism, and (5) the hexahedron. The following types of information are provided:

- Domain for the reference element.
- Expressions for the nodes (vertices) of the element.
- Definition of the edges of the element.
- Definition of the faces of the element should the element be a volume element.
- The lowest-order gradient-conforming basis functions that belong to the function space $H(\text{grad})$.
- The lowest-order curl-conforming basis functions that belong to the function space $H(\text{curl})$.
- The lowest-order divergence-conforming basis functions that belong to the function space $H(\text{div})$.
- The quadrature rule that can integrate quadratic polynomials on the reference element sufficiently well to yield an FEM with an error that is proportional to h^2.

Here, the function space $H(\text{grad})$ consists of all functions ϕ that satisfy

$$\int_{\Omega} |\nabla\phi|^2 + \phi^2 d\Omega < \infty, \tag{B.1}$$

where the computational domain Ω is one, two, or three dimensional. Similarly, the function space $H(\text{curl})$ consists of all functions $\boldsymbol{F}$ that satisfy

$$\int_{\Omega} |\nabla \times \boldsymbol{F}|^2 + |\boldsymbol{F}|^2 d\Omega < \infty, \tag{B.2}$$

T. Rylander et al., *Computational Electromagnetics*, Texts in Applied Mathematics 51, DOI 10.1007/978-1-4614-5351-2,

and the function space $H(\mathrm{div})$ consists of all functions $\boldsymbol{F}$ that satisfy

$$\int_{\Omega} |\nabla \cdot \boldsymbol{F}|^2 + |\boldsymbol{F}|^2 d\Omega < \infty. \tag{B.3}$$

Each gradient-conforming basis function φ_i (also referred to as nodal basis functions) is equal to unity at one node in the element and zero at all the other nodes. The node where the basis function evaluates to unity has the same index as the basis function itself. Consequently, there is exactly one nodal basis function for each node in the element. This can be expressed compactly as

$$\varphi_i(\boldsymbol{r}_j) = \delta_{ij}, \tag{B.4}$$

where δ_{ij} is the Kronecker delta that gives $\delta_{ij} = 1$ for $i = j$ and $\delta_{ij} = 0$ for $i \neq j$.

Each curl-conforming basis function $\boldsymbol{N}_i$ (also referred to as edge basis functions or edge elements) has a nonzero tangential component along one edge of the element and a zero tangential component along all the other edges of the element. Consequently, there is exactly one edge basis function associated with each edge in the element. This property of the curl-conforming basis functions listed in this appendix can be expressed compactly as

$$\hat{t}_j \cdot \boldsymbol{N}_i(\boldsymbol{r}_j) = \frac{1}{L_i}\delta_{ij} \;\; \forall\, \boldsymbol{r}_j \in \text{edge } j, \tag{B.5}$$

where $\hat{\boldsymbol{t}}_j$ is a unit vector tangential to the edge such that it points from the start node of the edge to the end node of edge i. Further, L_i is the length of edge i. Thus, it is easy to construct a basis function with a unit tangential component along its associated edge by the scaling $L_i \boldsymbol{N}_i$.

Each divergence-conforming basis function $\boldsymbol{M}_i$ (also referred to as a face basis function for volume elements) has a nonzero normal component on each edge (for surface elements) or face (for volume elements) and has a zero normal component on all the other edges/faces of the element. Consequently, there is exactly one divergence-conforming basis function for each edge/face of an element. This property of the divergence-conforming basis functions listed in this appendix can be expressed compactly as

$$\hat{\boldsymbol{n}}_j \cdot \boldsymbol{M}_i(\boldsymbol{r}_j) = \frac{1}{L_i}\delta_{ij} \;\; \forall\, \boldsymbol{r}_j \in \text{edge } j \tag{B.6}$$

for divergence-conforming basis functions on surface elements, i.e., the triangle and the quadrilateral. Here, $\hat{\boldsymbol{n}}_j$ is the outward unit normal to edge j. Similarly, this property of the divergence-conforming basis functions listed in this appendix can be expressed compactly as

$$\hat{\boldsymbol{n}}_j \cdot \boldsymbol{M}_i(\boldsymbol{r}_j) = \frac{1}{A_i}\delta_{ij} \;\; \forall\, \boldsymbol{r}_j \in \text{face } j \tag{B.7}$$

for divergence-conforming basis functions on the volume elements tetrahedron, prism, and hexahedron. Here, $\hat{\boldsymbol{n}}_j$ is the outward unit normal to face j. Furthermore, A_i is the area of face i. Thus, it is easy to construct a basis function with unit normal component by the scaling $L_i \boldsymbol{M}_i$ for the surface elements and $A_i \boldsymbol{M}_i$ for the volume elements.

The collection of MATLAB programs associated with this book also contains a program that exploits Symbolic Math Toolbox in MATLAB to calculate the basis functions listed in the following discussion, where this program can also visualize the basis functions on the reference elements.

B.1 2D Elements

B.1.1 Triangle

The linear reference triangle occupies the surface bounded by $0 \leq u \leq 1 - v$ and $0 \leq v \leq 1$. It has three vertices and three edges.

The nodes for the reference element are given by

$$\boldsymbol{r}_1 = [0, 0, 0],$$
$$\boldsymbol{r}_2 = [+1, 0, 0],$$
$$\boldsymbol{r}_3 = [0, +1, 0].$$

The edges for the reference element are listed in Table B.1.

Linear Basis Functions for $H(\text{grad})$

The node basis functions for the reference element are given by

$$\varphi_1 = 1 - u - v,$$
$$\varphi_2 = u,$$
$$\varphi_3 = v.$$

Table B.1 Definition of edges for triangular reference element: node 1—start node of edge; and node 2—end node of edge

Edge	Node #1	Node #2
1	1	2
2	2	3
3	3	1

Table B.2 Quadrature points for reference triangle

Point	u-coordinate	v-coordinate
1	$6.666666666666667 \times 10^{-1}$	$1.666666666666667 \times 10^{-1}$
2	$1.666666666666667 \times 10^{-1}$	$6.666666666666667 \times 10^{-1}$
3	$1.666666666666667 \times 10^{-1}$	$1.666666666666667 \times 10^{-1}$

Table B.3 Quadrature weights for reference triangle

Point	Weight
1	$1.666666666666667 \times 10^{-1}$
2	$1.666666666666667 \times 10^{-1}$
3	$1.666666666666667 \times 10^{-1}$

Linear Basis Functions for $H(\text{curl})$

The edge basis functions for the reference element are given by

$$\begin{aligned}
\boldsymbol{N}_1 &= \varphi_1 \tilde{\nabla} \varphi_2 - \varphi_2 \tilde{\nabla} \varphi_1 = (1-v)\hat{\boldsymbol{u}} + u\hat{\boldsymbol{v}}, \\
\boldsymbol{N}_2 &= \varphi_2 \tilde{\nabla} \varphi_3 - \varphi_3 \tilde{\nabla} \varphi_2 = -v\hat{\boldsymbol{u}} + u\hat{\boldsymbol{v}}, \\
\boldsymbol{N}_3 &= \varphi_3 \tilde{\nabla} \varphi_1 - \varphi_1 \tilde{\nabla} \varphi_3 = -v\hat{\boldsymbol{u}} + (u-1)\hat{\boldsymbol{v}}.
\end{aligned}$$

Linear Basis Functions for $H(\text{div})$

The face basis functions for the reference element are given by

$$\begin{aligned}
\boldsymbol{M}_1 &= \boldsymbol{N}_1 \times \hat{\boldsymbol{w}} = u\hat{\boldsymbol{u}} + (v-1)\hat{\boldsymbol{v}}, \\
\boldsymbol{M}_2 &= \boldsymbol{N}_2 \times \hat{\boldsymbol{w}} = u\hat{\boldsymbol{u}} + v\hat{\boldsymbol{v}}, \\
\boldsymbol{M}_3 &= \boldsymbol{N}_3 \times \hat{\boldsymbol{w}} = (u-1)\hat{\boldsymbol{u}} + v\hat{\boldsymbol{v}},
\end{aligned}$$

where $\hat{\boldsymbol{w}} = \hat{\boldsymbol{u}} \times \hat{\boldsymbol{v}}$.

Quadrature Rule

The quadrature points in Table B.2 and the corresponding weights in Table B.3 yield a quadrature rule that integrates quadratic polynomials exactly on the reference triangle.

Table B.4 Definition of edges for the quadrilateral reference element: Node #1—start node of the edge; and Node #2—end node of the edge

Edge	Node 1	Node 2
1	1	2
2	2	3
3	3	4
4	4	1

B.1.2 Quadrilateral

The linear reference quadrilateral occupies the surface bounded by $-1 \leq u \leq 1$ and $-1 \leq v \leq 1$. It has four vertices and four edges.

The nodes for the reference element are given by

$$\begin{aligned} \boldsymbol{r}_1 &= [-1, -1, 0], \\ \boldsymbol{r}_2 &= [+1, -1, 0], \\ \boldsymbol{r}_3 &= [+1, +1, 0], \\ \boldsymbol{r}_4 &= [-1, +1, 0]. \end{aligned}$$

The edges for the reference element are given in Table B.4.

Linear Basis Functions for H(grad)

The node basis functions for the reference element are given by

$$\begin{aligned} \varphi_1 &= \psi^-(u)\,\psi^-(v) = \frac{1}{4}(1-u)(1-v), \\ \varphi_2 &= \psi^+(u)\,\psi^-(v) = \frac{1}{4}(1+u)(1-v), \\ \varphi_3 &= \psi^+(u)\,\psi^+(v) = \frac{1}{4}(1+u)(1+v), \\ \varphi_4 &= \psi^-(u)\,\psi^+(v) = \frac{1}{4}(1-u)(1+v). \end{aligned}$$

Here, we use the basis functions

$$\begin{aligned} \psi^-(\xi) &= \frac{1}{2}(1-\xi), \\ \psi^+(\xi) &= \frac{1}{2}(1+\xi). \end{aligned}$$

Linear Basis Functions for $H(\mathbf{curl})$

The edge basis functions for the reference element are given by

$$\boldsymbol{N}_1 = \psi^-(v)\tilde{\nabla}\psi^+(u) = \frac{1}{4}(1-v)\hat{\boldsymbol{u}},$$

$$\boldsymbol{N}_2 = \psi^+(u)\tilde{\nabla}\psi^+(v) = \frac{1}{4}(1+u)\hat{\boldsymbol{v}},$$

$$\boldsymbol{N}_3 = \psi^+(v)\tilde{\nabla}\psi^-(u) = -\frac{1}{4}(1+v)\hat{\boldsymbol{u}},$$

$$\boldsymbol{N}_4 = \psi^-(u)\tilde{\nabla}\psi^-(v) = -\frac{1}{4}(1-u)\hat{\boldsymbol{v}}.$$

Linear Basis Functions for $H(\mathrm{div})$

The face basis functions for the reference element are given by

$$\boldsymbol{M}_1 = \boldsymbol{N}_1 \times \hat{\boldsymbol{w}} = -\frac{1}{4}(1-v)\hat{\boldsymbol{v}},$$

$$\boldsymbol{M}_2 = \boldsymbol{N}_2 \times \hat{\boldsymbol{w}} = \frac{1}{4}(1+u)\hat{\boldsymbol{u}},$$

$$\boldsymbol{M}_3 = \boldsymbol{N}_3 \times \hat{\boldsymbol{w}} = \frac{1}{4}(1+v)\hat{\boldsymbol{v}},$$

$$\boldsymbol{M}_4 = \boldsymbol{N}_4 \times \hat{\boldsymbol{w}} = -\frac{1}{4}(1-u)\hat{\boldsymbol{u}}.$$

Quadrature Rule

The quadrature points in Table B.5 and the corresponding weights in Table B.6 yield trapezoidal integration for the reference quadrilateral, by means of the product of two 1D trapezoidal integration rules. This quadrature rule provides mass lumping for rectangular elements and, despite the fact that it cannot integrate quadratic

Table B.5 Quadrature points for reference quadrilateral

Point	u-coordinate	v-coordinate
1	−1.000000000000000	−1.000000000000000
2	1.000000000000000	−1.000000000000000
3	1.000000000000000	1.000000000000000
4	−1.000000000000000	1.000000000000000

Table B.6 Quadrature weights for reference quadrilateral

Point	Weight
1	1.000000000000000
2	1.000000000000000
3	1.000000000000000
4	1.000000000000000

polynomials exactly, yields a, FEM with second order of convergence for a piecewise linear approximation of the field.

B.2 3D Elements

B.2.1 Tetrahedron

The linear reference tetrahedron occupies the volume bounded by $0 \leq u \leq 1-v-w$, $0 \leq v \leq 1-w$, and $0 \leq w \leq 1$. It has four vertices, six edges, and four triangular faces.

The nodes for the reference element are given by

$$\begin{aligned} \boldsymbol{r}_1 &= [0,0,0], \\ \boldsymbol{r}_2 &= [+1,0,0], \\ \boldsymbol{r}_3 &= [0,+1,0], \\ \boldsymbol{r}_4 &= [0,0,+1]. \end{aligned}$$

The edges for the reference element are given in Table B.7.

The faces for the reference element are given in Table B.8.

Linear Basis Functions for H(grad)

The node basis functions for the reference element are given by

$$\begin{aligned} \varphi_1 &= 1-u-v-w, \\ \varphi_2 &= u, \\ \varphi_3 &= v, \\ \varphi_4 &= w. \end{aligned}$$

Table B.7 Definition of edges for tetrahedral reference element: node 1—start node of edge; node 2—end node of edge

Edge	Node 1	Node 2
1	1	2
2	2	3
3	3	1
4	1	4
5	2	4
6	3	4

Table B.8 Definition of faces for tetrahedral reference element: node 1—first node of face; node 2—second node of face; and node 3—third node of face

Face	Node 1	Node 2	Node 3
1	3	2	1
2	1	2	4
3	2	3	4
4	3	1	4

Linear Basis Functions for $H(\mathrm{curl})$

The edge basis functions for the reference element are given by

$$\boldsymbol{N}_1 = \varphi_1 \tilde{\nabla}\varphi_2 - \varphi_2 \tilde{\nabla}\varphi_1 = (1 - w - v)\hat{\boldsymbol{u}} + u\hat{\boldsymbol{v}} + u\hat{\boldsymbol{w}},$$

$$\boldsymbol{N}_2 = \varphi_2 \tilde{\nabla}\varphi_3 - \varphi_3 \tilde{\nabla}\varphi_2 = -v\hat{\boldsymbol{u}} + u\hat{\boldsymbol{v}},$$

$$\boldsymbol{N}_3 = \varphi_3 \tilde{\nabla}\varphi_1 - \varphi_1 \tilde{\nabla}\varphi_3 = -v\hat{\boldsymbol{u}} + (u + w - 1)\hat{\boldsymbol{v}} - v\hat{\boldsymbol{w}},$$

$$\boldsymbol{N}_4 = \varphi_1 \tilde{\nabla}\varphi_4 - \varphi_4 \tilde{\nabla}\varphi_1 = w\hat{\boldsymbol{u}} + w\hat{\boldsymbol{v}} + (1 - v - u)\hat{\boldsymbol{w}},$$

$$\boldsymbol{N}_5 = \varphi_2 \tilde{\nabla}\varphi_4 - \varphi_4 \tilde{\nabla}\varphi_2 = -w\hat{\boldsymbol{u}} + u\hat{\boldsymbol{w}},$$

$$\boldsymbol{N}_6 = \varphi_3 \tilde{\nabla}\varphi_4 - \varphi_4 \tilde{\nabla}\varphi_3 = -w\hat{\boldsymbol{v}} + v\hat{\boldsymbol{w}}.$$

Linear Basis Functions for $H(\mathrm{div})$

The face basis functions for the reference element are given by

$$\begin{aligned}\boldsymbol{M}_1 &= 2(\varphi_3 \tilde{\nabla}\varphi_2 \times \tilde{\nabla}\varphi_1 + \varphi_2 \tilde{\nabla}\varphi_1 \times \tilde{\nabla}\varphi_3 + \varphi_1 \tilde{\nabla}\varphi_3 \times \tilde{\nabla}\varphi_2)\\ &= 2[u\hat{\boldsymbol{u}} + v\hat{\boldsymbol{v}} + (w - 1)\hat{\boldsymbol{w}}],\end{aligned}$$

$$\begin{aligned}\boldsymbol{M}_2 &= 2(\varphi_1 \tilde{\nabla}\varphi_2 \times \tilde{\nabla}\varphi_4 + \varphi_2 \tilde{\nabla}\varphi_4 \times \tilde{\nabla}\varphi_1 + \varphi_4 \tilde{\nabla}\varphi_1 \times \tilde{\nabla}\varphi_2)\\ &= 2[u\hat{\boldsymbol{u}} + (v - 1)\hat{\boldsymbol{v}} + w\hat{\boldsymbol{w}}],\end{aligned}$$

$$\begin{aligned}\boldsymbol{M}_3 &= 2(\varphi_2 \tilde{\nabla}\varphi_3 \times \tilde{\nabla}\varphi_4 + \varphi_3 \tilde{\nabla}\varphi_4 \times \tilde{\nabla}\varphi_2 + \varphi_4 \tilde{\nabla}\varphi_2 \times \tilde{\nabla}\varphi_3)\\ &= 2[u\hat{\boldsymbol{u}} + v\hat{\boldsymbol{v}} + w\hat{\boldsymbol{w}}],\end{aligned}$$

Table B.9 Quadrature points for reference tetrahedron

Point	u-coordinate	v-coordinate	w-coordinate
1	$5.854101966249685 \times 10^{-1}$	$1.381966011250105 \times 10^{-1}$	$1.381966011250105 \times 10^{-1}$
2	$1.381966011250105 \times 10^{-1}$	$5.854101966249685 \times 10^{-1}$	$1.381966011250105 \times 10^{-1}$
3	$1.381966011250105 \times 10^{-1}$	$1.381966011250105 \times 10^{-1}$	$5.854101966249685 \times 10^{-1}$
4	$1.381966011250105 \times 10^{-1}$	$1.381966011250105 \times 10^{-1}$	$1.381966011250105 \times 10^{-1}$

Table B.10 Quadrature weights for reference tetrahedron

Point	Weight
1	$4.166666666666666 \times 10^{-2}$
2	$4.166666666666666 \times 10^{-2}$
3	$4.166666666666666 \times 10^{-2}$
4	$4.166666666666666 \times 10^{-2}$

$$
\begin{aligned}
\boldsymbol{M}_4 &= 2(\varphi_3 \tilde{\nabla}\varphi_1 \times \tilde{\nabla}\varphi_4 + \varphi_1 \tilde{\nabla}\varphi_4 \times \tilde{\nabla}\varphi_3 + \varphi_4 \tilde{\nabla}\varphi_3 \times \tilde{\nabla}\varphi_1) \\
&= 2[(u-1)\hat{\boldsymbol{u}} + v\hat{\boldsymbol{v}} + w\hat{\boldsymbol{w}}].
\end{aligned}
$$

Quadrature Rule

The quadrature points in Table B.9 and the corresponding weights in Table B.10 yield a quadrature rule that integrates quadratic polynomials exactly on the reference tetrahedron.

B.2.2 Prism

The linear reference prism occupies the volume bounded by $0 \le u \le 1-v$, $0 \le v \le 1$, and $-1 \le w \le 1$. It has six vertices, nine edges, two triangular faces, and three quadrilateral faces.

The nodes for the reference element are given by

$$
\begin{aligned}
\boldsymbol{r}_1 &= [0, 0, -1], \\
\boldsymbol{r}_2 &= [+1, 0, -1], \\
\boldsymbol{r}_3 &= [0, +1, -1], \\
\boldsymbol{r}_4 &= [0, 0, +1], \\
\boldsymbol{r}_5 &= [+1, 0, +1], \\
\boldsymbol{r}_6 &= [0, +1, +1].
\end{aligned}
$$

The edges for the reference element are given in Table B.11.

The faces for the reference element are given in Table B.12.

Table B.11 Definition of edges for pyramidal reference element: node 1—start node of edge; node 2—end node of edge

Edge	Node 1	Node 2
1	1	2
2	2	3
3	3	1
4	1	4
5	2	5
6	3	6
7	4	5
8	5	6
9	6	4

Table B.12 Definition of faces for tetrahedral reference element: node 1—first node of face; node 2—second node of face; node 3—third node of face; node 4—fourth node of face (for three quadrilateral faces only)

Face	Node 1	Node 2	Node 3	Node 4
1	3	2	1	–
2	1	2	5	4
3	2	3	6	5
4	3	1	4	6
5	4	5	6	–

Linear Basis Functions for $H(\text{grad})$

The node basis functions for the reference element are given by

$$\varphi_1 = \phi_1(u,v)\,\psi^-(w) = \frac{1}{2}(1-u-v)(1-w),$$

$$\varphi_2 = \phi_2(u,v)\,\psi^-(w) = \frac{1}{2}u(1-w),$$

$$\varphi_3 = \phi_3(u,v)\,\psi^-(w) = \frac{1}{2}v(1-w),$$

$$\varphi_4 = \phi_1(u,v)\,\psi^+(w) = \frac{1}{2}(1-u-v)(1+w),$$

$$\varphi_5 = \phi_2(u,v)\,\psi^+(w) = \frac{1}{2}u(1+w),$$

$$\varphi_6 = \phi_3(u,v)\,\psi^+(w) = \frac{1}{2}v(1+w).$$

We have the basis functions

$$\phi_1 = 1-u-v,$$

$$\phi_2 = u,$$

$$\phi_3 = v$$

that vary in the plane perpendicular to the cylinder axis of the prism and, thus, depend only on u and v. Further, the two basis functions

$$\psi^- = \frac{1}{2}(1 - w),$$

$$\psi^+ = \frac{1}{2}(1 + w)$$

vary along the cylinder axis with the coordinate w.

Linear Basis Functions for $H(\text{curl})$

The edge basis functions for the reference element are given by

$$N_1 = \boldsymbol{n}_1(u, v)\,\psi^-(w) = \frac{1}{2}\left[(v-1)(w-1)\hat{\boldsymbol{u}} - u(w-1)\hat{\boldsymbol{v}}\right],$$

$$N_2 = \boldsymbol{n}_2(u, v)\,\psi^-(w) = \frac{1}{2}\left[v(w-1)\hat{\boldsymbol{u}} - u(w-1)\hat{\boldsymbol{v}}\right],$$

$$N_3 = \boldsymbol{n}_3(u, v)\,\psi^-(w) = \frac{1}{2}\left[v(w-1)\hat{\boldsymbol{u}} - (u-1)(w-1)\hat{\boldsymbol{v}}\right],$$

$$N_4 = \phi_1(u, v)\,\tilde{\nabla}\psi^+(w) = \frac{1}{2}(1 - v - u)\hat{\boldsymbol{w}},$$

$$N_5 = \phi_2(u, v)\,\tilde{\nabla}\psi^+(w) = \frac{1}{2}u\hat{\boldsymbol{w}},$$

$$N_6 = \phi_3(u, v)\,\tilde{\nabla}\psi^+(w) = \frac{1}{2}v\hat{\boldsymbol{w}},$$

$$N_7 = \boldsymbol{n}_1(u, v)\,\psi^+(w) = \frac{1}{2}\left[-(v-1)(w+1)\hat{\boldsymbol{u}} + u(w+1)\hat{\boldsymbol{v}}\right],$$

$$N_8 = \boldsymbol{n}_2(u, v)\,\psi^+(w) = \frac{1}{2}\left[-v(w+1)\hat{\boldsymbol{u}} + u(w+1)\hat{\boldsymbol{v}}\right],$$

$$N_9 = \boldsymbol{n}_3(u, v)\,\psi^+(w) = \frac{1}{2}\left[-v(w+1)\hat{\boldsymbol{u}} + (u-1)(w+1)\hat{\boldsymbol{v}}\right],$$

where

$$\boldsymbol{n}_1 = \phi_1\tilde{\nabla}\phi_2 - \phi_2\tilde{\nabla}\phi_1,$$

$$\boldsymbol{n}_2 = \phi_2\tilde{\nabla}\phi_3 - \phi_3\tilde{\nabla}\phi_2,$$

$$\boldsymbol{n}_3 = \phi_3\tilde{\nabla}\phi_1 - \phi_1\tilde{\nabla}\phi_3$$

are the edge elements on the triangular cross section of the prism perpendicular to the cylinder axis.

Linear Basis Functions for $H(\text{div})$

The face basis functions for the reference element are given by

$$\begin{aligned}
\boldsymbol{M}_1 &= 4\psi^-(w)\tilde{\nabla}\psi^-(w) = (w-1)\hat{\boldsymbol{w}},\\
\boldsymbol{M}_2 &= \frac{1}{2}\boldsymbol{m}_1(u,v) = \frac{1}{2}\left[u\hat{\boldsymbol{u}} + (v-1)\hat{\boldsymbol{v}}\right],\\
\boldsymbol{M}_3 &= \frac{1}{2}\boldsymbol{m}_2(u,v) = \frac{1}{2}\left[u\hat{\boldsymbol{u}} + v\hat{\boldsymbol{v}}\right],\\
\boldsymbol{M}_4 &= \frac{1}{2}\boldsymbol{m}_3(u,v) = \frac{1}{2}\left[(u-1)\hat{\boldsymbol{u}} + v\hat{\boldsymbol{v}}\right],\\
\boldsymbol{M}_5 &= 4\psi^+(w)\tilde{\nabla}\psi^+(w) = (w+1)\hat{\boldsymbol{w}},
\end{aligned}$$

where

$$\begin{aligned}
\boldsymbol{m}_1 &= \boldsymbol{n}_1 \times \hat{\boldsymbol{w}},\\
\boldsymbol{m}_2 &= \boldsymbol{n}_2 \times \hat{\boldsymbol{w}},\\
\boldsymbol{m}_3 &= \boldsymbol{n}_3 \times \hat{\boldsymbol{w}}.
\end{aligned}$$

Quadrature Rule

The quadrature points in Table B.13 and the corresponding weights in Table B.14 yield a quadrature rule that integrates quadratic polynomials exactly in the plane perpendicular to the axis of the prism. However, the quadrature rule exploits trapezoidal integration along the prism axis and, consequently, provides mass lumping along the cylinder axis for straight prisms. Despite the fact that this quadrature rule for a prism cannot integrate quadratic polynomials exactly, it yields an FEM with second order of convergence for a piecewise linear approximation of the field.

B.2.3 Hexahedron

The linear reference hexahedron occupies the volume bounded by $-1 \le u \le 1$, $-1 \le v \le 1$, and $-1 \le w \le 1$. It has 8 vertices, 12 edges, and 6 quadrilateral faces.

The nodes for the reference element are given by

Table B.13 Quadrature points for reference prism

Point	u-coordinate	v-coordinate	w-coordinate
1	$6.666666666666667 \times 10^{-1}$	$1.666666666666667 \times 10^{-1}$	-1.000000000000000
2	$1.666666666666667 \times 10^{-1}$	$6.666666666666667 \times 10^{-1}$	-1.000000000000000
3	$1.666666666666667 \times 10^{-1}$	$1.666666666666667 \times 10^{-1}$	-1.000000000000000
4	$6.666666666666667 \times 10^{-1}$	$1.666666666666667 \times 10^{-1}$	1.000000000000000
5	$1.666666666666667 \times 10^{-1}$	$6.666666666666667 \times 10^{-1}$	1.000000000000000
6	$1.666666666666667 \times 10^{-1}$	$1.666666666666667 \times 10^{-1}$	1.000000000000000

Table B.14 Quadrature weights for reference prism

Point	Weight
1	$1.666666666666667 \times 10^{-1}$
2	$1.666666666666667 \times 10^{-1}$
3	$1.666666666666667 \times 10^{-1}$
4	$1.666666666666667 \times 10^{-1}$
5	$1.666666666666667 \times 10^{-1}$
6	$1.666666666666667 \times 10^{-1}$

Table B.15 Definition of edges for hexahedral reference element: node 1—start node of edge; node 2—end node of edge

Edge	Node 1	Node 2
1	1	2
2	2	3
3	3	4
4	4	1
5	1	5
6	2	6
7	3	7
8	4	8
9	5	6
10	6	7
11	7	8
12	8	5

$$
\begin{aligned}
\boldsymbol{r}_1 &= [-1, -1, -1], \\
\boldsymbol{r}_2 &= [+1, -1, -1], \\
\boldsymbol{r}_3 &= [+1, +1, -1], \\
\boldsymbol{r}_4 &= [-1, +1, -1], \\
\boldsymbol{r}_5 &= [-1, -1, +1], \\
\boldsymbol{r}_6 &= [+1, -1, +1], \\
\boldsymbol{r}_7 &= [+1, +1, +1], \\
\boldsymbol{r}_8 &= [-1, +1, +1].
\end{aligned}
$$

The edges for the reference element are given in Table B.15.
The faces for the reference element are given in Table B.16.

Table B.16 Definition of faces for tetrahedral reference element: node 1—first node of face; node 2—second node of face; node 3—third node of face; node 4—fourth node of face

Face	Node 1	Node 2	Node 3	Node 4
1	4	3	2	1
2	1	2	6	5
3	2	3	7	6
4	3	4	8	7
5	4	1	5	8
6	5	6	7	8

Linear Basis Functions for *H*(grad)

The node basis functions for the reference element are given by

$$\varphi_1 = \psi^-(u)\,\psi^-(v)\,\psi^-(w) = \frac{1}{8}(1-u)(1-v)(1-w),$$

$$\varphi_2 = \psi^+(u)\,\psi^-(v)\,\psi^-(w) = \frac{1}{8}(1+u)(1-v)(1-w),$$

$$\varphi_3 = \psi^+(u)\,\psi^+(v)\,\psi^-(w) = \frac{1}{8}(1+u)(1+v)(1-w),$$

$$\varphi_4 = \psi^-(u)\,\psi^+(v)\,\psi^-(w) = \frac{1}{8}(1-u)(1+v)(1-w),$$

$$\varphi_5 = \psi^-(u)\,\psi^-(v)\,\psi^+(w) = \frac{1}{8}(1-u)(1-v)(1+w),$$

$$\varphi_6 = \psi^+(u)\,\psi^-(v)\,\psi^+(w) = \frac{1}{8}(1+u)(1-v)(1+w),$$

$$\varphi_7 = \psi^+(u)\,\psi^+(v)\,\psi^+(w) = \frac{1}{8}(1+u)(1+v)(1+w),$$

$$\varphi_8 = \psi^-(u)\,\psi^+(v)\,\psi^+(w) = \frac{1}{8}(1-u)(1+v)(1+w).$$

Here, we use the basis functions

$$\psi^-(\xi) = \frac{1}{2}(1-\xi),$$

$$\psi^+(\xi) = \frac{1}{2}(1+\xi).$$

Linear Basis Functions for *H*(curl)

The edge basis functions for the reference element are given by

$$\boldsymbol{N}_1 = \psi^-(v)\,\psi^-(w)\tilde{\nabla}\psi^+(u) = \frac{1}{8}(1-v)(1-w)\hat{\boldsymbol{u}},$$

$$\boldsymbol{N}_2 = \psi^+(u)\,\psi^-(w)\tilde{\nabla}\psi^+(v) = \frac{1}{8}(1+u)(1-w)\hat{\boldsymbol{v}},$$

$$\boldsymbol{N}_3 = \psi^+(v)\,\psi^-(w)\tilde{\nabla}\psi^-(u) = -\frac{1}{8}(1+v)(1-w)\hat{\boldsymbol{u}},$$

$$\boldsymbol{N}_4 = \psi^-(u)\,\psi^-(w)\tilde{\nabla}\psi^-(v) = -\frac{1}{8}(1-u)(1-w)\hat{\boldsymbol{v}},$$

$$\boldsymbol{N}_5 = \psi^-(u)\,\psi^-(v)\tilde{\nabla}\psi^+(w) = \frac{1}{8}(1-u)(1-v)\hat{\boldsymbol{w}},$$

$$\boldsymbol{N}_6 = \psi^+(u)\,\psi^-(v)\tilde{\nabla}\psi^+(w) = \frac{1}{8}(1+u)(1-v)\hat{\boldsymbol{w}},$$

$$\boldsymbol{N}_7 = \psi^+(u)\,\psi^+(v)\tilde{\nabla}\psi^+(w) = \frac{1}{8}(1+u)(1+v)\hat{\boldsymbol{w}},$$

$$\boldsymbol{N}_8 = \psi^-(u)\,\psi^+(v)\tilde{\nabla}\psi^+(w) = \frac{1}{8}(1-u)(1+v)\hat{\boldsymbol{w}},$$

$$\boldsymbol{N}_9 = \psi^-(v)\,\psi^+(w)\tilde{\nabla}\psi^+(u) = \frac{1}{8}(1-v)(1+w)\hat{\boldsymbol{u}},$$

$$\boldsymbol{N}_{10} = \psi^+(u)\,\psi^+(w)\tilde{\nabla}\psi^+(v) = \frac{1}{8}(1+u)(1+w)\hat{\boldsymbol{v}},$$

$$\boldsymbol{N}_{11} = \psi^+(v)\,\psi^+(w)\tilde{\nabla}\psi^-(u) = -\frac{1}{8}(1+v)(1+w)\hat{\boldsymbol{u}},$$

$$\boldsymbol{N}_{12} = \psi^-(u)\,\psi^+(w)\tilde{\nabla}\psi^-(v) = -\frac{1}{8}(1-u)(1+w)\hat{\boldsymbol{v}}.$$

Linear Basis Functions for $\boldsymbol{H}(\mathrm{div})$

The face basis functions for the reference element are given by

$$\boldsymbol{M}_1 = \frac{1}{2}\psi^-(w)\tilde{\nabla}\psi^-(w) = -\frac{1}{8}(1-w)\hat{\boldsymbol{w}},$$

$$\boldsymbol{M}_2 = \frac{1}{2}\psi^-(v)\tilde{\nabla}\psi^-(v) = -\frac{1}{8}(1-v)\hat{\boldsymbol{v}},$$

$$\boldsymbol{M}_3 = \frac{1}{2}\psi^+(u)\tilde{\nabla}\psi^+(u) = +\frac{1}{8}(1+u)\hat{\boldsymbol{u}},$$

$$\boldsymbol{M}_4 = \frac{1}{2}\psi^+(v)\tilde{\nabla}\psi^+(v) = +\frac{1}{8}(1+v)\hat{\boldsymbol{v}},$$

Table B.17 Quadrature points for reference hexahedron

Point	u-coordinate	v-coordinate	w-coordinate
1	−1.000000000000000	−1.000000000000000	−1.000000000000000
2	1.000000000000000	−1.000000000000000	−1.000000000000000
3	1.000000000000000	1.000000000000000	−1.000000000000000
4	−1.000000000000000	1.000000000000000	−1.000000000000000
5	−1.000000000000000	−1.000000000000000	1.000000000000000
6	1.000000000000000	−1.000000000000000	1.000000000000000
7	1.000000000000000	1.000000000000000	1.000000000000000
8	−1.000000000000000	1.000000000000000	1.000000000000000

Table B.18 Quadrature weights for reference hexahedron

Point	Weight
1	1.000000000000000
2	1.000000000000000
3	1.000000000000000
4	1.000000000000000
5	1.000000000000000
6	1.000000000000000
7	1.000000000000000
8	1.000000000000000

$$\boldsymbol{M}_5 = \frac{1}{2}\psi^-(u)\tilde{\nabla}\psi^-(u) = -\frac{1}{8}(1-u)\hat{\boldsymbol{u}},$$

$$\boldsymbol{M}_6 = \frac{1}{2}\psi^+(w)\tilde{\nabla}\psi^+(w) = +\frac{1}{8}(1-w)\hat{\boldsymbol{w}}.$$

Quadrature Rule

The quadrature points in Table B.17 and the corresponding weights in Table B.18 yield trapezoidal integration for the reference hexahedron by means of the product of three 1D trapezoidal integration rules. This quadrature rule provides mass lumping for brick-shaped elements and, despite the fact that it cannot integrate quadratic polynomials exactly, yields an FEM with second order of convergence for a piecewise linear approximation of the field.

Appendix C
Large Linear Systems

C.1 Sparse Matrices

Many CEM problems require the solution of large linear systems of equations. This is generally the case for the finite element method (FEM), both for frequency- and time-domain applications. In realistic 3D applications, the number of unknowns can be in the range of tens of thousands to several millions. For the largest systems, direct inversion is seldom possible, and iterative methods are needed. Here, we will introduce some routines for large linear systems.

Below, we give a MATLAB function that assembles the sparse system that we solved using Gauss–Seidel iterations in the capacitance calculation in Chap. 3. The study was then limited to a 50×50 grid. With the assembled system we can use more efficient methods and therefore use higher resolutions. For this 2D problem, the direct solver invoked by "\" in MATLAB performs very well.

We write the discretized problem as $\mathbf{Af} = \mathbf{s}$ and use the MATLAB function `setAs` listed below to set $\mathbf{A}$ and $\mathbf{s}$. Note that this script was written so as to make very few references to the sparse matrix. This is faster than referencing the individual elements in the sparse matrix, because each reference requires a function call, which is quite slow.

```
% ------------------------------------------------------------------
% Set up matrix A and right-hand side s
% ------------------------------------------------------------------
function [A, s] = setAs(a, b, c, d, n, m)

% Arguments:
%    a    =  width of inner conductor
%    b    = height of inner conductor
%    c    =  width of outer conductor
%    d    = height of outer conductor
%    n    = number of points in the x-direction (horizontal)
%    m    = number of points in the y-direction (vertical)
% Returns:
%    A    = matrix on sparse storage format
%    s    = right-hand side on sparse storage format
```

T. Rylander et al., *Computational Electromagnetics*, Texts in Applied Mathematics 51, DOI 10.1007/978-1-4614-5351-2,

```
hx = 0.5*c/n;                % Grid size in x-direction
na = round(0.5*a/hx);        % Number of cells for half width of
                             % inner conductor
hy = 0.5*d/m;                % Grid size in y-direction
m  = round(0.5*d/hy);        % Number of cells for half height of
                             % outer conductor
mb = round(0.5*b/hy);        % Number of cells for half height of
                             % inner conductor
p  = 1;                      % Potential on inner conductor

%          The upper right corner is discretized
%
%                    --------------------+
%                            c/2         |
%                                        |
%                    ------------+       | d/2
%                        a/2     |       |
%                                | b/2   |
%                                |       |
%
%                  (Dimensions)
%
%
%              The nodes are numbered like this
%          y
%          ^
%          |
% (m-1)hy| (m-1)n+1 (m-1)n+2 (m-1)n+3   ...   mn
%                :        :        :                :
%   2hy  |    2n+1     2n+2     2n+3     ...   3n
%    hy  |     n+1      n+2      n+3     ...   2n
%     0  |      1        2        3      ...    n
%          -----------------------------------------> x
%           0       hx       2 hx          (n-1)hx
%
%                  (Discretization)

N  = n * m;       % Total number of unknowns.
cx = hx^-2;
cy = hy^-2;

% Generate a matrix with N = m*n rows (-> nodes on the grid),
% and five columns, one for each nonzero diagonal of A.
% The first column gives contribution from nodes beneath.
% The second column gives contribution from nodes to the left.
% The third column gives self-contribution.
% The fourth column gives contribution from nodes to the right.
% The fifth column gives contribution from nodes above.

% The following lines assume some knowledge of MATLAB.  If you
% feel uncertain, insert the 'keyboard' command.  This causes
% MATLAB to stop.  Then execute lines by 'dbstep' and examine
% the result.
```

```
C = repmat([cy cx -2*(cx+cy) cx cy], N, 1);

% Find indices of nodes that are not surrounded by four interior
% nodes.
idx0R = n:n:N-n;             % Nodes with     V = 0 to the right
idxNB = na+2:n;              % Nodes with dV/dy = 0 beneath
idxNL = 1+n*(mb+1):n:N;      % Nodes with dV/dx = 0 to the left

idx1C = repmat((1:na+1)', 1, mb+1) + repmat((0:n:n*mb),na+1,1);
                             % 'x-index + n*(y-index-1)' for all
idx1C = idx1C(:)';           % nodes on (or inside) the inner
                             % conductor where V = 1
                             % and convert to row vector

C(idx1C,[1 2 4 5]) = 0;
C(idx1C, 3) = 1;
C(idx0R, 4) = 0;
C(idxNB, 5) = 2*cy;
C(idxNL, 4) = 2*cx;
C(idxNL, 2) = 0;

% Find the nonzero elements (si) of each column and the
% corresponding row indices (ii).  Do not include elements
% corresponding to nodes outside the grid.
[i1,j,s1] = find(C(n+1:end,   1)); % The first 'nc' nodes have no
                                   % neighbors beneath
[i2,j,s2] = find(C(1+1:end,   2)); % The first node has no
                                   % neighbor to the left
[i3,j,s3] = find(C(  1:end,   3));
[i4,j,s4] = find(C(  1:end-1, 4)); % The last node has no
                                   % neighbor to the right
[i5,j,s5] = find(C(  1:end-n, 5)); % The last 'nc' nodes have no
                                   % neighbors above

% Put the elements (si) into a sparse matrix. The first input
% are row indices, the second is column indices and the third
% is the elements.
A = sparse([i1+n; i2+1; i3; i4; i5], ...
           [i1; i2; i3; i4+1; i5+n], ...
           [s1; s2; s3; s4; s5], N, N);
s = sparse(idx1C', 1, p, N, 1);
```

C.2 Solvers for Large Sparse Systems of Equations

As we already mentioned, the 2D discretized Laplace equation can be solved in MATLAB by direct inversion `f = A\s`. For 2D problems, direct methods are generally very competitive, unless the problems are very large. However, for 3D problems, iterative solvers are often more efficient. We will here give a brief overview of solvers for sparse linear systems of equations that are used in CEM.

C.2.1 Direct Solvers

In direct methods, a complete factorization (e.g., an LU decomposition) of the matrix $\mathbf{A}$ is done. Clever reordering of the rows and the columns of $\mathbf{A}$ plays an important role; a good reordering scheme can reduce the operation count and the memory requirements for the factorization by more than an order of magnitude. In MATLAB, one can, for example, use column approximate minimum degree permutation, `colamd` (for nonsymmetric matrices), or symmetric approximate minimum degree permutation, `symamd` (for symmetric matrices), to reorder matrices. However, when the backslash operator "\" is invoked, this is done automatically.

A major advantage of direct methods compared to iterative methods is that since a complete factorization is done, additional right-hand sides can be solved for with low additional cost. Another advantage is that direct methods generally are less sensitive to ill conditioning and can be used where many iterative methods fail to converge.

However, both time and memory requirements scale unfavorably with problem size; hence direct methods become prohibitively expensive for very large problems. Often the memory requirements are the limiting factor.

Efficient, freely available algorithms for direct factorization and reordering of sparse matrices include UMFPACK [23], SuperLU [24], TAUCS [84], and METIS [44].

C.2.2 Iterative Solvers

The matrices that result from finite element discretizations of Poisson's equation (1.3) or the time-domain version of the curl-curl equation (6.86) are symmetric and positive definite. For such systems, iterative so-called Krylov methods (see Appendix D) generally work very well.

However, to speed up the convergence of the iterative algorithm, it is very useful to precondition the matrix. The idea of preconditioning is to find an approximate inverse of $\mathbf{A}$, say $\mathbf{M}^{-1}$, and multiply $\mathbf{Af} = \mathbf{s}$ by the approximate inverse from the left. If $\mathbf{M}^{-1}\mathbf{A} \approx \mathbf{I}$, the iterative solver will converge much faster. The choice of preconditioner generally has a much stronger effect on the speed of convergence than the choice of Krylov method. A choice that often works well is the so-called incomplete LU decomposition, in which $\mathbf{M} = \mathbf{LU} \approx \mathbf{A}$, with $\mathbf{L}$ a lower triangular and $\mathbf{U}$ an upper triangular matrix. Then $\mathbf{M}^{-1} = \mathbf{U}^{-1}\mathbf{L}^{-1}$, which is inexpensive to apply if $\mathbf{L}$ and $\mathbf{U}$ are sparse. When $\mathbf{A}$ is symmetric, the factorization can be made such that $\mathbf{U} = \mathbf{L}^T$, and this is called incomplete Cholesky decomposition. The degree of incompleteness can be specified by how much fill-in is allowed in $\mathbf{L}$ and $\mathbf{U}$, that is, how many extra nonzero elements $\mathbf{L}$ and $\mathbf{U}$ have in comparison with $\mathbf{A}$. In MATLAB, this is controlled by setting a relative tolerance below which elements in **L** and **U** are dropped. This tolerance is chosen as a compromise between good accuracy of the decomposition (favored by a small tolerance) and minimizing memory and CPU time for a matrix multiplication (which is favored by a high tolerance).

Also in the case with incomplete factorizations, it is strongly recommended to reorder the rows and columns of **A** before the incomplete factorization is computed.

Another, less complicated, preconditioner is symmetric successive overrelaxation (SSOR) [7], in which the preconditioning matrix **M** never is stored explicitly. Hence the memory requirements are smaller when SSOR is used as a preconditioner instead of some incomplete factorization of **A**.

An important note is that for the time-harmonic version of the curl-curl equation, and for low-frequency eddy current computations (Sect. 6.8.3), the null-space of the curl operator causes problems for the Krylov methods, and therefore more advanced preconditioners [26, 27, 46] are required.

Reliable implementations of Krylov methods and preconditioners are available, e.g., in the PETSc library [6]. Also MATLAB provides implementations of many popular Krylov methods.

C.2.3 Multigrid Methods

The multigrid (MG) method [33, 91] was introduced about four decades ago, but has only very recently been applied to Maxwell's equations [36]. The MG method can be used either as an iterative solver on its own, or as a very efficient preconditioner for iterative Krylov methods. It greatly improves the convergence rate of iterative solvers for large sparse matrices that occur in differential equation formulations. In fact, the convergence rate can be made independent of the cell size h, rather than to scale as some power of h.

The underlying principle is the observation that for the Laplace equation, the "short-wavelength error" (which varies on the scale of the grid) is reduced quickly by local operations (known as smoothers) such as Jacobi or Gauss–Seidel iterations; see Sect. 3.1.1. However, the long-wavelength error is reduced much more slowly by the smoothers. Since such error has short wavelength with respect to a *coarser* grid, one expects that this error can be reduced more rapidly on a coarse grid. Therefore, the basic idea of MG is to introduce a hierarchy of grids, starting from the finest one, and try to improve the solution on the finer grid by looking for a correction from the coarser grid. Optimally, the coarsest grid has only a small number of cells, and a direct solver can be used at a low computational cost.

So far, MG is used mostly for electrostatic and magnetostatic problems [59, 70] and transient eddy current problems. Generally, MG is among the most efficient solvers [33, 59] for Laplace-type equations. However, little research on MG has been devoted to fully electromagnetic problems, such as time-harmonic problems for eddy current computations [8, 35]. Certain difficulties (due to the null-space of the curl-curl operator) are encountered when this method is applied to the full

Table C.1 Capacitance vs. cell size for finite difference solution on larger grids

n	$h \times 10^2$	C [pF/m]
50	2.000	90.78080 583
100	1.000	90.68006 976
200	0.500	90.64044 979
300	0.333	90.62961 567
400	0.250	90.62481 230

Maxwell's equations. For wave problems, another complicating aspect is that the coarsest grid must resolve the wavelength $\lambda \propto 1/f$, which limits the hierarchy of grids and therefore the recursive MG algorithm.

C.3 Capacitance Calculation on Larger Grids

With the more efficient solvers we can extend the capacitance calculation of Sect. 3.1 to much larger grids. Results for grids up to 400 by 400 are shown in Table C.1.

One can estimate the order of convergence from formula (2.4) for 100, 200, and 400 points, and the order of convergence in h comes out as 1.341. This is close to the asymptotic result 4/3, which occurs for the 270° corners. If we do polynomial fits to $h^{4/3}$, the extrapolated value is 90.6145 pF/m. It should be pointed out that a *higher-order* fit to noninteger powers of h, such as $h^{4/3}$, is not an optimal representation, because the regular parts of the solution contribute errors that scale as h^2. Nevertheless, the extrapolation has added three figures of accuracy. If we tried to achieve this accuracy by a single calculation with uniform refinement of the grid, we would have to decrease h by more than a factor of 100, and the execution time would increase by at least 100^3, that is, one million times. Evidently, extrapolation can be a very efficient way of increasing the accuracy. In the chapter on finite elements, we show that the accuracy can also be improved by adaptive grid refinement, which aims at increasing the resolution in regions where the solution varies rapidly.

Appendix D
Krylov Methods

Here, we will discuss some iterative methods for solving large linear systems of equations

$$\mathbf{Ax} = \mathbf{b}. \tag{D.1}$$

For large 3D problems, it is generally too demanding to use a direct solver. Iterative, so-called Krylov methods are often a much better choice for these problems. Multigrid methods, which we discussed very briefly in Sect. C.2.3, have proven even more efficient for many problems but will not be discussed here.

D.1 Projection Methods

In projection methods, one minimizes the residual

$$\mathbf{r} = \mathbf{b} - \mathbf{Ax} \tag{D.2}$$

by an approach similar to the Galerkin and Petrov–Galerkin methods for finite elements. The vector $\mathbf{x}$ will be constructed as a sum of basis vectors $\mathbf{v}$; $\mathbf{x} = \mathbf{x}_0 + \sum_{i=1}^{m} \mathbf{v}_i y_i$, and y is an array of coefficients. This can be written compactly by introducing the matrix $\mathbf{V} = (\mathbf{v}_1, \mathbf{v}_2, \ldots, \mathbf{v}_m)$ and the column vector $\mathbf{y} = (y_1, y_2, \ldots, y_m)^T$:

$$\mathbf{x} = \mathbf{x}_0 + \mathbf{Vy}. \tag{D.3}$$

The vectors $\mathbf{v}_1, \mathbf{v}_2, \ldots, \mathbf{v}_m$ span a space K_m of "basis" vectors. Similarly, one chooses a space L_m of "test" vectors $\mathbf{w}_1, \mathbf{w}_2, \ldots, \mathbf{w}_m$ and demands that on the mth step of the iteration the residual $\mathbf{r}_m$ be orthogonal to all vectors in L_m. If $K_m = L_m$, this is Galerkin's method; otherwise, it is a Petrov–Galerkin method.

The most important part of the iteration is the choice of the search directions $\mathbf{v}_1, \mathbf{v}_2, \ldots, \mathbf{v}_m$. The simplest case is that in which $\mathbf{A}$ is real and symmetric. The old-fashioned "steepest descent" method chooses the increment directions $\mathbf{v}_i$ in

T. Rylander et al., *Computational Electromagnetics*, Texts in Applied Mathematics 51, DOI 10.1007/978-1-4614-5351-2,

the gradient direction of the error functional $(\mathbf{x} - \mathbf{x}_{\text{exact}})^T \mathbf{A}(\mathbf{x} - \mathbf{x}_{\text{exact}})$, on every step of the iteration. It turns out that this is a bad strategy. When the matrix $\mathbf{A}$ is positive definite and symmetric, the number of iterations for the steepest descent method scales as the condition number of $\mathbf{A}$, that is, the ratio of largest to smallest eigenvalues, $\kappa = \lambda_{\max}/\lambda_{\min}$.

D.2 Krylov Methods

A better strategy is to generate the increment directions as $\mathbf{r}_0$, $\mathbf{A}\mathbf{r}_0$, $\mathbf{A}^2\mathbf{r}_0$, ..., $\mathbf{A}^{m-1}\mathbf{r}_0$, where $\mathbf{r}_0$ is the first residual. Then K is called a *Krylov space*. The Arnoldi algorithm does exactly this and projects out components of the new $\mathbf{v}$'s to keep them orthonormal.

1. *Choose a vector* $\mathbf{v}_1$ *of norm 1*
2. *For* $j = 1, 2, \ldots, m$, *Do:*
3. $\quad h_{ij} = (\mathbf{A}\mathbf{v}_j, \mathbf{v}_i)$ *for* $i = 1, 2, \ldots, j$
4. $\quad \mathbf{w}_j = \mathbf{A}\mathbf{v}_j - \sum_{i=1}^{j} h_{ij}\mathbf{v}_i$
5. $\quad h_{j+1,j} = (\mathbf{w}_j, \mathbf{w}_j)^{1/2}$
6. $\quad$ *If* $h_{j+1,j} = 0$ *then Stop*
7. $\quad \mathbf{v}_{j+1} = \mathbf{w}_j / h_{j+1,j}$
8. *EndDo*

GMRES is Arnoldi's method followed by a minimization of $(\mathbf{r}, \mathbf{r})$. This is a reliable method, and it has the nice property that the error decreases monotonically with the iteration number. The disadvantage of GMRES is that one needs to store all the incremental directions $\mathbf{v}_1, \ldots, \mathbf{v}_m$ to do the minimizations. Therefore, it can become very memory-demanding if the number of iterations is large. To circumvent the memory problem, one can restart GMRES after a certain number of iterations (typically 5 to 50). However, at the restart, orthogonality is lost.

There are cleverer ways of generating the incremental directions $\mathbf{v}$. The standard method, which assumes that $\mathbf{A}$ is symmetric, is the Lanczos method. Here it suffices to save three increment directions.

1. *Choose a start vector* $\mathbf{v}_1$ *of norm 1.*
2. *Set* $\beta_1 = 0$, $\mathbf{v}_0 = 0$
3. *For* $j = 1, 2, \ldots, m$, *Do:*
4. $\quad \mathbf{w}_j = \mathbf{A}\mathbf{v}_j - \beta_j \mathbf{v}_{j-1}$
5. $\quad \alpha_j = (\mathbf{w}_j, \mathbf{v}_j)$
6. $\quad \mathbf{w}_j = \mathbf{w}_j - \alpha_j \mathbf{v}_j$
7. $\quad \beta_{j+1} = (\mathbf{w}_j, \mathbf{w}_j)^{1/2}$. *If* $\beta_{j+1} = 0$ *then Stop*
8. $\quad \mathbf{v}_{j+1} = \mathbf{w}_j / \beta_{j+1}$
9. *EndDo*

This makes all the vectors $\mathbf{v}_i, i = 1, 2, \ldots,$ orthogonal (in infinite-precision arithmetic). With finite precision, orthogonality may be lost if the iteration runs many steps. Consequently, the iteration may have to be restarted.

A method that is related to the Lanczos method is the conjugate gradient (CG) method, where one keeps going in orthogonal directions. At least with infinite-precision arithmetic, this method can guarantee convergence when the number of steps equals the number of unknowns. The CG method for a symmetric $\mathbf{A}$ can be written as follows:

1. *Compute* $\mathbf{r}_0 = \mathbf{b} - \mathbf{A}\mathbf{x}_0$, $\mathbf{p}_0 = \mathbf{r}_0$
2. *For* $j = 0, 1, \ldots,$ *until convergence, Do:*
3. $\quad \alpha_j = (\mathbf{r}_j, \mathbf{r}_j)/(\mathbf{A}\mathbf{p}_j, \mathbf{p}_j)$
4. $\quad \mathbf{x}_{j+1} = \mathbf{x}_j + \alpha_j \mathbf{p}_j$
5. $\quad \mathbf{r}_{j+1} = \mathbf{r}_j - \alpha_j \mathbf{A}\mathbf{p}_j$
6. $\quad \beta_j = (\mathbf{r}_{j+1}, \mathbf{r}_{j+1})/(\mathbf{r}_j, \mathbf{r}_j)$
7. $\quad \mathbf{p}_{j+1} = \mathbf{r}_{j+1} + \beta_j \mathbf{p}_j$
8. *EndDo*

An advantage of the CG method is that one does not store the whole history of incremental directions. For positive definite symmetric matrices, the required number of iterations for CG is proportional to the square root of the condition number of the matrix.

D.3 Nonsymmetric A

Lanczos Biorthogonalization

The symmetric Lanczos algorithm can be extended to nonsymmetric matrices. The biorthogonal Lanczos algorithm constructs a pair of biorthogonal bases

$$\mathbf{v}_1, \mathbf{A}\mathbf{v}_1, \ldots, \mathbf{A}^{m-1}\mathbf{v}_1,$$

$$\mathbf{w}_1, \mathbf{A}^T\mathbf{w}_1, \ldots, (\mathbf{A}^T)^{m-1}\mathbf{w}_1,$$

with the orthogonality property $(\mathbf{v}_i, \mathbf{w}_j) = \delta_{ij}$. The procedure can be written as follows:

1. *Choose two vectors* $\mathbf{v}_1, \mathbf{w}_1$ *such that* $(\mathbf{v}_1, \mathbf{w}_1) = 1$.
2. *Set* $\beta_1 = \delta_1 = 0$, $\mathbf{v}_0 = \mathbf{w}_0 = 0$
3. *For* $j = 1, 2, \ldots, m$, *Do:*
4. $\quad \alpha_j = (\mathbf{A}\mathbf{v}_j, \mathbf{w}_j)$
5. $\quad \hat{\mathbf{v}}_{j+1} = \mathbf{A}\mathbf{v}_j - \alpha_j \mathbf{v}_j - \beta_j \mathbf{v}_{j-1}$
6. $\quad \hat{\mathbf{w}}_{j+1} = \mathbf{A}^T\mathbf{v}_j - \alpha_j \mathbf{w}_j - \delta_j \mathbf{w}_{j-1}$
7. $\quad \delta_{j+1} = |(\hat{\mathbf{v}}_{j+1}, \hat{\mathbf{w}}_{j+1})|^{1/2}$. *If* $\delta_{j+1} = 0$ *Stop*
8. $\quad \beta_{j+1} = (\hat{\mathbf{v}}_{j+1}, \hat{\mathbf{w}}_{j+1})/\delta_{j+1}$

9. $\mathbf{w}_{j+1} = \hat{\mathbf{w}}_{j+1}/\beta_{j+1}, \mathbf{v}_{j+1} = \hat{\mathbf{v}}_{j+1}/\delta_{j+1}$
10. EndDo

BICG and QMR
Relatively new methods are the biconjugate gradient (BICG) and quasi-minimal residual (QMR) algorithms. BICG is a generalization of CG to nonsymmetric matrices. BICG generates the space of test vectors from powers of $\mathbf{A}^T$ rather than of $\mathbf{A}$, so this is a Petrov–Galerkin method. The BICG method works as follows ($\mathbf{x}^*$ denotes the complex conjugate of $\mathbf{x}$):

1. *Set* $\mathbf{r}_0 = \mathbf{b} - \mathbf{A}\mathbf{x}_0$. *Choose* $\mathbf{r}_0^*$ *so that* $(\mathbf{r}_0, \mathbf{r}_0^*) \neq 0$
2. *Set* $\mathbf{p}_0 = \mathbf{r}_0$, $\mathbf{p}_0^* = \mathbf{r}_0^*$
3. *For* $j = 0, 1, \ldots,$ *until convergence, Do:*
4. $\alpha_j = (\mathbf{r}_j, \mathbf{r}_j^*)/(\mathbf{A}\mathbf{p}_j, \mathbf{p}_j^*)$
5. $\mathbf{x}_{j+1} = \mathbf{x}_j + \alpha_j \mathbf{p}_j$
6. $\mathbf{r}_{j+1} = \mathbf{r}_j - \alpha_j \mathbf{A}\mathbf{p}_j, \quad \mathbf{r}_{j+1}^* = \mathbf{r}_j^* - \alpha_j \mathbf{A}\mathbf{p}_j^*$
7. $\beta_j = (\mathbf{r}_{j+1}, \mathbf{r}_{j+1}^*)/(\mathbf{r}_j, \mathbf{r}_j^*)$
8. $\mathbf{p}_{j+1} = \mathbf{r}_{j+1} + \beta_j \mathbf{p}_j, \quad \mathbf{p}_{j+1}^* = \mathbf{r}_{j+1}^* + \beta_j \mathbf{p}_j^*$
9. *EndDo*

QMR uses the Lanczos procedure to generate the incremental directions but still manages to avoid saving the $\mathbf{v}$'s. Finally, QMR minimizes a quantity that is related to (but not quite the same as) the residual. Hence the name "quasi." QMR does not require storage of the $\mathbf{v}$ vectors. As long as it does not lose orthogonality, it is probably the most useful of the iterative schemes for nonsymmetric matrices. In case the method loses orthogonality, QMR can be restarted using the last $\mathbf{x}$ as a starting point.

A disadvantage of both BICG and QMR is that they also use the transpose of the matrix $\mathbf{A}$. Improvements in which $\mathbf{A}^T$ is eliminated are called BICGSTAB and TFQMR (transpose-free QMR).

D.4 Preconditioning

For good efficiency, the iterative solver must in general be combined with a preconditioner; i.e., (D.1) is multiplied by some approximate inverse of $\mathbf{A}$ from the left. This can strongly improve the convergence. A preconditioner that often works, and is commonly used for eddy current calculations, is the incomplete LU decomposition; see Appendix C.2.2. Iterative methods are described in [4, 6, 7, 34, 67].

References

1. T Abboud, J C Nédélec, and J Volakis. Stable solution of the retarded potential equations. *17th Annual Review of Progress in Applied Computational Electromagnetics, Monterey, CA*, pages 146–151, 2001.
2. M Abramowitz and I A Stegun. *Handbook of Mathematical Functions*. National Bureau of Standards, 1965.
3. F Alimenti, P Mezzanotte, L Roselli, and R Sorrentino. A revised formulation of model absorbing and matched modal source boundary conditions for the efficient FDTD analysis of waveguide structures. *IEEE Trans. Microwave Theory Tech.*, 48(1):50–59, January 2000.
4. O Axelsson. *Iterative Solution Methods*. New York, NY: Cambridge University Press, 1994.
5. C A Balanis. *Advanced Engineering Electromagnetics*. New York, NY: John Wiley & Sons, 1989.
6. S Balay, W Gropp, L Curfman McInnes, and B Smith. The portable, extensible toolkit for scientific computation. http://www-unix.mcs.anl.gov/petsc/petsc-2/, 2005.
7. R Barret, M Berry, T F Chan, J Demmel, J Donato, J Dongarra, V Eijkhout, R Pozo, C Romine, and H Van der Vorst. *Templates for the Solution of Linear Systems: Building Blocks for Iterative Methods*. SIAM, Philadelphia, PA, 1994. available at: `ftp://ftp.netlib.org/templates/templates.ps`.
8. R Beck and R Hiptmair. Multilevel solution of the time-harmonic Maxwell's equations based on edge elements. *Int. J. Numer. Meth. Engng.*, 45(7):901–920, 1999.
9. J P Bérenger. A perfectly matched layer for the absorption of electromagnetic waves. *J. Comput. Phys.*, 114(2):185–200, October 1994.
10. J Bey. Tetrahedral grid refinement. *Computing*, 55(4):355–378, 1995.
11. M J Bluck and S P Walker. Time-domain BIE analysis of large three-dimensional electromagnetic scattering problems. *IEEE Trans. Antennas Propagat.*, 45(5):894–901, May 1997.
12. Alain Bossavit. *Computational Electromagnetism*. Boston, MA: Academic Press, 1998.
13. M M Botha and J M Jin. On the variational formulation of hybrid finite element–boundary integral techniques for electromagnetic analysis. *IEEE Trans. Antennas Propagat.*, 52(11):3037–3047, November 2004.
14. A C Cangellaris and D B Wright. Analysis of the numerical error caused by the stair-stepped approximation of a conducting boundary in FDTD simulations of electromagnetic phenomena. *IEEE Trans. Antennas Propagat.*, 39(10):1518–1525, October 1991.
15. F X Canning. Improved impedance matrix localization method. *IEEE Trans. Antennas Propagat.*, 41(5):659–667, May 1993.
16. F X Canning and K Rogovin. Fast direct solution of standard moment-method matrices. *IEEE Antennas Propagat. Mag.*, 40(3):15–26, June 1998.

T. Rylander et al., *Computational Electromagnetics*, Texts in Applied Mathematics 51, DOI 10.1007/978-1-4614-5351-2,

17. M Celuch-Marcysiak and W K Gwarek. Generalized TLM algorithms with controlled stability margin and their equivalence with finite-difference formulations for modified grids. *IEEE Trans. Microwave Theory Tech.*, 43(9):2081–2089, September 1995.
18. Z Chen, M M Ney, and W J R Hoefer. A new finite-difference time-domain formulation and its equivalence with the TLM symmetrical condensed node. *IEEE Trans. Microwave Theory Tech.*, 39(12):2160–2169, December 1991.
19. D K Cheng. *Fundamentals of Engineering Electromagnetics*. Reading, MA: Addison-Wesley, 1993.
20. W C Chew, J M Jin, E Michielssen, and J Song. *Fast and Efficient Algorithms in Computational Electromagnetics*. Norwood, MA: Artech House, 2001.
21. R Coifman, V Rokhlin, and S Wandzura. The fast multipole method for the wave equation: A pedestrian prescription. *IEEE Antennas Propagat. Mag.*, 35(3):7–12, June 1993.
22. D B Davidson. *Computational Electromagnetics for RF and Microwave Engineering*. Cambridge: Cambridge University Press, second edition, 2011.
23. T Davis. UMFPACK. http://www.cise.ufl.edu/research/sparse/umfpack/, 2005.
24. J W Demmel, J R Gilbert, and X S Li. SuperLU. http://crd.lbl.gov/~xiaoye/SuperLU/, 2005.
25. S J Dodson, S P Walker, and M J Bluck. Costs and cost scaling in time-domain integral-equation analysis of electromagnetic scattering. *IEEE Antennas Propagat. Mag.*, 40(4):12–21, August 1998.
26. R Dyczij-Edlinger and O Biro. A joint vector and scalar potential formulation for driven high frequency problems using hybrid edge and nodal finite elements. *IEEE Trans. Microwave Theory Tech.*, 44(1):15–23, 1996.
27. R Dyczij-Edlinger, G Peng, and J F Lee. A fast vector-potential method using tangentially continuous vector finite elements. *IEEE Trans. Microwave Theory Tech.*, 46(6):863–868, 1998.
28. K Eriksson, D Estep, P Hansbo, and C Johnson. *Computational Differential Equations*. New York, NY: Cambridge University Press, 1996.
29. R Garg. *Analytical and Computational Methods in Electromagnetics*. Norwood, MA: Artech House, 2008.
30. W L Golik. Wavelet packets for fast solution of electromagnetic integral equations. *IEEE Trans. Antennas Propagat.*, 46(5):618–624, May 1998.
31. R D Graglia. On the numerical integration of the linear shape functions times the 3-D Green's function or its gradient on a plane triangle. *IEEE Trans. Antennas Propagat.*, 41(10):1448–1455, October 1993.
32. D J Griffiths. *Introduction to Electrodynamics*. Upper Saddle River, NJ: Prentice-Hall, third edition, 1999.
33. W Hackbusch. *Multi-Grid Methods and Application*. Berlin: Springer-Verlag, 1985.
34. W Hackbush. *Iterative Solution of Large Sparse Linear Systems of Equations*. New York, NY: Springer-Verlag, 1994.
35. V Hill, O Farle, and R Dyczij-Edlinger. A stabilized multilevel vector finite-element solver for time-harmonic electromagnetic waves. *IEEE Trans. Magn.*, 39(3):1203–1206, 2003.
36. R Hiptmair. Multigrid method for Maxwell's equations. *SIAM J. Numer. Anal.*, 36(1):204–225, 1998.
37. W J R Hoefer. The transmission-line method – theory and applications. *IEEE Trans. Microwave Theory Tech.*, 33(10):882–893, October 1985.
38. T J R Hughes. *The finite element method: linear static and dynamic finite element analysis*. Englewood Cliffs, NJ: Prentice-Hall, 1987.
39. P Ingelström. *Higher Order Finite Elements and Adaptivity in Computational Electromagnetics*. PhD thesis, Chalmers University of Technology, Göteborg, Sweden, 2004.
40. J M Jin. *The Finite Element Method in Electromagnetics*. New York, NY: John Wiley & Sons, 1993.
41. J M Jin. *The Finite Element Method in Electromagnetics*. New York, NY: John Wiley & Sons, second edition, 2002.
42. J M Jin. *Theory and Computation of Electromagnetic Fields*. New York, NY: John Wiley & Sons, 2010.

43. P B Johns. A symmetrical condensed node for the TLM method. *IEEE Trans. Microwave Theory Tech.*, 35(4):370–377, April 1987.
44. G Karypis. METIS. http://www-users.cs.umn.edu/~karypis/metis/, 2005.
45. P S Kildal, S Rengarajan, and A Moldsvor. Analysis of nearly cylindrical antennas and scattering problems using a spectrum of two-dimensional solutions. *IEEE Trans. Antennas Propagat.*, 44(8):1183–1192, August 1996.
46. Y Q Liu, A Bondeson, R Bergström, C Johnson, M G Larson, and K Samuelsson. Eddy-current computations using adaptive grids and edge elements. *IEEE Trans. Magn.*, 38(2):449–452, March 2002.
47. N K Madsen and R W Ziolkowski. A three-dimensional modified finite volume technique for maxwell's equations. *Electromagnetics*, 10(1-2):147–161, January-June 1990.
48. P Monk. *Finite Element Methods for Maxwell's Equations*. Oxford: Clarendon Press, 2003.
49. P B Monk. A comparison of three mixed methods for the time dependent Maxwell equations. *SIAM Journal on Scientific and Statistical Computing*, 13(5):1097–1122, September 1992.
50. A Monorchio and R Mittra. A hybrid finite-element finite-difference time-domain (FE/FDTD) technique for solving complex electromagnetic problems. *IEEE Microw. Guided Wave Lett.*, 8(2):93–95, February 1998.
51. J C Nédélec. Mixed finite elements in R^3. *Numer. Math.*, 35(3):315–341, 1980.
52. N M Newmark. A method of computation for structural dynamics. *J. Eng. Mech. Div., Proc. Am. Soc. Civil Eng.*, 85(EM 3):67–94, July 1959.
53. S Owen. Meshing Research Corner. http://www.andrew.cmu.edu/user/sowen/mesh.html, 2005.
54. A F Peterson, S L Ray, and R Mittra. *Computational Methods for Electromagnetics*. New York, NY: IEEE Press, 1997.
55. P G Petropoulos, L Zhao, and A C Cangellaris. A reflectionless sponge layer absorbing boundary condition for the solution of Maxwell's equations with high-order staggered finite difference schemes. *J. Comput. Phys.*, 139(1):184–208, January 1998.
56. A J Poggio and E K Miller. Integral equation solutions of three-dimensional scattering problems. *Computer Techniques for Electromagnetics*, Oxford: Pergamon:159–264, 1973.
57. S M Rao and D R Wilton. Transient scattering by conducting surfaces of arbitrary shape. *IEEE Trans. Antennas Propagat.*, 39(1):56–61, January 1991.
58. S M Rao, D R Wilton, and A W Glisson. Electromagnetic scattering by surfaces of arbitrary shape. *IEEE Trans. Antennas Propagat.*, AP-30(3):409–418, May 1982.
59. S Reitzinger and M Kaltenbacher. Algebraic multigrid methods for magnetostatic field problems. *IEEE Trans. Magn.*, 38(2):477–480, 2002.
60. D J Riley and C D Turner. VOLMAX: A solid-model-based, transient volumetric Maxwell solver using hybrid grids. *IEEE Antennas Propagat. Mag.*, 39(1):20–33, February 1997.
61. V Rokhlin. Rapid solution of integral equations of classical potential theory. *J. Comput. Phys.*, 60(2):187–207, 1985.
62. V Rokhlin. Rapid solution of integral equations of scattering theory in two dimensions. *J. Comput. Phys.*, 86(2):414–439, 1990.
63. T Rylander and A Bondeson. Stability of explicit-implicit hybrid time-stepping schemes for Maxwell's equations. *J. Comput. Phys.*, 179(2):426–438, July 2002.
64. T Rylander and J M Jin. Perfectly matched layer for the time domain finite element method. *J. of Comput. Phys.*, 200(1):238–250, October 2004.
65. T Rylander, T McKelvey, and M Viberg. Estimation of resonant frequencies and quality factors from time domain computations. *J. of Comput. Phys.*, 192(2):523–545, December 2003.
66. B P Rynne. Instabilities in time marching methods for scattering problems. *Electromagnetics*, 6(2):129–144, 1986.
67. Y Saad. *Iterative methods for sparse linear systems*. Boston, MA: PWS Publishing, 1996.
68. M N O Sadiku. *Numerical Techniques in Electromagnetics with MATLAB*. Boca Raton, FL: CRC Press, third edition, 2009.
69. M Salazar-Palma, T K Sarkar, L E Garcia-Castillo, T Roy, and A Djordjevic. *Iterative and Self-Adaptive Finite-Elements in Electromagnetic Modeling*. Norwood, MA: Artech House, 1998.

70. M Schinnerl, J Schöberl, and M Kaltenbacher. Nested multigrid methods for the fast numerical computation of 3D magnetic fields. *IEEE Trans. Magn.*, 36(4):1557–1560, 2000.
71. R Schuhmann and T Weiland. Stability of the FDTD algorithm on nonorthogonal grids related to the spatial interpolation scheme. *IEEE Trans. Magn.*, 34(5):2751–2754, September 1998.
72. X Q Sheng and W Song. *Essentials of Computational Electromagnetics*. Singapore: John Wiley & Sons, 2012.
73. J R Shewchuk. Trianlge – a two-dimensional quality mesh generator and delaunay triangulator. http://www.cs.cmu.edu/~quake/triangle.html.
74. J R Shewchuk. Triangle: Engineering a 2D Quality Mesh Generator and Delaunay Triangulator. *Lecture Notes in Computer Science*, 1148:203–222, May 1996.
75. P P Silvester and R L Ferrari. *Finite Elements for Electrical Engineers*. New York, NY: Cambridge University Press, second edition, 1990.
76. P D Smith. Instabilities in time marching methods for scattering: cause and rectification. *Electromagnetics*, 10(4):439–451, October–December 1990.
77. J M Song and W C Chew. The fast Illinois solver code: requirements and scaling properties. *IEEE Comput. Sci. Eng.*, 5(3):19–23, July–September 1998.
78. J M Song, C C Lu, and W C Chew. Multilevel fast multipole algorithm for electromagnetic scattering by large complex objects. *IEEE Trans. Antennas Propagat.*, 45(10):1488–1493, October 1997.
79. J M Song, C C Lu, W C Chew, and S W Lee. Fast Illinois solver code (FISC). *IEEE Antennas Propagat. Mag.*, 40(3):27–34, June 1998.
80. A Taflove. *Computational Electrodynamics: The Finite-Difference Time-Domain Method.* Norwood, MA: Artech House, 1995.
81. A Taflove, editor. *Advances in Computational Electrodynamics: The Finite-Difference Time-Domain Method.* Norwood, MA: Artech House, 1998.
82. A Taflove and S C Hagness. *Computational Electrodynamics: The Finite-Difference Time-Domain Method.* Norwood, MA: Artech House, second edition, 2000.
83. P Thoma and T Weiland. Numerical stability of finite difference time domain methods. *IEEE Trans. Magn.*, 34(5):2740–2743, September 1998.
84. S Toledo, D Chen, and V Rotkin. TAUCS, A Library of Sparse Linear Solvers. http://www.tau.ac.il/~stoledo/taucs/, 2005.
85. D A Vechinski and S M Rao. A stable procedure to calculate the transient scattering by conducting surfaces of arbitrary shape. *IEEE Trans. Antennas Propagat.*, 40(6):661–665, June 1992.
86. R L Wagner and W C Chew. A study of wavelets for the solution of electromagnetic integral equations. *IEEE Trans. Antennas Propagat.*, 43(8):802–810, August 1995.
87. J J H Wang. *Generalized Moment Methods in Electromagnetics*. New York, NY: John Wiley & Sons, 1991.
88. K F Warnick. *NUMERICAL METHODS FOR ENGINEERING - An Introduction Using MATLAB and Computational Electromagnetics Examples*. Raleigh, NC: SciTech Publishing, 2011.
89. J P Webb. Hierarchal vector basis functions of arbitrary order for triangular and tetrahedral finite elements. *IEEE. Trans. Antennas Propagat.*, 47(8):1244–1253, 1999.
90. T Weiland. Time domain electromagnetic field computation with finite difference methods. *Int. J. Numer. Model. El.*, 9(4):295–319, July-August 1996.
91. P Wesseling. *An Introduction to Multigrid Methods*. Chichester: John Wiley & Sons, 1992.
92. R B Wu and T Itoh. Hybrid finite-difference time-domain modeling of curved surfaces using tetrahedral edge elements. *IEEE Trans. Antennas Propagat.*, 45(8):1302–1309, August 1997.
93. K S Yee. Numerical solution of initial boundary value problems involving Maxwell's equations in isotropic media. *IEEE Trans. Antennas Propagat.*, AP-14(3):302–307, May 1966.
94. K S Yee and J S Chen. The finite-difference time-domain (FDTD) and the finite-volume time-domain (FVTD) methods in solving Maxwell's equations. *IEEE Trans. Antennas Propagat.*, 45(3):354–363, March 1997.
95. K S Yee, J S Chen, and A H Chang. Numerical experiments on PEC boundary condition and late time growth involving the FDTD/FDTD and FDTD/FVTD hybrid. *IEEE Antennas Propagat. Soc. Int. Symp.*, 1:624–627, 1995.

Index

T. Rylander et al., *Computational Electromagnetics*, Texts in Applied Mathematics 51, DOI 10.1007/978-1-4614-5351-2,